International Review of
Cytology

A Survey of
Cell Biology

VOLUME 163

International Review of Cytology

A Survey of Cell Biology

Edited by

Kwang W. Jeon
Department of Zoology
University of Tennessee
Knoxville, Tennessee

Jonathan Jarvik
Department of Biological Sciences
Carnegie Mellon University
Pittsburgh, Pennsylvania

VOLUME 163

ACADEMIC PRESS
San Diego   New York   Boston   London   Sydney   Tokyo   Toronto

Copyright © 1995 by ACADEMIC PRESS, INC.

Academic Press, Inc.
A Division of Harcourt Brace & Company
525 B Street, Suite 1900, San Diego, California 92101-4495

United Kingdom Edition published by
Academic Press Limited
24-28 Oval Road, London NW1 7DX

International Standard Serial Number: 0074-7696

International Standard Book Number: 0-12-364567-0

PRINTED IN THE UNITED STATES OF AMERICA
95 96 97 98 99 00 EB 9 8 7 6 5 4 3 2 1

CONTENTS

Programmed Cell Death in Development

E. J. Sanders and M. A. Wride

Biochemistry and Molecular Biology of Chromoplast Development

Bilal Camara, Philippe Hugueney, Florence Bouvier, Marcel Kuntz, and René Monéger

Sperm-Binding Proteins

Kathleen R. Foltz

The Ultrastructure of Epithelial and Fiber Cells in the Crystalline Lens

J. R. Kuszak

CONTRIBUTORS

Numbers in parentheses indicate the pages on which the authors' contributions begin.

Florence Bouvier (175), *Institut de Biologie Moléculaire des Plantes du Centre National de la Recherche Scientifique and Université Louis Pasteur, Strasbourg 67087 France*

Bilal Camara (175), *Institut de Biologie Moléculaire des Plantes du Centre National de la Recherche Scientifique and Université Louis Pasteur, Strasbourg 67087 France*

Kathleen R. Foltz (249), *Division of Molecular, Cell, and Developmental Biology and the Marine Science Institute, University of California at Santa Barbara, Santa Barbara, California 93106*

Adrian E. Hill (1), *Physiological Laboratory, University of Cambridge, Cambridge CB2 3EG, United Kingdom*

Philippe Hugueney (175), *Institut de Biologie Moléculaire des Plantes du Centre National de la Recherche Scientifique and Université Louis Pasteur, Strasbourg 67087 France*

Marcel Kuntz (175), *Institut de Biologie Moléculaire des Plantes du Centre National de la Recherche Scientifique and Université Louis Pasteur, Strasbourg 67087 France*

J. R. Kuszak (305), *Departments of Pathology and Ophthalmology, Rush-Presbyterian-St. Luke's Medical Center, Chicago, Illinois 60612*

René Monéger (175), *Laboratoire de Pathologie et Biochimie Végétales, Université Pierre et Marie Curie, 75025 Paris, France*

E. J. Sanders (105), *Department of Physiology, University of Alberta, Edmonton, Alberta, Canada T6G 2M7*

David R. Soll (43), *Department of Biological Sciences, University of Iowa, Iowa City, Iowa 52242*

M. A. Wride (105), *Department of Physiology, University of Alberta, Edmonton, Alberta, Canada T6G 2M7*

Osmotic Flow in Membrane Pores

Adrian E. Hill
The Physiological Laboratory, University of Cambridge, Cambridge CB2 3EG
United Kingdom

A comparison is made between the conventional macroscopic pore theory, the single-file (no-pass) theory, and the bimodal theory in their ability to predict the values of the unit osmotic permeability P_{os} of single pores. In larger pores osmosis is thought to be a viscous (bulk) flow, while in molecular-sized pores only diffusive flow is considered possible. The physical assumptions underlying these theories are examined and compared with bimodal theory in which (i) viscous flow is impossible in any pore region that can be permeated by the driving osmolyte, and (ii) a distinction between diffusive and viscous flow can still be present in no-pass pores. Experimental values for the osmotic permeability of channels formed by the antibiotics amphotericin, nystatin, and gramicidin and the cellular aquaporin CHIP28 determined with different osmolytes are compared with theoretical expressions for P_{os} as a function of osmolyte radius. Aquaporins are probably pores of variable internal cross section and bimodal theory predicts that they can be probed by osmolytes of different radius to give different osmotic flows, although the overall permeability to each molecule is apparently zero. Such information can be used to construct a model of the pore channel. Conversely, if the pore structure is known, the unit osmotic permeability to any osmolyte can be calculated.

KEY WORDS: Osmosis, Pores, Water channels, Reflexion coefficient, Bimodel theory, Drag coefficients.

I. Introduction

After more than a hundred years of theory and experiment, osmosis is still an imperfectly understood phenomenon. Apart from the interest attached to describing such a fundamental process, it has been given new impetus with the more recent characterization of cell membranes as containing

1

Symbols Used in the Text

a_i	activity
A_i	available filtration area
c_i	concentration
D_i	diffusion coefficient
f	molecular frictional coefficient
F	force
F_i	diffusive drag factor
G_i	viscous drag factor
J_i	flux
k	Boltzmann's constant
K_s	partition ratio of solute between bath and pore
l	length of pore section
L	pore length
n	filing number
n_s, n_w	moles of salt or water
n_2	mole fraction of solute in the pore
N	number of water molecules in a pore
p	pore permeability per unit length
P	hydrostatic pressure
P_d	diffusional water permeability
P_d*	tracer water permeability
P_f	hydraulic permeability
P_{os}	osmotic permeability
r, r_i	radius of pore or molecule
v	fluid velocity
$\bar{v}$	molecular volume of water in the pore
$\bar{V}_s^p$	molar volume of the pore water containing osmolyte
$\bar{V}_w^p$	molar volume of water in the pore
$\bar{V}_i$	molar volume of component in solution
R	gas constant
T	temperature in degrees Kelvin
μ_i	chemical potential
σ_i	reflexion coefficient
η_w	viscosity of water
Π	osmotic pressure

specific pore proteins that mediate osmotic flow through an aqueous core. This will serve to focus attention on whether any particular osmotic model can provide workable expressions, i.e., formulae in which osmotic parameters, principally the osmotic permeability, can be calculated with any degree of accuracy from the properties of the pore and osmolyte, and which have predictive power.

This review does not cover the large body of theoretical and experimental work on membrane osmosis of the last four decades, which has been presented in several monographs (Dick, 1966; House, 1974; Finkelstein, 1987); instead, I concentrate here on the mechanism of osmotic flow and its relevance for understanding flow in relation to structure in pores of biological origin, both antibiotic and eukaryotic.

Osmosis has traditionally been treated as a hydrodynamic problem (Solomon, 1968), but more recently osmotic flow through molecular-sized pores has come to be regarded as a diffusive or quasi-diffusive flow (Finkelstein, 1987). It becomes clear that experimental data can be used to provide information about pore structure and function only if the mechanism of osmosis is decided upon beforehand; this involves essentially a description of osmotic flow at the molecular level. In this review the essential biophysics of these two approaches is examined. I shall argue that the (macroscopic) hydrodynamic approach, scaled down as it is to molecular pores, overlooks the role of water diffusion in osmotic flow, while the (microscopic) theory of transport in single-file pores by a diffusive queuing process ignores the viscous aspects of the problem. Obviously, if one takes the view that there cannot be any real distinction between diffusion and viscous flow in micropores then macroscopic physical concepts cannot be used and a new approach has to be sought. Although various authors have strongly advocated this, no such theory has yet been constructed, let alone fitted to the present slender data for pore osmotic permeabilities.

These two approaches are contrasted with a different one, the bimodal theory, which involves either a viscous or a diffusive flow of water depending upon the radius of the driving osmolyte in relation to that of the pore. This theory can be applied to pores of irregular or tapering cross section in which both modes of transport can even occur in series within the same pore, depending on the osmolyte radius; this represents the most general case and often the most realistic situation for a cell membrane pore.

II. Unit Osmotic Conductances

A. Antibiotic Pores

Of the pores formed in membranes by the two polyene antibiotics nystatin and amphotericin, and by the short polypeptide gramicidin, we have osmotic data that are unique (Holz and Finkelstein, 1970; Rosenberg and Finkelstein, 1978; Dani and Levitt, 1981). Most data on osmosis have been obtained from fiber or neutron-etched membranes in which the pores are polydisperse and of irregular shape. After insertion into an artificial bilayer membrane it is possible to measure unit electrical conductances of these antibiotic pores (Siemens per pore); measurement of the total conductance of an antibiotic-doped bilayer then gives the number of pores in the membrane, and subsequent measurement of the osmotic permeability of the same bilayer yields the unit osmotic conductance P_{os} (cm³/s). There are as yet no comparable data for any abiotic structure of known dimensions. The values are shown in Table I. It should be noted that for nystatin and amphotericin some of the values are calculated from reflexion coefficients given in the papers, which are *relative* values, i.e., relative to another osmolyte (here sucrose) which is assumed to give the maximum osmotic flow; this is an important point which is dealt with below in the discussion on bimodal theory (Section V).

The geometry of the antibiotic pores has been determined independently of the osmotic studies by X-ray diffraction and NMR in the case of gramicidin (Killian, 1992) and by kinetic analysis backed by model building for the polyene antibiotic pores (Finkelstein and Holz, 1973) which are assem-

TABLE I

Unit Osmotic Conductances of Some Antibiotics and the Aquaporin CHIP28

Pore	Osmolyte	$P_{os}10^{-14}$ cm³/s	Refs.
Gramicidin	Urea	0.95	Rosenberg and Finkelstein (1978)
		6–7	Wang *et al.* (1995)
	Mannitol	6.0	Dani and Levitt (1981)
Amphotericin[a]	NaCl	4.5	Holz and Finkelstein (1970), Finkelstein (1987)
	Urea	2.6	Holz and Finkelstein (1970), Finkelstein (1987)
	Glucose	4.5	Holz and Finkelstein (1970), Finkelstein (1987)
Nystatin[a]	NaCl	1.5	Holz and Finkelstein (1970), Finkelstein (1987)
	Ethandiol	1.01	Holz and Finkelstein (1970), Finkelstein (1987)
	Urea	0.83	Holz and Finkelstein (1970), Finkelstein (1987)
	Gycerol	1.17	Holz and Finkelstein (1970), Finkelstein (1987)
	Glucose	1.5	Holz and Finkelstein (1970), Finkelstein (1987)
	Sucrose	1.5	Holz and Finkelstein (1970), Finkelstein (1987)
CHIP28	NaCl	1	Preston *et al.* (1992)
	Mannitol	6	van Hoek and Verkman (1992), van Hoek *et al.* (1993)
	Sucrose	11.7	Zeidel *et al.* (1992)

[a] The P_{os} values for this pore are obtained by multiplying the values for the reflexion coefficient σ given in Holz and Finkelstein (1970) by the P_{os} values for glucose and sucrose ($\sigma = 1$) given in Finkelstein (1987).

blies in the lipid environment of several molecules. Fortunately for the simplest theoretical models, they all form right cylindrical pores.

B. Cellular Aquapores

Water permeability can be attributed to protein components of the membrane (Verkman, 1992). Three criteria are recognized as constituting evidence for aqueous pores: (i) measured P_{os} values of the membrane are found to be higher than expected from diffusion through the lipid phase of the membrane; it should be noted, however, that the upper limit of water permeability in lipid membranes can be as high as is found for most cell membranes (2×10^{-2} cm/s) (Fettiplace and Haydon, 1980) and it is a difficult task to apply this test with any clear result. (ii) The inhibition of water transport by organomercurials such as pCMBS which covalently modify protein structure containing labile sulfhydryl groups implies that water transfer is under protein control. (iii) Lower activation energies for water transfer through the membrane are measured, of the order of those found for diffusion or viscous flow in bulk water, and these activation energies usually rise to values expected for lipid membrane transfer when agents such as organomercurials are applied.

CHIP28, a member of the protein family called "aquaporins," is found in the membranes of many tissues (Preston and Agre, 1991; Agre *et al.*, 1993) and has been extracted from red cells where it appears to be responsible for most if not all of the osmotic conductance of the membranes. In addition there have been direct measurements showing other membrane proteins such as glucose transporters (Fischbarg *et al.*, 1990) and the cystic fibrosis transporter, CFTR (Hasegawa *et al.*, 1992), acting as pores. The general properties of transmembrane proteins in this regard are considered below (Section VII.B).

The unit osmotic conductance of the naturally occurring membrane protein CHIP28 has been measured with various osmolytes and the values are given in Table I, with the antibiotics. The P_{os} of the protein pore monomer has been determined in two ways: (i) by expression in *Xenopus* oocytes with subsequent measurement of the incremental osmotic swelling of the egg membrane; estimates of the number of copies expressed then enable the unit P_{os} to be calculated (Preston *et al.*, 1992). (ii) The protein is extracted and incorporated into a bilayer system, whose osmotic permeability is measured; estimates of the number of copies by an independent assay again lead to an estimate of the unitary P_{os}. Now that the protein has been extracted, fractionated, and crystallized it is possible to add virtually pure CHIP28 to the bilayer and calculate the number of molecules per area of membrane with some accuracy (Zeidel *et al.*, 1992).

The data are tantalizingly sparse, as is also the case for the antibiotics. In Table I three values are shown, determined with the osmolytes NaCl, mannitol, and sucrose. As yet no attention has been given to the possibility that the reflexion coefficient σ may be varying. Since

$$P_{os} = \sigma P_f, \tag{1}$$

where P_f is the hydraulic permeability (i.e., viscous flow permeability) of the pore as would be measured with a totally excluded osmolyte or equivalent pressure difference, we can see that these measurements assume that $P_{os} = P_f$.

III. Hydrodynamic Theory

A. Semipermeable Pores

The original theory of membrane pores regarded osmosis as a viscous flow phenomenon. In a cylindrical pore viscous flow is proportional to the fourth power of the radius, while diffusive flow is proportional to the square of the radius. This distinction underlies the method for estimating the pore radius in red cell membranes exploited by Solomon and his co-workers over many years (Solomon, 1968). In essence, if measurement of P_f is made by osmosis (assuming that osmosis is a wholly viscous phenomenon) and the diffusional water permeability P_d* by tracer water diffusion, their ratio is given by

$$\frac{P_f}{P_d^*} = \frac{RT}{8\eta_w D_w \overline{V}_w} r^2, \tag{2}$$

from which the pore radius r can be calculated. D_w is the diffusion coefficient, η_w the viscosity, and $\overline{V}_w$ the molar volume, of water. Equation (2) is derived from the equations of Poiseuille and Fick and the perfect gas law. Although, as it will be argued below, this formula is unrealistic (or rather, incomplete) due to a neglect of drag coefficients, the essence is that P_f/P_d^* cannot be equal to 1.0 in any pore larger than a water molecule.

To avoid confusion between the permeability coefficient for water diffusion (as might be measured osmotically if transfer through a pore were only diffusive) and the value determined with tracer water (which is subject to a filing effect) the former is designated P_d and the latter P_d* throughout this review. No distinction between the two is normally made in the literature but here it is essential because, as will become clear, P_f/P_d and P_f/P_d^* are not the same in many circumstances.

In red cell membranes the most reliable tracer values for P_d^* are those obtained from NMR measurements, i.e., spin labeling using extracellular spin relaxation agents (manganese) in which external unstirred layers outside the cell membrane are virtually absent (Fabry and Eisenstadt, 1975). This is because the spin label is quickly destroyed outside the cell, although the relaxation agent is not used up in this process. Recently obtained P_d^* values by either tracer diffusion or NMR are now very similar and the P_f/P_d^* ratio has settled at a value of about 10. It can easily be calculated from Eq. (2) that the corresponding pore radius is about 1 nm. Unfortunately, it is very difficult to reconcile this value with what we know of the permeability of the cell membrane for it would be massively permeable to ions and most small solutes (Galey and Brahm, 1985).

A possible explanation of this ratio (one which calls the whole interpretation given by Eq. (2) into question) is that the water permeability as measured with labeled water is reduced due to molecular queuing in pores where molecules cannot pass each other and are confined to single files ("no-pass" pores). Thus the P_f/P_d^* ratio is an underestimate due to the fact that the tracer and unlabeled water fluxes are not independent. In other words, the pore, or a substantial part of it, is a no-pass structure. This is dealt with below (Section VI.A.1), but here we may note that the simple hydrodynamic approach faces a problem.

The mechanism proposed for viscous flow driven by osmosis involves a "translation" of the osmotic driving force into a hydrostatic pressure difference of a thermodynamically equivalent magnitude (Vegard, 1908; Mauro, 1957). It is brought about by the fact that osmolyte molecules are excluded from the pore, but water at the pore extremities must be virtually in equilibrium with the baths, i.e.,

$$\overline{V}_w (\Delta P - \Delta \Pi) = 0, \tag{3}$$

where ΔP and $\Delta \Pi$ are the hydrostatic and osmotic pressure differences between pore and bath. The pore core therefore experiences a pressure gradient $(P_1 - P_2)$ between baths 1 and 2 equal to $(\Pi_1 - \Pi_2)$, the osmotic difference between the baths. This mechanism has been extensively discussed and need not be further presented here. It explains osmotic water flow across a semipermeable pore, which occurs at the same rate as viscous flow, by invoking a pressure gradient within the pore. If the pore is only water-filled there is nothing except a pressure gradient that could drive the flow. Thermodynamically this is quite sound, built on the fundamental principle that when one component of a mixture (here the solute) is excluded from a compartment, a compensatory pressure difference is set up to preserve equality of chemical potential of the other (the solvent)

(Guggenheim, 1957). The complexity and interest lie in the situation where the osmolyte is not sterically excluded by being larger than the pore.

B. Leaky Pores

The hydrodynamic theory which underlies the theory of osmotic flow in pores permeable to the osmolyte is unclear. The pressure gradient must be present, as in the semipermeable case, but somehow it has to be attenuated (by an increasing osmolyte permeability of the pore) until the driving force is very small or even zero. Any such theory must result in an equation for the reflexion coefficient as defined by Eq. (1). This coefficient can be seen conveniently to summarize the theory and provide a quantitative test for how well it fits the experimental results. Several treatments have been applied, none of them really compatible with each other.

1. Ultrafiltration Theory

The expression for the reflexion coefficient most commonly met with is

$$\sigma = (1 - A_s/A_w), \tag{4}$$

which gives σ in terms of the "available filtration areas" for solute and water, A_s and A_w. These areas are equal to the pore cross-sectional area multiplied by a frictional factor, i.e., $A_s = \pi r^2 \times F_s$. They are derived not from any model of osmosis, but rather from a consideration of solute and water flow through a pore driven by pressure, i.e., from ultrafiltration (Durbin, 1960). The reflexion coefficient of Eq. (4) is therefore for ultrafiltration (σ_f) and it can be shown that it is only equal to the reflexion coefficient for osmosis (σ_s) under certain conditions, those being that water flow during osmosis is wholly viscous, as it is for ultrafiltration. Reciprocity is thus assumed between the reflexion coefficient for ultrafiltration and osmosis, i.e., that $\sigma_f = \sigma_s$. The areas A_s and A_w have traditionally been derived by applying drag theory to molecules in the pore using a diffusive drag coefficient F (Renkin, 1954) where a convective drag coefficient G should be used; these coefficients (F and G) must play an important role in microscopic pores and are dealt with below (Section V.A), where they are shown graphically in Fig. 6.b. Moreover, diffusion of solute, which occurs in leaky pore osmosis, is not considered in any way; because it plays no part in a simplified treatment of ultrafiltration, Eq. (4) is independent of the intrinsic diffusion coefficients of solute and solvent, D_s and D_w. It is difficult, therefore, to ascertain what physical model of osmosis underlies this treatment.

2. Surface Forces

In a more specific model of leaky pore osmosis the interaction of osmolyte with the pore wall has been considered. A simple version of the theory (Manning, 1968) makes use of a one-dimensional simplification, i.e., it uses averaged values for the concentration of osmolyte in the pore interior rather than a detailed three-dimensional treatment of the forces at the pore wall which vary with radius. The forces at the wall change the osmolyte concentration in the pore from a bulk solution value, c_{bulk}, to an average internal value, c_{pore}. This averaging leads to the expression

$$\sigma = 1 - \frac{c_{pore}}{c_{bulk}} = 1 - K_s \tag{5}$$

for the reflexion coefficient, where K_s is the partition ratio between pore and bath.

In a more detailed treatment (Anderson and Malone, 1974) the forces which may be present at the internal surface of the pore and which cause changes of osmolyte concentration at that surface are considered. These create differences in hydrostatic pressure between the pore fluid and the wall according to Eq. (3) (the water is considered to be in local equilibrium); this pressure is proportional to the concentration along the pore. There is thus a pressure gradient along the pore which impels the fluid along the pore axis from bath to bath.

This theory suffers from the serious defect that all forces emanating from a surface cause accumulations of solutes: electric fields cause net accumulation of salt, independently of the sign of the charge, while van der Waals forces are always attractive. As a consequence of this, the pressure gradient occurs in the same direction as the concentration gradient, and must therefore drive fluid from a higher to a lower concentration. This effect is hardly ever observed and is considered to be anomalous—osmotic flow is always from the lower to the higher concentration. To rectify this situation there must be a repulsive wall field to offset the attractive fields which would always be present. The theory thus postulates that the mass centers are repelled by being held a solute radius away from the wall, i.e., there is effectively a square-well potential of width r_s which constitutes a repulsive "steric effect" at the surface. However, pressure on the wall is caused by exchange of momentum mediated only by contacts between molecules and the surface, not by the behavior of mass centers. If this were not so, the molecules of a gas confined to a vessel would exert no pressure on the walls. It is impossible not to conclude that the infinite square-well potential applied to the mass centers is irrelevant.

An inconsistency in this theory is that although the solute is considered particulate, the solvent is regarded as a continuum. In reality, of course, the same "wall repulsion" effect should apply to the solvent too. More interesting is the assumption that pore wall fields, even if they were repulsive, would create hydrostatic gradients within the pore. It is questionable whether this is true, and there are physical arguments against there being any hydrostatic pressure differences between the bath and the pore when the latter can freely enter (Hill, 1989a), but a detailed consideration of this point forms the basis of the bimodal theory (Section V) and will be considered there.

3. Summary

The hydrodynamic theory proposes that fluid is driven through semipermeable pores during osmosis by hydrostatic forces. The physics of this is generally accepted; indeed, it is hard to see what other forces could be acting on water in a purely water-filled core. It follows from the equation of state for a pure liquid that a gradient of water concentration along the pore is equivalent to a pressure gradient

$$\frac{dP}{dx} = \frac{dc_w}{dx}RT, \tag{6}$$

which is a situation that can only lead to poiseuillean flow.

When the pore is leaky to osmolyte the only mechanism that has been proposed is that of modified pressure gradients created by pore wall interactions with osmolyte. Apart from the dubious physics, it will be seen below that this cannot be correct because osmosis induced by permeant osmolytes in small pores occurs at rates even lower than those expected for the diffusive transfer of water.

IV. Single-File Theory

This theory argues that flow in pores where molecules cannot pass each other must be radically different from flow in wide tubes, and therefore that simple extrapolation of flow concepts such as poiseuillean and fickian flow to molecular-sized pores is unrealistic if not impossible. New concepts are introduced and simplifications are made. Some treatments (e.g., Finkelstein, 1987) make a clear distinction between macroscopic and single-file pore theory, and it is quite clear that the concepts employed are in fact different. To what extent these changes in treatment are really relevant to the no-pass condition of single-file pores or merely additions which may be unnecessary is an important point which must be settled.

The phenomenon of tracer water filing, the dependence of the flux of labeled water molecules through the pore on the number of molecules in the pore due to the fact that molecules cannot pass each other, is not dealt with here because it does not affect the basic physics of osmosis, i.e., osmotic volume transfer. However, it is essential to distinguish between the measurement of the water diffusion coefficient by tracer in the absence of an applied osmotic gradient (P_d^*) and the osmotic flow coefficient P_{os} which may occur by diffusion under certain conditions (in which case $P_{os} = P_d$). This is because P_d plays an important role in osmotic theory independently of the tracer permeability. The two are related by the expression

$$P_d = nP_d^*, \tag{7}$$

where n may be conveniently called the filing number. The original demonstration of potassium filing (Hodgkin and Keynes, 1955) showed both theoretically and experimentally that tracer fluxes can be smaller by a factor $1/n$ than the true unidirectional fluxes due to queuing phenomena. The theory was extended to water molecules initially by Lea (1963) and has since received quite a detailed treatment (Kohler and Heckmann, 1979a,b; Hernandez and Fischbarg, 1992). The debate about the value of n as being equal to N, $N + 1$, or $N - 1$, where N is the number of water molecules in the pore, is due to the precise model of the process in respect of vacancies created when water molecules enter or leave the pore (see review in Finkelstein, 1987).

A. Semipermeable Pores

In the semipermeable pore the osmotic permeability P_{os} must be equivalent to the viscous permeability P_f. An influential treatment of very narrow pores in which the osmotic transport coefficient is determined from statistical mechanics is a convenient starting point (Longuet-Higgins and Austin, 1966). The authors conclude that osmotic flow in single-file channels which exclude solute is best described as diffusional in nature, and obtain a final expression for the flux J_1 of water molecules

$$J_1 = ND_1 \Delta n_2 / L^2, \tag{8}$$

where N is the number of solvent molecules in the pore, Δn_2 is the difference in mole fraction of solute across the pore, L is the pore length, and D_1 is the "self-diffusion" coefficient of solvent (op. cit. Eq. (4.9)). This may be readily converted to the fickian expression

$$J = \pi r^2 D_w \Delta c_w / L \tag{9}$$

for a water-filled pore.

However, the argument is not so clear cut. If the pore excludes solute, the water in the core must be under reduced pressure, and there will be a gradient of pressure from one bath toward that of higher osmotic pressure. This is the basis of the Vegard-Mauro mechanism that is generally accepted for macroscopic pores, and there is no reason why it should be absent in the case of single-file pores. The authors wrongly assume that because there is no external pressure gradient, there cannot be an internal one. A generalized transport coefficient is developed as a correlation integral and finally assigned to a "diffusion coefficient." Thus it is not surprising that the final expression turns out to be a steady-state fickian one. Of course, there is a sense in which pressure-driven flow in one dimension can be loosely regarded as diffusional: in a pure solvent the equation of state demands that a pressure gradient is commensurate with a concentration gradient or increase in vacancies. In Fig.1 this situation is shown together with another situation in which a concentration gradient is created by a gradient of permeant solute. Both involve concentration gradients of solvent, but it is only the gradient of momentum exchange in the first which gives rise to viscous flow.

It is widely supposed that in a single-file or no-pass pore viscous flow is impossible because in such a flow there is a velocity distribution which is parabolic, i.e., streamlines parallel to the axis are moving at different speeds. In poiseuillean flow the hydraulic permeability of a cylindrical pore of radius r and length L is given by

$$P_f = \pi r^4 kT/8\eta_w L\bar{v}, \tag{10}$$

and the velocity v_x at a distance x from the axis is given by

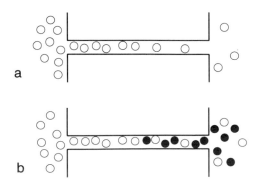

FIG. 1 (a) Gradient of vacancies caused by a pressure gradient in a solvent-filled pore leading to viscous flow. (b) Gradient of solvent concentration created by a gradient of osmolyte leading to a diffusive flow of solvent.

$$v_x = v_0 \left(1 - x^2/r^2\right) \tag{11}$$

where v_0 is the axial (maximum) velocity. This is shown to the right in Fig.2. Longuet-Higgins and Austin express a commonly held view when they state, "If the tube is so small that molecules cannot overtake one another inside it, then one can scarcely speak of the variation of velocity across the tube. So, for pores which have radii nearly as small as the molecular radius of the solvent, equation [10] loses all physical justification."

This statement would seem to be correct, but in fact it shows a misunderstanding of the nature of the velocity profile. It is not individual molecules which follow stream lines with the velocities given by Eq. (11). If this were true, in pores of any size two molecules on stream lines less than a molecular diameter apart would never be able to flow past each other, and this would apply to all stream lines. What occurs is that the time-average of the vector components of all molecules moving in different directions across a line parallel to the axis at a radial distance x behaves as v_x in Eq. (11) above. A diagram of this averaging is shown in Fig.2. It must be clear that this process can occur in a pore of any size and does not suddenly change at the transition to a no-pass pore. The only restrictions on Eq. (10) are due to the fact that Poiseuille's equation must incorporate a drag term so that at a radius equal to that of the water molecule $P_f = 0$ (discussed below). There are therefore no physical reasons why viscous flow cannot occur in molecular-sized pores that do not allow water molecules physically to pass one another.

It has been claimed that water passage during osmosis in the semipermeable membrane is diffusive by a different argument. This applies Newton's third law in the steady state to the force acting upon a molecule in a pore to produce a velocity v

$$F = fv, \tag{12}$$

where f is the molecular frictional coefficient for water in the pore. The

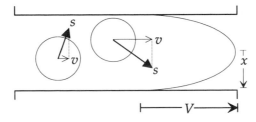

FIG. 2 The distribution of velocities in a "no-pass" pore. The familiar velocity profile is shown to the right. In the pore, molecules have different instantaneous velocity vectors s with components v parallel to the axis. It is the time-average of v at any radial position that constitutes the velocity profile, but molecules do not have to pass each other on "stream lines."

treatment here is to be found in somewhat greater detail elsewhere (Levitt, 1973; Finkelstein and Rosenberg, 1979; Finkelstein, 1987; Hernandez and Fischbarg, 1992); the situation is shown in Fig.3.

The volume flux J_v, where N is the number of molecules in the pore of length L and $\bar{v}$ is the molecular volume of water in the pore, is

$$J_v = vN\bar{v}/L = -(P_f\bar{v}/kT)\Delta\Pi. \tag{13}$$

In the semipermeable pore the osmotic difference is physically translated into a pressure gradient by the Vegard-Mauro mechanism, so that the work of transferring a water molecule becomes

$$\Delta P\bar{v} = FL. \tag{14}$$

Remembering that the molecular volume of water $\bar{v}$ is equivalent to, $\pi r^2L/N$, we obtain from Eqs. (12)–(14)

$$P_f = \frac{\bar{v}NkT}{fL^2} = \frac{\pi r^2kT}{fL}. \tag{15}$$

This expression appears to be very general and is actually similar to that of Longuet-Higgins and Austin (1966). As such it gives little indication of the nature of the osmotic flow process. However, it has been interpreted in the following way: using the Einstein relation

$$D_w = \frac{kT}{f}, \tag{16}$$

where D_w is the diffusion coefficient of water within the confines of the pore at a particular solvent density, Eq. (15) becomes

$$P_f = \frac{\pi r^2 D_w}{L}, \tag{17}$$

which is a fickian diffusion equation (Finkelstein, 1987). We have, therefore, the paradoxical situation in which water is driven along the pore by a pressure gradient but the flow seems to be by diffusion. This would of

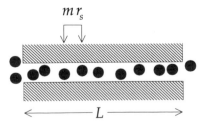

FIG. 3   Molecules in a "no-pass" pore of length L. The average spacing is equal to mr_s.

course fit the preconception that the pore is so narrow that diffusion is the only means of transfer, but we have shown that this need not be so.

Equation (15) can be interpreted quite differently, as I shall now show. The basis of the Einstein equation for diffusion in solution is the stokesian hydraulic drag force

$$f_s = A\pi\eta r_s \tag{18}$$

required to push a spherical particle of radius r_s through a medium of viscosity η. The numerical value of the constant A is 6 for a free solution but larger when the sphere is confined between walls, due to viscous drag forces (Wang and Skalak, 1969). Consider a single water molecule in the pore. If the radius of the pore is comparable to that of the sphere ($r_s \to r$), then the drag force f_s is equivalent to the molecular frictional force f. Furthermore, the molecular volume in the pore is given by

$$\bar{v} = \pi r^2 \cdot mr_s, \tag{19}$$

where m is the mean free path in the pore in units of solute radius ($m = 2$ for close packing). Equation (18) can therefore be written

$$f = \frac{A\eta\bar{v}}{mr^2}, \tag{20}$$

which when substituted into Eq. (15) yields

$$P_f = \frac{\pi r^4}{B\eta L} \cdot \frac{kT}{\bar{v}}, \tag{21}$$

which is a poiseuillean expression for the permeability in terms of viscous flow, with the constant B proportional to the drag coefficient on the molecule ($B = A/m$). This can now be seen to be a more self-consistent situation: a pressure gradient operating together with a viscous flow coefficient.

It should by now be apparent that there are no grounds for assuming that the transfer of water in small pores is diffusive rather than viscous in semipermeable single-file pores of the sort encountered in cell membranes. The expressions derived above are based on a simplified model in which the magnitude of the permeability in Eq. (15) is governed by the frictional parameter f. If f is given a diffusive interpretation by setting it equal to kT/D_w then the fickian expression Eq. (17) results. If, however, f is interpreted purely hydrodynamically and set equal to $A\pi\eta r_s$ then the result is the poiseuillean Eq. (21). Although Eq. (18), the hydrodynamic friction on a sphere, underlies both, the difference between the two approaches is that they are determined by the nature of the force driving the flow. The choice of driving force under any particular set of conditions stems from the use of a suitable physical model which precedes the mathematical description, not vice versa. In either case, the diffusion coefficient D_w (which may play a role in leaky pores) or the viscous parameter B is subject to drag corrections, which in the restricted case of small pores as shown in Fig.3 must be very substantial indeed. The conversion of such a description into an accurate physical model requires a calculation of these drag corrections.

The magnitude of the coefficient f is very different for diffusive and viscous flow. The situation is not unlike that arising in the irreversible thermodynamic approach to membrane transport (Kedem and Katchalsky, 1961; Dainty and Ginzburg, 1963; Ginzburg and Katchalsky, 1963). There the generality of the approach actually masked the fact that the same frictional coefficient f_{wm} between water and membrane in porous systems was used for situations in which either diffusive flow or viscous flow was being considered. It is easy to see that this is incorrect, for example, in the case of macroscopic pores because the coefficients of tracer water diffusion and viscous water flow are so different. The physical mechanism lying behind this is simply that water transfer is due to friction with both the pore structure and other water molecules, creating internal liquid friction. In viscous flow this latter component is much reduced due to the cooperative movement of water molecules.

For no-pass pores the single-file theory maintains that there is no longer any meaningful distinction between diffusive and viscous flow, and that there is only one mechanism of water flux—a "quasi-diffusive" one. The ratio P_f/P_d* should tend to 1.0 at small radii and any departures from this are envisaged as being due to tracer filing effects, which tend to N. However, when the pore is small enough to become a no-pass channel, but not small enough for water molecules to have lost all their radial freedom of movement, this distinction can still apply. Surprisingly, as will become apparent in the discussion of bimodal theory, the distinction gets greater when the pore becomes very small.

The problem of the drag effects, and how they determine certain aspects of the osmotic flow rate (and the P_f/P_d* ratio), is dealt with in Section V. If they can be assessed to the best of our current ability, then Eq. (21) can be regarded as the basis of osmosis, by viscous flow, in semipermeable pores of any radius, and it will become apparent that such a "corrected" equation gives quite a good account of certain experimental values.

B. Leaky Pores

In leaky pores where osmolyte molecules can enter the pore, single-file theory regards the mechanism of flow to be essentially the same as that in semipermeable pores. When a pore does not contain an osmolyte molecule, the driving force is considered to be ΔP ($\Delta\Pi$ and ΔP are regarded as physically interchangeable), i.e., the mechanism of flow in an osmolyte-free pore is similar to that of a semipermeable pore. When an osmolyte does enter the pore a mechanism is envisaged in which a solute molecule drives out the content of the pore as a "plug" containing all the water molecules before it (Levitt, 1974). Thus for every osmolyte molecule traversing the pore there is a volume of water accompanying it in the same direction equal to $\overline{V}_s^p$ which is the volume of the pore containing the osmolyte (when translated into solution). For this mechanism, in pores where osmolyte and solvent cannot pass, the reflexion coefficient is given as a result of the hydraulic plug-flow mechanism by

$$\sigma = 1 - \frac{P_s \overline{V}_s^p}{P_f V_w}, \tag{22}$$

where P_s is the solute permeability coefficient of the pore. A similar equation has been derived for a pore in which osmolyte and solvent cannot pass, and which in addition is no-pass for solvent,

$$\eta = 1 - \frac{P_s \overline{V}_s^p}{P_f V_s^p}. \tag{23}$$

$\overline{V}_s^p$ here is the molar volume of water in the pore. The derivations are quite straightforward, but the original paper should be consulted for details (Levitt, 1974).

It is debatable that solvent in pores that (momentarily) contain no osmolyte is flowing at the same rate as if it were impelled by a hydrostatic pressure gradient—again the conceptual model is that diffusive and viscous flows are identical in no-pass pores. For reasons outlined above for semipermeable pores this cannot be proven. Quite apart from this there is a difficulty with the concept of coupled osmolyte and water flow. It is doubtful whether the pore contents can ever be physically defined as such a plug that can be "displaced," and it is possible to envisage a situation in which an osmolyte molecule traverses a pore containing a fluctuating population of water molecules without actually being accompanied by any of them. Nor is there any reason why the osmolyte should be in such a privileged position. On grounds of symmetry it should also be postulated that a water molecule traversing the pore flushes out with it a similar volume as does the solute, but this would probably not lead to Eqs. (22) or (23).

Enough has been presented here to outline the nature of this theory of osmosis in single-file leaky pores and to discuss the assumptions. The separate resulting expressions for osmosis in different pore types (depending on what can pass what) and the overall approach make no distinction at all between osmotic and pressure forces or between diffusive and hydraulic flow mechanisms with their different permeability coefficients. By contrast, however, instead of regarding a leaky pore as a time-averaged entity in which there are gradients of solute and solvent molecules in opposite directions, this approach depends uniquely upon treating pores containing water only, and those containing water with a single osmolyte molecule, as distinct entities which can be regarded mechanically as piston-driven cylinders. The formulae for osmotic flow that are given above are compared with experimental data in Section VI.

V. Bimodal Theory

In this theory the term bimodal refers to the fact that there are two modes of water transfer, each of which may be operating under different conditions within a pore section (Hill, 1982). The approach is based upon an analysis of the forces operative inside a pore when an osmotic gradient is applied across the membrane, and in this respect it is no different from some other theories. It also makes a distinction between osmotic flow by a viscous mechanism and osmotic flow by a diffusive mechanism. The fact that volume can be transferred by diffusion has to be distinguished from the measurement of the diffusive permeability of a pore by tracer ($P_d{}^*$); it implies the transfer of volume by diffusion as opposed to viscous (bulk) flow, when groups of molecules begin to translate with a common drift velocity. Osmosis was regarded traditionally as a wholly diffusive phenomenon up until the 1950s—Vegard's paper of 1908 was little known. It was taught as being due to the unequal water activities on two sides of a membrane, and in some circles it was even called the "diffusion pressure deficit" or DPD. It is probable that most physicists and chemists still subscribe to this view.

The mechanism of transfer, whether diffusive or viscous, is dependent on the nature of the driving force operative within the pore: there is nothing inherent in the pore structure or the molecular constitution of liquids which dictates the transfer coefficient. The two coefficients have different magnitudes, which depend ultimately upon the pore size, but there is no radius at which single-file or no-pass conditions begin to impose special characteristics on the osmotic flow (as opposed to the tracer fluxes). The same formulae for viscous or diffusive transfer apply to any pore size and hold right down to the point where the water molecules jam in the pore at about 0.15 nm radius.

What then determines whether the driving force in the pore is diffusive or hydrostatic? The theory postulates that if the osmolyte can enter a pore section then there is only a gradient of solvent (and osmolyte) concentration and no hydrostatic pressure is set up. The concentration gradient sets up a diffusive flow of water and osmolyte in opposing directions. If, however, the osmolyte is too large to enter the pore section (the semipermeable case), then the osmotic pressure difference between the baths and the pore section creates a gradient of pressure across the section which drives viscous flow. In any pore section osmotic flow driven by an osmolyte is either diffusive or viscous, and the important determining factor of the osmotic flow rate is whether or not the osmolyte is sterically excluded.

A. Semipermeable Pores

Clearly, when the osmolyte is excluded the reflexion coefficients for such species are then equal to 1.0 and the maximum osmotic flow is seen. This much is common to all theories; although $\sigma = 1$ the question of the magnitude of the osmotic permeability coefficient depends upon the mechanism of flow. Hydrodynamic theory applied to macroscopic pores envisages that the Vegard-Mauro mechanism operates and pressure-driven flow occurs. However, we have seen that single-file pore theory considers the flow to occur by a mechanism akin to diffusion. The bimodal theory makes no distinction between small and large pores where the mechanism of water transfer is concerned and therefore considers all flow driven by impermeable osmolytes to be viscous in nature. This leads to simple predictions, namely, that the osmotic permeability can only be appropriately calculated using a poiseuillean expression dependent upon the viscosity of water and not by a fickian formula involving the diffusion coefficient.

It is an important element of such a theory that compares viscous and diffusive flow that they can be computed to a reasonable degree of accuracy. This requires the use of drag coefficients. These coefficients apply to both viscous and diffusive flow and take the form of factors which range from 1 (when the pore is considerably larger than the molecular radius) to 0 (when the two are the same). Viscous flow demands a consideration of drag coefficients because obviously poiseuillean formulae cannot hold down to zero radius, but only to the radius of water; diffusive flow involves a drag coefficient for the same reasons but the two take a different form. It is convenient to deal with them here because they cannot be ignored, as has been the case, and must play an important role in osmotic theory. It is very difficult to understand this neglect because such effects on solutes causing flow were made the basis of the well-known expression for the reflexion coefficient in terms of "available areas" of solute and water (Eq. (4)). The viscous drag factor, G, represents the decrease in velocity of a solute driven through a pore by moving solvent due to wall interactions, while the diffusive factor F represents the decrease in mobility of a solute in a pore when the solvent in the pore is stationary. These have both been calculated for spheres in tubes filled with fluid using macroscopic continuum theory (Haberman and Sayre, 1958; Wang and Skalak, 1969), though better solutions could almost certainly be derived and may occur in the future. Development of both the diffusive (F) and convective (G) factors for circular channels in terms of λ_s (the ratio of molecule to pore radius, r_s/r) has been used here, where

$$F_w = 1 - 2.1019\lambda_w + 2.0829\lambda_w^3 - 1.6764\lambda_w^5 + 0.6772\lambda_w^6$$

and

$$G_w = 1 - 0.8341\lambda_w^2 + 0.8977\lambda_w^3 - 1.0586\lambda_w^4$$

are polynomials fitted to the numerical solutions (Faxen, 1922; Wang and Skalak, 1969) for the water molecule. These solutions are made for spherical molecules traversing the axis of the pore and the friction is lowest at that position. No other solutions are available at the moment but solutions for a "radially distributed" coefficient could be made. Brownian motion knocks the sphere (i.e., the solute molecule in question) about over all radial positions very quickly: for flows close to the Reynolds number the molecular velocity is about 10^5 times the flow speed. Both F and G are usually multiplied by a steric factor S. This has been called the "entrance" factor (Renkin and Curry, 1979; Renkin, 1954; Ferry, 1936). Modern ideas do not support the concept of restricted entrance and it may rather be interpreted as a partition coefficient between pore and bath, so bringing it into line with the theory of diffusion through any generalized membrane. In this paper S is given by the expression $(1 - \lambda_s)^2/(1 - \lambda_w)^2$ representing the mole fraction ratio between pore and bath related to the reduced area coefficient $(1 - \lambda)^2$ for a cylindrical pore. This expression fits experimental data well (Beck and Schultz, 1972). When the species under consideration is water itself, the expression is equal to 1.0, because the mole fraction of water is unchanged by the pore in dilute solutions. In reality, if proper "radially averaged" drag coefficients were to be derived, the steric factor would merely appear as the limits of a radial integral where J_s is either the diffusive flow given by

$$J_s = dc/dx \int_0^{(r_p - r_s)} D_0 F_r 2\pi r \, dr$$

or the solvent drag flux caused by a fluid moving with velocity v_r through the pore between baths of mean concentration c_0

$$J_s = c_0 \int_0^{(r_p - r_s)} G_r v_r 2\pi r \, dr.$$

The situation is rather unsatisfactory from a theoretical point of view, although we have reasonable approximations.

Setting the osmotic coefficient P_{os} equal to the filtration coefficient (Eq. (10)) which now incorporates the viscous coefficient G_w for water,

$$P_{os} = P_f = \frac{\pi r^4 kT}{8\eta_w L v_w} G_w, \qquad (24)$$

we obtain a viscous interpretation of osmotic permeability. This expression can be compared with Eq. (9).

The ratio P_f/P_d is also dependent upon the drag coefficients. Incorporating these factors we obtain (Hill, 1994)

$$\frac{P_f}{P_d} = \frac{P_f}{nP_d^*} = \frac{RTG_w}{8\eta_w D_w^o V_w F_w} r^2. \qquad (25)$$

The shape of this curve is rather unexpected due to the effects of the drag and can be compared with that of the hydrodynamic theory (Eq. (2)) in Section VI.

B. Leaky Pores

The mechanism by which different driving forces are created by osmolyte exclusion is fundamental to the theory. The concept that when osmolyte enters a pore no pressure drop is created relies on an analysis of the interaction between a solute molecule and the pore interior, which is in fact an interaction with the wall forces. This has already been discussed in connection with leaky pores in Section III.B.2. The problem therefore presents itself in the simplest form as an analysis of the effect of a polar surface on a solution in contact with it. It is convenient to present the arguments under three heads: a basic thermodynamic one, a molecular kinetic one, and one based on experimental observation.

1. Basic Physical Mechanism

a. Thermodynamics   The Gibbs-Duhem equation is often used here, but is expressed in different forms. Applied to a polar surface in equilibrium with a solution (containing one solute, the osmolyte, for simplicity) it may be written

$$n_s \delta\mu_s + n_w \delta\mu_w + V\delta P = 0, \qquad (26)$$

where n_s and n_w are the moles of solute and solvent in the system of total volume V and pressure P, and μ is the chemical potential. This equation tells us, in physical terms, that a slight perturbation in (say) one of the μ will cause compensatory perturbations in any or all of the other intensives,

μ or P. This is a thermodynamic equivalent of the principle of virtual work. In its differential form Eq. (26) becomes

$$c_s\frac{d\mu_s}{dx} + c_w\frac{d\mu_w}{dx} + \frac{dP}{dx} = 0. \tag{27}$$

The chemical potentials are, however, constant with approach to the surface because we are in equilibrium, and so there is no change in pressure on approaching the surface, i.e., dP/dx is zero. The upshot of this is that the pore wall may cause a decrease in concentration of osmolyte in the pore but there is no change in pressure. The osmotic pressure difference is not translated into a difference of hydrostatic pressure either within the pore or between the pore space and the baths.

It is not infrequent to find that pressure gradients in pores due to wall forces are theoretically underpinned by an incorrect form of the Gibbs-Duhem equation. This is usually presented in the form $n_s\delta\mu_s + n_w\delta\mu_w + V\delta P + n_s\delta\Phi = 0$, where Φ is a surface potential in J/mol acting on the solute. Obviously, such an equation leads directly to a pressure difference $\delta P = c_s\delta\Phi$. It is sufficient to remark that all the terms in the Gibbs-Duhem equation are conjugate pairs of system variables ($[V\ P]$, $[S\ T]$, $[n, \mu]$) and n_s, Φ is not such a pair. The contribution of Φ is contained within the chemical potential μ_s and the additional term $n_s\delta\Phi$ is superfluous. The form of the equation given above cannot be derived from any basic thermodynamics.

If the pore can alter the concentration of osmolyte in the pore, is there not a difference of osmotic pressure between pore and baths according to the well-known expression $\Delta\Pi = RT\Delta c_s$? Why does not equilibrium at the pore mouths between pore and bath lead to a pressure difference of this magnitude? The theory answers this by arguing that the effect of the pore surface field is to change the activity coefficient γ such that the activity of osmolyte (given by $a_s = \gamma c_s$) remains unchanged. There is therefore no activity difference between the bath and the adjacent pore interior. The osmotic pressure is actually a function of the activity (not the concentration, although this is frequently assumed for dilute solutions), i.e., $\Delta\Pi = f(\Delta a_s) = 0$ so there are no pressures generated by this mechanism. An expression such as Eq. (5) should therefore be amended to contain activities in which the ratio c_{pore}/c_{bulk} becomes a_{pore}/a_{bulk} which effectively equals 1.0, rendering the relation meaningless. It is activity gradients within the pore, the theory argues, which drive volume flow.

In the semipermeable case where the osmolyte is sterically excluded from the pore, the situation is completely altered. Whether or not there is a pore field is irrelevant because the osmolyte can never enter the pore and the activity within the pore is zero. We return, therefore, to the Vegard-Mauro mechanism.

b. Kinetics The thermodynamic argument above can be seen entirely from a viewpoint of molecular dynamics. In a binary solution the two components are reflecting from the wall and the pressure there is the sum of the rates of momentum exchange of the solute and solvent. This is the basis of Dalton's law. If the wall surface is polar (attractive to solute, say) then on approaching it the solute molecules will be accelerated and strike the wall with greater momentum. Concomitantly, they pull a molecule of the wall toward them, which also acquires acceleration. After impact the reflected solute molecule withdraws the same momentum from the wall and decelerates, while the attractive force between them also decelerates the wall molecule. Some time later after collision when the solute molecule is its initial distance away, the wall molecule is again stationary. The situation is the same as if there had been no attractive wall force (Tabor, 1991). This argument is not always found in textbooks, probably because it is taken for granted that Newton's third law is always obeyed at the wall in such a manner.

There can be no change in solute pressure (osmotic pressure) due to the pore wall forces, a situation that accords with the basic thermodynamics presented above, and there can be no changes in hydrostatic pressure in a leaky pore for the same reasons. Water is thus envisaged as flowing only by diffusion in response to activity gradients set up between the baths.

c. Empirical If there were pressures set up by polar surfaces interacting with solutes in a bath, it should be possible to measure them relatively easily. The surface of a pressure probe is just such a surface. It is common experience that if such a probe is immersed in a solution (of constant density and at constant height) it does not respond to changes in the nature or strength of the solution. The surface may be made of any material, with different polarity of surface, and the solution may contain any solutes, but changes in composition of the bath produce no detectable change in pressure. There may of course be detectable changes in concentration in the region adjacent to the probe surface, but the absence of hydrostatic pressure changes indicates that there is no change in activity or osmotic pressure in this surface region in accord with the thermodynamics and kinetics outlined previously. Indeed, if there were changes in pressure the use of such probes in physiology and medicine, not to mention soil science or industry, would render them useless because they would be continually varying either with composition of the medium or with surface contamination of the probe which would change the polarity of the surface. Nor can it be argued that a pressure drop at the surface is present, but that the deflection of the probe surface is compensated by the presence of other forces—whatever they might be. These other forces would represent a pressure component themselves, and so this is equivalent to saying that the pressure changes are always "compensated" by another pressure change—therefore there is no pressure change at all.

2. Reflexion Coefficient and P_{os}

The volume flow J_v during osmosis in leaky pores is considered to be wholly diffusive in nature and due to the separate contributions from water and salt. It is given by

$$J_v = J_w \overline{V}_s + J_s \overline{V}_s = (D_w \Delta c_w \overline{V}_w + D_s \Delta c_s \overline{V}_s) A/L, \qquad (28)$$

which leads to an expression for the reflexion coefficient in terms of the diffusion coefficients within the pore (Hill, 1994; 1989b).

$$\sigma = \frac{P_d}{P_f}\left[1 - \frac{D_s}{D_w}\right] \qquad (29)$$

These diffusion coefficients are equal to the free solution values $D°$ modified by the diffusive drag and steric coefficients so that D_s/D_w is equal to $D_s°F_sS/D_w°F_wS$. The importance of having reasonable values of the drag coefficients is therefore apparent: the value of P_{os}, equal to σP_f, by which the theory can be compared with experimental data, is crucially dependent upon them. At small radii these factors are substantially smaller than 1.0 and their ratio, used in Eq. (25), can be quite large.

If Eq. (4) is a widely held model for the relationship in a porous system between osmolyte size and the rate of osmosis, then the essential difference in physical approach between Eq. (29) and Eq. (4) (bearing in mind that Eq. (4) has not been derived from an osmotic model) can be seen to be the factor P_d/P_f. In the hydrodynamic approach a molecule that has difficulty traversing the pore gives rise to virtually the maximum osmotic flow rate ($\sigma \to 1$), as does one which cannot enter the pore at all. Equation (29) means that the former situation, which entails that $F_s \to 0$, leads to a state where osmosis is occurring with the maximum diffusive rate, i.e., $\sigma = P_d/P_f$ and $P_{os} = \sigma P_f = P_d$. This maximum rate is given by

$$P_d = \pi r^2 D_w F_w/L. \qquad (30)$$

Therefore an important difference exists between a quasi-impermeable pore and a truly impermeable one; that factor is the ratio P_d/P_f. When the osmolyte radius exceeds the pore radius then viscous flow sets in with the consequence that $\sigma = 1$. The situation demanded by this theory is shown diagrammatically in Fig.4.

Thus there is no single formula which covers both the semipermeable and the leaky pore because osmosis is a bimodal phenomenon; i.e., there are two modes of water transfer that can occur in response to different

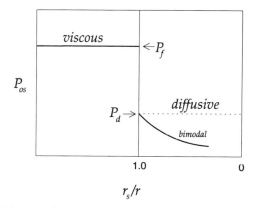

FIG. 4 Behavior of the osmotic permeability P_{os} as a function of osmolyte size for a pore of given radius. There is a discontinuity at $r_s/r = 1.0$ when the osmolyte is excluded because diffusive flow takes over from viscous flow. As the osmolyte size becomes smaller the value of P_{os} falls according to σP_f (Eqs. (1) and (29)).

driving forces, by different mechanisms. Consequently there is a discontinuity or jump in osmotic flow rate when the osmolyte radius crosses the size barrier from being smaller to larger than that of the pore. It may be argued that abrupt jumps are unlikely without a transition occurring as the flow regime switches from one flow to the other, but this probably does exist. The pore is a molecular structure of fluctuating radius rather than a "hard cylinder" and as the osmolyte radius approaches the critical size there must be pressure fluctuations within the pore which grade into a mean decrease of pressure. Thus the discontinuity is smoothed over, depending on the constitution of the pore.

It is possible with this approach to treat the pore as being composed of separate sections with differing radii, the smaller of which may be impermeable to the osmolyte. The pore would then present both a semipermeable and a leaky section in series with each other; the theory predicts that both viscous and diffusive flow would occur within the same pore, though sequentially (Fig.5). If the semipermeable and leaky sections are of lengths l_1 and l_2 and radii r_1 and r_2 and we define a diffusivity p for each section such that $P = p/l$, the overall osmotic permeability of the pore becomes

$$\frac{1}{P_{os}} = \frac{1}{P_{f1}} + \frac{1}{P_{d2}} = \frac{l_1}{p_{f1}} + \frac{l_2}{p_{d2}}. \tag{31}$$

This has direct implications for analysis of the pore by osmotic measurements. For example, by measuring P_{os} with differently sized osmolytes it

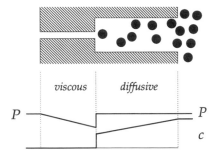

FIG. 5   The entry of osmolyte into a pore section (right) with exclusion from the semipermeable section (left). The osmotic pressure difference draws water through the left section by viscous flow and through the right section by diffusion, as shown in the profiles of pressure (P) and concentration (c) below.

should be possible to determine what models of the pore (in this particular case l_1, l_2, r_1, r_2) are compatible with the data; p_f and σ are functions of pore radius r as given by the theory behind Eqs. (24), (25), and (29). Several such determinations reduce the options proportionately. Conversely, if the pore shape were known it would be possible to divide it into axial sections, calculate p_{os} for each, and assemble P_{os} from the finite elements. For a symmetrical pore with a semipermeable central region there are essentially three pore sections, but the principle is similar.

VI. Interpretation of Existing Data

A. Fixed-Section Pores

1. P_f/P_d Ratios

For these pores, here considered as straight cylindrical throughout, the ratio P_f/P_d, which plays such an important part in determining the pore radius, is a good starting point. Plots of Eq. (2), the classic hydrodynamic expression on which the determination of the "equivalent pore" radius was originally based, are shown in Fig. 6a. Equation (25), the bimodal version of this expression incorporating drag coefficients, is also shown. It is clear that the necessary incorporation of the two drag coefficients completely alters the picture. It has come to be realized that conventional theory fails to give an adequate account of the pore radius because with a current value of 10 for the red cell the equivalent cylindrical radius from Eq. (2) is 0.9–1.0 nm, large enough to be massively permeable to small ions and metabo-

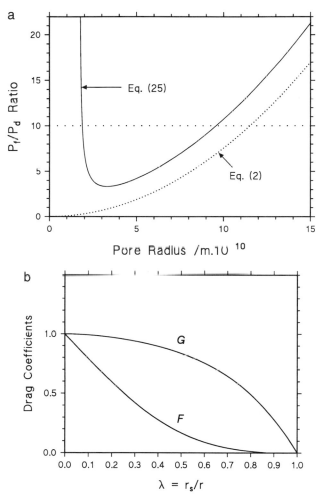

FIG. 6 (a) The P_f/P_d ratio for hydrodynamic theory Eq. (2) and bimodal theory Eq. (25). The current ratio for the red cell is shown as a horizontal line (dotted). (b) The two drag coefficients G and F as a function of molecular size relative to pore radius. The ratio G/F becomes very large in the limit when $r_s/r \to 1$.

lites within the cell (Galey and Brahm, 1985); however, Eq. (25) predicts a value close to 0.2 nm, which is similar to that expected from the solute permeability of the membrane and from components that have been extracted from it. This is gratifying because it shows that the basic idea that pore radii could be determined from the ratio of hydrodynamic to diffusive flows (Solomon, 1968) is essentially correct and does not need to be jettisoned as has been suggested. It is also true, though, that the idea is incomplete without the drag concept.

In Fig.6a the rising limb at small pore radii is an unexpected feature that comes from the relative behavior of the drag coefficients. As the pore radius decreases viscous flow is choked off relative to diffusive flow, but then a minimum is reached, in contradistinction to what might have been supposed. In Fig.6b plots of G and F are shown where it is clear that the ratio G/F is responsible for the subsequent steep rise. It does seem that the use of these coefficients leads to realistic results, however difficult it may be to visualize the molecular processes in detail.

We may also contrast the curves in Fig.6a with the prediction of the single-file theory which should apply to pores below about 0.3 nm, or twice the radius of the water molecule. Although there is no specific expression for this, it can be deduced from the theory and has been represented in diagrammatic form (Finkelstein, 1987). The form of the curve will be that of Eq. (2) ("macroscopic" pores) until the pore becomes no-pass at 0.3 nm. Then, according to the theory, there will be no difference between P_f and P_d, i.e., the ratio is equal to 1.0, but tracer filing takes over and the ratio becomes n (i.e., close to N, the number of water molecules in the pore). It is for this reason that a unique curve cannot be drawn because n is a function of the pore length L. The theory would interpret a P_f/P_d ratio of 10 for the red cell as solely due to filing. We can calculate what this would mean in terms of pore length. Assuming that water has the same density in the pore as the baths, N will be given by $\pi r^2 L/\bar{v}$ which yields a value of $L = 4.2$ nm. This is a very long pore section, and much longer than it would seem could be accommodated by a protein such as CHIP28 which is thought to dominate P_{os} of the red cell (van Hoek and Verkman, 1992; Zeidel et al., 1992). In addition, it is highly probable that the intermolecular distance increases when a liquid or gas is confined to one dimension, although this must depend upon the pliancy of the wall, and so the no-pass section would have to be even longer.

It appears to be true on theoretical grounds that when a gas is confined to one (or two) dimensions the equation of state requires that the intermolecular spacing increase (Hirsch, 1967; Gürsey, 1950) but this is for confinement within totally reflecting walls. If the walls are composed of fluid molecules that can freely exchange energy and momentum with the pore water, then there should be little effect on the mean free path. I do not know whether any simulation of this phenomenon has ever been done or to what extent it applies to proteins that form pores. These are presumably more rigid than the semiliquid membrane in which they are embedded.

An alternative explanation is that the pore, which is thought to be about 0.2 nm in radius at its smallest section only, may have an average radius of (say) 0.3 nm, equivalent to a P_f/P_d ratio of about 3. In the smallest no-pass section there is single-file transfer with a filing number n of about 3. Thus by Eq. (7) the P_f/P_d* ratio, which is what has been measured in experiments, is about 9–10 ($P_f/P_d \times n$). This is the consequence of a pore which is not a straight cylinder but rather of differing cross-section with a narrow constriction; such a pore is discussed in the next section.

There are other values for P_f/P_d* which are even higher than the red cell and which it is customary to consider as due to abnormally low P_d* values caused by unstirred-layer effects during the tracer measurement. A value of 18 has been recorded for the basolateral membrane of the proximal tubule (Whittembury and Carpi-Medina, 1988). This does not have to be so, however, and the result is still within the limits of a small pore of about 0.19 nm radius; if there is some filing effect, too, the true ratio P_f/P_d could represent that of a pore radius of about 0.2 nm, in common with other cells. The point is that it does not indicate a filing of 18 water molecules along a pore.

An interesting result that flows from the bimodal treatment is that there is in principle no need to use tracer solvent fluxes to determine the apparent pore radius. This is because both diffusive and viscous fluxes are an inherent part of the treatment, and so under the right conditions it should be possible to extract a value of P_f/P_d from osmotic measurements and use this to calculate r directly. Measured values of P_{os}/P_d* by themselves give no measure of either r or n unless the mode of water transfer is also known. The advantage of this procedure would be clear: in single-file pores the phenomenon of filing introduces an additional unknown parameter into the ratio P_f/P_d* due to the distinction between P_d and P_d* introduced by the tracer measurements. Accordingly, the bimodal theory demands that the ratio must incorporate both the true viscous:diffusive ratio and the effects of any possible filing,

$$\frac{P_{os}}{P_d*} = n\left[\frac{P_f}{P_d}\right], \tag{32}$$

i.e., the product of two unknowns quantities. The filing number n may be measured independently by measuring the osmotic permeability with a driving osmolyte that enters the pore but is slightly permeant ($D_s \rightarrow 0$; Eq. (29)), which gives the upper diffusive limit for P_{os} equivalent to P_d. This estimate of P_d is hard to obtain without any complication due to tracer filing. The value may then be compared to the measured tracer permeability P_d* when n is given by the ratio P_d/P_d* according to Eq.(7). The pore radius r can be determined without complications of filing by measuring the osmotic flow first with an impermeant osmolyte (which gives P_f) and then with a poorly permeant one as described above, which gives P_d. The ratio of the two is then close to the true ratio P_f/P_d; r may then be determined from Eq. (25).

Gramicidin is the only pore whose radius is known with reasonable accuracy to approximate internally a right cylinder, and for which all three permeabilities (P_f, P_d, and P_d^*) have been measured. P_f/P_d is equal to 6, given by the ratio of osmotic permeabilities determined with mannitol (impermeant) and urea (poorly permeant) which from Eq. (25) gives a pore radius of 0.21 nm, close to the accepted value of about 0.24 nm (see Fig.6a). If the value of P_{os} for gramicidin measured with urea is the diffusive value (P_d), its ratio to the tracer water permeability (P_d^*) is ~5, a value of n of the same size as that extracted from streaming potential measurements (Levitt et al., 1978; Finkelstein, 1987).

The ability to predict the pore radius fairly accurately from osmotic measurements alone and to dissect it from an estimate of n must be considered as unique to the bimodal approach.

2. P_{os} Values

We now come to a central issue—to what extent can different theories account for the osmotic permeabilities found in different molecular-sized pores in terms of what is known of their structure ? There are values, though precious few in number, for antibiotic pores whose structure is relatively well characterized, and one or two measurements for a natural cellular pore CHIP28, a member of the MIP family of aquaporins, whose structure is emerging (Table I). Although there are now several other biological pore systems known that transport ions across membrane systems, the osmotic conductances of these pores have not been measured yet.

In Fig.7 values are plotted in a way which corresponds with the diagram of Fig.4. In Figs. 7a and 7b the values for amphotericin and nystatin , which are larger pores of 0.4 nm radius, are shown, and in Figs. 7c and 7d the values for gramicidin and CHIP28, which are smaller are shown. Gramicidin has a known structure: it shows quite a regular internal cylindrical radius of about 0.24 nm. CHIP28 will be considered below as an example of a pore of variable cross-section: its overall structure may be common to other pores such as ion channels (Unwin, 1995) and porins (Cowan et al., 1992) in possessing wider "entrance" sections with a central restricted section which is about the size of the gramicidin pore.

If we consider first the polyene antibiotics of Figs.7a and 7b, it is immediately apparent that the P_{os} values are below the diffusive upper limit for transfer. That is, water cannot traverse the pores by any sort of viscous mechanism. In this case all the solutes can permeate the pores, which are estimated to be about 0.4 nm in radius. The largest osmolyte, sucrose, can permeate this pore end-on, i.e., by its shortest molecular dimension; in addition, permeability studies show that the larger solutes do indeed have small but measurable permeabilities (Andreoli et al., 1969).

All the osmolytes used are smaller than the pore, or rather, can enter the pore by their smallest molecular radius, which is a strong correlate of permeation (Soll, 1967). Bimodal theory therefore predicts that water transfer in these cases is diffusive and cannot be greater than P_d. It can be seen from Eq. (29) that the calculated diffusive limit is reached for poorly permeable osmolytes when $D_s \rightarrow 0$, in which case $P_{os} = \sigma P_f = P_d$. This is due to the fact that the volume transfer by the osmolyte is effectively zero and all the osmosis is solvent transfer; the pore, it must be remembered, is not completely impermeable for then it would be a semipermeable system and viscous flow would occur. We do not know the exact constitution of the polyene pore when it is assembled, and it is clear that it cannot be just a 0.4-nm radius cylinder (as assumed here for the purposes of calculation) because there are differences in P_{os} and solute permeability between amphotericin and nystatin which have yet to be elucidated (Finkelstein, 1987). Furthermore, the water permeability of these pores (P_w), which are not small enough to show tracer filing, is smaller than would be expected for diffusion through such a pore. It can be seen from Eq. (29) that a reduction in the diffusion coefficient of water with respect to osmolyte would decrease P_{os}. For these reasons it is virtually impossible to make an exact calculation of P_{os} but it is clear that the values lie well below the diffusive maximum in accordance with the theory. The values also fall off with decreasing molecular size, as predicted.

This is also true of formulae for the reflexion coefficient based upon hydrodynamic and single-file theories (Eqs. (4) and (22)), but it can be seen in these expressions that when the permeability of the osmolyte (P_s or A_s) falls to near zero the reflexion coefficient approaches 1.0 and the osmotic permeability is given as $P_{os} = \sigma P_f = P_f$, i.e., the osmotic flow rate is given by a viscous flow expression such as Eq.(10). It can be seen from the data that this is not the case, and that the flow is clearly not viscous in nature.

Figure.7.c shows a similar plot for gramicidin. There are two P_{os} values at present, one measured with a permeant osmolyte (urea), another with an impermeant one (mannitol). These values are close to those expected for diffusive and viscous flow, respectively. Again, however, Eqs. (4) and (22) predict that the value with urea as osmolyte should be close to the viscous value because urea is barely permeable but can enter the pore. In the case of the protein CHIP28 the situation is more complex; this is dealt with in more detail in the following section. The precise shape of the internal pore section is unknown, although the region spanning the membrane may be compared to that of gramicidin: a length of about 2.5 nm and a diameter of 0.2–0.25 nm. It is shown on a similar pore diagram for comparison with those of gramicidin in Fig.7d.

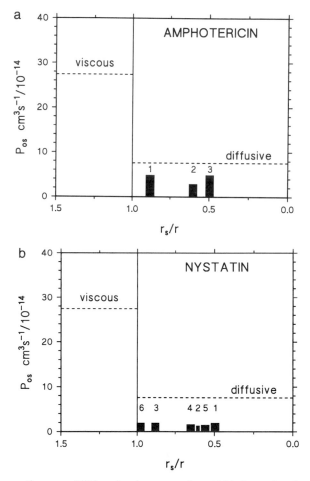

FIG. 7   The osmotic permeabilities of various pores from Table I as a function of r_s/r, shown in a similar way to Fig. 4. The numbers on the bars indicate the osmolytes used which are NaCl (1), urea (2), glucose (3), glycerol (4), ethandiol (5), sucrose (6), and mannitol (7). Shown on each graph are the rates for viscous flow (Eq. (24)) or the maximum diffusive flow P_d (Eq. (30)). (a) Amphotericin (double length). $L = 4.5$ nm, $r = 0.4$ nm. (b) Nystatin (double length). $L = 4.5$ nm, $r = 0.4$ nm. (c) Gramicidin. $L = 2.5$ nm, $r = 0.24$ nm. Note that with the lipid length used by Wang *et al.* (1995) the radius of gramicidin is probably at a minimum equal to that of urea (see Boehler *et al.* (1978). (d) Chip28. The diffusive and viscous maxima for gramicidin are shown here for comparison. It is clear that these osmotic permeabilities cannot be accommodated by a single gramicidin-like pore of approximate length 2.5 nm but require at least two pore sections of different radii.

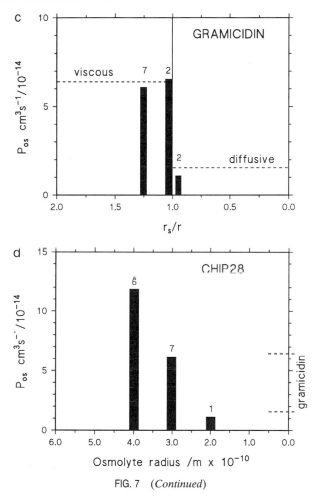

FIG. 7   (*Continued*)

The lowest value is determined with an osmolyte gradient of saline (Preston *et al.,* 1992), which represents the ions Na^+ and Cl^-; these would be expected to enter much of the pore section, although the trans-pore permeability would be very small. The intermediate value is determined with mannitol (van Hoek and Verkman, 1992) which would not be expected to enter the pore and thus to initiate viscous flow, which is apparently the case. The highest value of P_{os}, also shown in Fig.7d, was obtained with an even larger osmolyte, sucrose, and this requires the participation of an even larger pore section that would flank the main core section considered so far. In addition, it is observed that the red cell P_{os} is apparently the same whether it is measured with sucrose or NaCl, and it is doubtful whether CHIP28 has any measurable ion conductance. This means that there must be a semipermeable pore section some-where along the length, but it can be very short.

If these results are taken at face value they point to the fact that different osmolytes lead to different values of osmotic permeability; the possibility then opens up of using such measurements to probe pore structure in pores of variable cross section.

B. Variable-Section Pores

It is very unlikely that the internal cross sections of natural membrane pores are right cylinders. We know that this is not the case for other channels such as porins and ion channels. We have seen in Section V.B.2 how it should be possible to determine the overall osmotic permeability of a variable-section pore by essentially dividing it up into several sections, and having decided whether each one is large enough to admit a particular osmolyte or not, to calculate P_{os} for each using either a diffusive or viscous expression, respectively, and summing them up in series. If this may be called a synthetic approach, in which the pore geometry has been deter-mined independently by structural methods, there is also an analytical approach which can yield information on the internal regions of the pore. It also may provide an understanding of why two osmolytes, each virtually (or entirely) impermeant, can produce substantially different osmotic flows by probing different internal regions of the pore, which is inexplicable in terms of the models underlying either Eq. (4) or Eq. (22).

It is instructive to analyze the data for osmotic flow through the CHIP28 channel; it is scanty, comprising only the three values that have so far been obtained and which are shown in Fig.7d. These values may subsequently be emended, but a bimodal analysis does seem to provide a starting point that shows how such a pore may behave, which is more realistic than that of a straight cylinder.

Starting from the fact that three osmolytes were used, the pore is shown in Fig.8a divided into three sections. NaCl can enter the middle section of length l_1 (though its permeability may be exceptionally small such that $\sigma \to 1$), mannitol enters the outer sections which have a similar radius and lengths l_2' and l_2'', and sucrose is excluded completely. The basic assumptions are (i) that the smallest pore section has a radius of about 0.24 nm, similar to that of gramicidin with which it shares a very low permeability to urea, and (ii) that the outer radius is less than 0.38 nm, which is the maximum required to exclude the sucrose molecule end-on. The equations describe the following situations: with NaCl as osmolyte there must be diffusive flow throughout the pore; with mannitol there is viscous flow through the middle section and diffusive flow through the outer sections; with sucrose

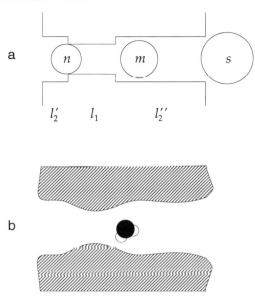

FIG. 8   The internal channel of the aquaporin CHIP28. (a) A simple reconstruction of the pore from osmotic data (Table I) using the solutions of Eqs. (33). The osmolyte (Stokes–Einstein) radii shown here are Na,Cl (0.2 nm), mannitol (0.3 nm), and sucrose (0.42 nm). (b) An impression of the pore approximating to the dimensions in (a). From such a pore with known contours the value of P_{os} for any osmolyte (in either direction) could be calculated from equations similar to Eq. (33).

there is viscous flow throughout. In terms of the specific diffusivities p and lengths l of Eq. (31),

$$\frac{1}{P_{os(Na,Cl)}} = \frac{l_1}{p_{d1}} + \frac{l_2' + l_2''}{p_{d2}}$$

$$\frac{1}{P_{os(mann)}} = \frac{l_1}{p_{f1}} + \frac{l_2' + l_2''}{p_{d2}} \tag{33}$$

$$\frac{1}{P_{os(suc)}} = \frac{l_1}{p_{f1}} + \frac{l_2' + l_2''}{p_{f2}}$$

A solution of these equations yields the values $l_1 = 0.76$, $(l_2' + l_2'') = 1.84$, and $r_2 = 0.35$ nm; a pore with these dimensions is shown to scale in Fig.8a. It can be seen from Eq. (33) that the flanking sections need not be of equal length and l_2' has been arbitrarily made much smaller than l_2''. An interesting aspect of the analysis is that no constraints were put upon the pore length, but the overall value comes out to 2.6 nm. An impression of such a pore is shown in Fig.8b.

This is no more than an exercise with preliminary results which shows what could be done with a fuller data set. If there were more values for P_{os} with different osmolytes there would be proportionately more equations and the analysis could be refined with more pore sections. Another aspect of the theory shows that if the pore is internally asymmetric, as must be in the most general case, then it will be a rectifying channel for osmotic flow. If we consider the elementary situation shown in Fig.5 it becomes apparent that if the gradient of osmolyte were reversed then there would be viscous flow throughout the entire pore and the flow would therefore be greater in this direction. If the pores formed by CHIP28 are not symmetrical about their long axis then different values of P_{os} should be found on reversing the gradient of any particular osmolyte, thereby generating another set of equations, similar in general form to Eqs. (33), which would (in principle) cover the lengths of all the sections. There are no experimental results that provide any information about this as yet.

A final point to note is that filing effects may be quite small in variable section pores, being confined to the narrow parts which may represent only a small fraction of the total pore depth.

VII. Physiological Considerations

The upper value of osmotic permeability obtainable from lipid bilayers is close to that observed in most cell membranes (2×10^{-2} cm/s) and so pores are not absolutely required to achieve the magnitude of natural P_{os} values. It is clear, though, that pores are indeed responsible for most of the observed values because they are found embedded in a lipid membrane of lower permeability. This has to be born in mind when discussing their possible physiological function because it is sometimes presented as though only water pores could confer the "high" osmotic permeabilities observed in certain tissues.

Here two aspects of pore function in cell membranes are dealt with: the insertion and removal of pores to modulate the water permeability of the membrane in response to cellular signals, and the role of membrane-spanning proteins in general in providing pathways between the aqueous phases of cytoplasm and medium.

A. ADH-Sensitive Epithelia

The insertion of a water channel into the apical membrane of renal tubular collecting duct in response to a rise in the antidiuretic hormone levels in the circulating plasma seems to occur by recruitment of pore proteins in membrane vesicles as originally suggested (Bourguet *et al.,* 1976; Wade *et al.,* 1981). There is a shift in the balance between exocytosis and endocytosis of membrane-containing water channels, leading to a raised population of these pore proteins in the apical membrane when the hormone levels rise. The water channel concerned is not CHIP28 but WCH–CD (Fushimi *et al.,* 1993) and seems to be a member of the MIP family whose general function is not clear (if there is any common role).

What is interesting about this mechanism is that it seems to be very complicated compared to that of simply modulating the activity of channel proteins already present in the membrane. There is as yet no indication that any water pore can exist in two states (open or closed, for example) as is the case with voltage- or ligand-gated ion channels. One explanation for why water channelling is not easily gated may lie in the fact that the changes in pore structure required would be much more extensive than those that occur in ion channels. In these it seems, in theory at least, that small changes in radius at the part where the pore is very constricted, and where the electric field is very high, can control the passage of ions very effectively. In other words, the unit conductance is a very steep function of pore radius in certain parts of the channel. We know relatively little as yet about the structure and functioning of the water channel, in particular its unit conductance, and until we do there is little that can be known with precision. However, osmotic theory does provide some clues.

The problem with osmotic channels is that there is both solvent and solute to be considered. To increase the osmotic permeability by widening the pore, i.e., by increasing P_f, would make the pore very leaky to ions and reduce the reflexion coefficient, so the radius cannot be increased very much. It would seem easier to insert pores which exert a discriminatory effect on water and ion fluxes over most of their length into a lipid membrane which lacks them, rather than to try to modulate the P_{os} of existing pores.

B. Transmembrane Proteins and "Designer" Pores

Quite a high proportion of membrane-located proteins span the membrane; this involves several alpha-helical domains in a parallel cluster. Translocation or pore-mediated transfers of molecules or ions between the cytoplasm and medium may well involve structural pathways through the core of such proteins (or the monomers if they are composed of subunits) that are both continuous and permeable to water molecules. Band 3 of red cells has been described as a water pore, although the evidence is conflicting (reviewed by Verkman, 1992). The Na-independent glucose transporter is a water channel (Fischbarg *et al.,* 1990; Zhang *et al.,* 1991) and the CFTR protein also appears to form an osmotically active channel (Hasegawa *et al.,* 1992). It is important to make a distinction between the possible water channel activity of a protein and its relative contribution to the overall osmotic permeability of any particular membrane. It may emerge that many proteins have water channel activity by virtue of spanning the membrane, but some of these have been selected for mediating most of the osmotic exchange between a cell and its environment. In a similar way various proteins have been selected for providing the crystalline structure of the lens (the MIP family) when in fact they have other functions, or have evolved from them. To what extent the aquaporins with their hydrophilic channel regions have been "designed" by selection for this task or are performing another function in addition to water exchange remains to be discovered.

VIII. Conclusions

Any description of osmosis results in a formula that should be testable by comparison with experimental data. The fundamental test is that the theory should be able to predict the osmotic permeability, P_{os}, of a pore of defined structure. Originally, data were confined to results obtained from synthetic membranes, which are basically of two kinds. They are made of either randomly oriented polymers fibers, with the spaces between these representing continuous but oddly shaped spaces, or neutron-etched membranes with a large spread of pore radii, the smaller ones being rhomboidal in cross section. Data resulting from the volume changes of cells, the membranes of which are heterogeneous and of uncertain composition, were used analytically; i.e., they were used to calculate "equivalent" pore radii. With experiments on antibiotic pores it has finally become possible to measure unit osmotic permeabilities of single channels and compare them with theory; these pores are regular in cross section, basically cylindrical. Finally, the emergence of membrane proteins with measurable unit permeabilities has raised the question of whether the osmotic permeability of these pores, whose shape is being determined by structural methods, can be predicted from osmotic theory.

Osmotic theory has been based on the idea that water traverses pores by some sort of viscous (bulk) flow down to very small radii when they become single-file or no-pass pores. It has been claimed that when this point is reached water transfer is diffusional in nature because the pore is so narrow. Thus viscous flow grades into diffusive flow in the transition from macroscopic to molecular-sized pores. An alternative approach that has been developed, the bimodal theory, recognizes that water within the same pore can flow in two modes of transfer, viscous or diffusive, depending on the size of the osmolyte relative to that of the pore. It is steric exclusion from the pore that determines the mode of transfer. In this review I have tried to show that the physics of this theory is sounder and that there is no pore radius at which viscous flow must give way to diffusion. Either can occur at small radii, even though no-pass conditions hold, the determining factor being the osmolyte radius.

Experimental values of P_{os} for molecular pores are not easy to obtain and the number of measurements is still very small. What data there are, however, taken at their face value, are hard to interpret with the expressions derived from the older hydrodynamic approach; it is not clear at all what values of P_{os} could be extracted from the theory for no-pass pores. This is mainly because of unclarity regarding the transfer coefficients for molecules, particularly water, in the pore, and how these coefficients could be calculated. With bimodal theory the diffusion and viscous flows can be calculated by fickian and poiseuillean expressions as for macroscopic pores, but they have to be modified by drag coefficients to take into account the smaller radii. The fits, though not perfect, are quite encouraging. It is clear that the size of the driving osmolyte cannot be ignored when comparing different P_{os} values. In addition, it should be possible to calculate the osmotic permeability of pores of varying cross-section where the osmolyte can probe only parts of the channel, which is probably the natural biological situation.

Acknowledgment

I thank Dr. Bruria Shachar-Hill for the considerable time and work given to critically reading the drafts of this review at all stages of its production.

References

Agre, P., Preston, G. M., Smith, B. L., Jung, J. S., Raina, S., Moon, C., Guggino, W. B., and Nielsen, S. (1993). Aquaporin CHIP: The archetypal molecular water channel. *Am. J. Physiol.* **265**, F463–F476.

Anderson, J. L., and Malone, D. M. (1974). Mechanism of osmotic flow in porous membranes. *Biophys. J.* **14**, 957–982.

Andreoli, T. E., Dennis, V. W., and Weigl, A. M. (1969). The effect of Amphotericin B on the water and non-electrolyte permeability of thin lipid membranes. *J. Gen. Physiol.* **53**, 133–156.

Beck, R. E., and Schultz, J. S. (1972). Hindrance of solute diffusion within membranes as measured with microporous membranes of known pore geometry. *Biochim. Biophys. Acta* **255**, 273–303.

Boehler, B. A., De Gier, J., and Van Deenen, L. L. M. (1978). The effect of gramicidin A on the temperature dependence of water permeation through liposomal membranes prepared from phosphatidylcholines with different chain lengths. *Biochim. Biophys. Acta* **512**, 480–488.

Bourguet, J., Chevalier, J., and Hugon, J. S. (1976). Alterations in membrane-associated particle distribution during antidiuretic challenge in frog urinary bladder epithelium. *Biophys. J.* **16**, 627–639.

Cowan, S. W., Schirmer, T., Rummel, G., Steiert, M., Ghosh, R., Pauptit, R. A., Jansonius, J. N., and Rosenbusch, J. P. (1992). Crystal structures explain functional properties of two *E.coli* porins. *Nature (London)* **358**, 727–733.

Dainty, J., and Ginzburg, B. Z. (1963). Irreversible thermodynamics and frictional models of membrane processes. *J. Theor. Biol.* **5**, 256–265.

Dani, J. A., and Levitt, D. G. (1981). Binding constants for Li, K, and Tl, in the gramicidin channel determined from water permeability measurements. *Biophys. J.* **35**, 485–500.

Dick, D. A. T. (1966). "Cell Water." Butterworth, London.

Durbin, R. P. (1960). Osmotic flow of water across permeable cellulose membranes. *J. Gen. Physiol.* **44**, 315–326.

Fabry, M. E., and Eisenstadt, M. (1975). Water exchange between red cells and plasma. Measurements by nuclear magnetic relaxation. *Biophys. J.* 15, 1101–1110.

Faxen, H. (1922). Die Bewegung einer starren kugel längs der Achse eines mit zaher Flussigkeit gefullten Rohres. *Arch. Mat., Astron. Fys.* **17**(27), 1–28.

Ferry, J. D. (1936). Statistical evaluation of sieve constants in ultrafiltration. *J. Gen. Physiol.* **20**, 95–104.

Fettiplace, R., and Haydon, D. A. (1980). Water permeability of lipid membranes. *Physiol. Rev.* **60**, 510–550.

Finkelstein, A. (1987). "Water Movement Through Lipid Bilayers, Pores and Plasma Membranes." Wiley, New York.

Finkelstein, A., and Holz, R. (1973). Aqueous pores created in thin lipid membranes by the polyene antibiotics nystatin and amphotericin B. *In* "Membranes, A Series of Advances" (G. Eisenman, ed.), vol. 2, pp. 377–408. Dekker, New York.

Finkelstein, A., and Rosenberg, P. A. (1979). Single file transport: Implications for ion and water movement through gramicidin A channels. *In* "Membrane Transport Processes" (C.F. Stevens and R.W. Tsien, eds.), Vol. 3, pp. 73–88. Raven Press, New York.

Fischbarg, J., Kuang, K. Y., Vera, J. C., Arant, S., Silverstein, S. C., Loike, J., and Rosen, O. M. (1990). Glucose transporters serve as water channels. *Proc. Natl. Acad. Sci. U.S.A.* **87**, 3244–3247.

Fushimi, K., Uchida, S., Hara, Y., Hirata, Y., Marumo, F., and Sasaki, S. (1993). Cloning and expression of apical membrane water channel of rat kidney collecting tubule. *Nature (London)* **361**, 549–552.

Galey, W. R., and Brahm, J. (1985). The failure of hydrodynamic analysis to define pore size in cell membranes. *Biochim. Biophys. Acta* **818**, 425–428.

Ginzburg, B. Z., and Katchalsky, A. (1963). The frictional coefficients of the flows of nonelectrolytes through artificial membranes. *J. Gen. Physiol.* **47**, 403–418.

Guggenheim, E. A. (1957). "Thermodynamics," p. 41. North-Holland Publ., Amsterdam.

Gürsey, F. (1950). Classical statistical mechanics of a rectilinear assembly. *Proc. Cambridge Philos. Soc.* **46**, 182–194.

Haberman, W. L., and Sayre, R. M. (1958). "Motions of Rigid and Fluid Spheres in Stationary and Moving Liquids Inside Cylindrical Tubes," U.S. Dep. Navy Rep. No. 1143. David Taylor Model Basin.

Hasegawa, H., Skach, W., Baker, O., Calayag, M. C., and Verkman, A. S. (1992). A multifunctional aqueous channel formed by CTFR. *Science* **258**, 1477–1479.

Hernandez, J. A., and Fischbarg, J. (1992). Kinetic analysis of water transport through a single-file pore. *J. Gen. Physiol.* **99**, 645–662.

Hill, A. E. (1982). Osmosis: A bimodal theory with implications for symmetry. *Proc. R. Soc. London, Ser. B* **215**, 155–174.

Hill, A. E. (1989a). Osmosis in leaky pores: The role of pressure. *Proc. R. Soc. London, Ser. B* **237**, 363–367.

Hill, A. E. (1989b). Osmotic flow equations for leaky porous membranes. *Proc. R. Soc. London, Ser. B* **237**, 369–377.

Hill, A. E. (1994). Osmotic flow in membrane pores of molecular size. *J. Membr. Biol.* **137**, 197–203.

Hirsch, H. (1967). Relevance of the single-file model to water flow through porous cell membranes. *Curr. Mod. Biol.* **1**, 139–142.

Hodgkin, A. L., and Keynes, R. D. (1955). Active transport of cations in giant axons from *Sepia* and *Loligo. J. Physiol. (London)* **128**, 28–60.

Holz, R., and Finkelstein, A. (1970). The water and nonelectrolyte permeability induced in thin lipid membranes by the polyene antibiotics Nystatin and Amphotericin B. *J. Gen. Physiol.* **56**, 125–145

House, C. R. (1974). "Water Transport in Cells and Tissues." Edward Arnold, London.

Kedem, O., and Katchalsky, A. (1961). A physical interpretation of the phenomenological coefficients of membrane permeability. *J. Gen. Physiol.* **45**, 143–179.

Killian, J. A. (1992). Gramicidin and gramicidin-lipid interactions. *Biochim. Biophys. Acta* **1113**, 391–425.

Kohler, H.-H., and Heckmann, K. (1979a). Unidirectional fluxes in saturated single-file pores of biological and artificial membranes. I. Pores containing no more than one vacancy. *J. Theor. Biol.* **79**, 381–401.

Kohler, H.-H., and Heckmann, K. (1979b). Unidirectional fluxes in saturated single-file pores of biological and artificial membranes. II. Asymptotic behavior at high degrees of saturation. *J.Theor. Biol.* **85**, 575–595.

Lea, E. J. A. (1963). Permeation through long narrow pores. *J.Theor. Biol.* **5**, 102–107.

Levitt, D. G. (1973). Kinetics of diffusion and convection in 3.2A pores. *Biophys. J.* **13**, 186–206.

Levitt, D. G. (1974). A new theory of transport for cell membrane pores. 1. General theory and application to red cell. *Biochim. Biophys. Acta* **373**, 115–131.

Levitt, D. G., Elias, S. R., and Hautman, J. M. (1978). Number of water molecules coupled to the transport of sodium, potassium and hydrogen ions via gramicidin, nonactin or valinomycin. *Biochim. Biophys. Acta* **512**, 436–451.

Longuet-Higgins, H. C., and Austin, G. (1966). The kinetics of osmotic transport through pores of molecular dimensions. *Biophys. J.* **6**, 217–224.

Manning, G. S. (1968). Binary diffusion and bulk flow through a potential energy profile: a kinetic basis for the thermodynamic equations of flow through membranes. *J. Chem. Phys.* **49**, 2668–2675.

Mauro, A. (1957). Nature of solvent transfer in osmosis. *Science* **126**, 252–253.

Preston, G. M., and Agre, P. (1991). Isolation of the cDNA for erythrocyte integral membrane protein of 28 kilodaltons: Member of an ancient channel family. *Proc. Natl. Acad. Sci. U.S.A.* **88**, 11110–11114.

Preston, G. M., Carroll, T. P., Guggino, W. B., and Agre, P. (1992). Appearance of water channels in Xenopus oocytes expressing red cell CHIP28 protein. *Science* **256**, 385–387.

Renkin, E. M. (1954). Filtration, diffusion and molecular sieving through porous cellular membranes. *J. Gen. Physiol.* **38**, 225–243.

Renkin, E. M., and Curry, F. E. (1979). Transport of water and solutes across capillary endothelium. *In* "Membrane Transport in Biology" (G. Giebisch, D.C., Tosteson, and H.H. Ussing, Eds.), Vol. 4 Chapter 1, pp. 1–45. Springer, New York.

Rosenberg, P. A., and Finkelstein, A. (1978). Water permeability of gramicidin A-treated lipid bilayer membranes. *J. Gen. Physiol.* **72**, 341–350.

Soll, A. H. (1967). A new approach to molecular configuration applied to aqueous pore transport. *J. Gen. Physiol.* **50**, 2565–2578.

Solomon, A. K. (1968). Characterization of biological membranes by equivalent pores. *J. Gen. Physiol.* **51**, 335s–364s.

Tabor, D. (1991). "Gases, Liquids and Solids," 3rd ed., pp. 133–134. Cambridge Univ. Press, Cambridge, UK.

Unwin, N. (1995). Acetylcholine-receptor channel imaged in the open state. *Nature (London)* **373**, 37–43.

van Hoek, A. N., and Verkman, A. S. (1992). Functional reconstitution of the isolated erythrocyte water channel CHIP28. *J. Biol. Chem.* **267**, 18267–18269

van Hoek, A. N., Wiener, M., Bicknese, S., Miercke, L., Biwersi, J., and Verkman, A. S. (1993). Secondary structure analysis of purified CHIP28 water channels by CD and FTIR spectroscopy. *Biochemistry* **32**, 11847–11856.

Vegard, L. (1908). On the free pressure in osmosis. *Proc. Cambridge Philos. Soc.* **15**, 13–23.

Verkman, A. S. (1992). Water channels in cell membranes. *Annu. Rev. Physiol.* **54**, 97–108.

Wade, J. B., Stetson, D. L. and Lewis, S. A. (1981). ADH action: Evidence for a membrane shuttle hypothesis. *Ann. N.Y. Acad. Sci.* **372**, 106–117.

Wang, K.-W., Tripathi, S. and Hladky, S. B. (1995). Ion binding constants for gramicidin A obtained from water permeability measurements. *J. Membrane Biol.* **143**, 247–257.

Wang, H., and Skalak, R. (1969). Viscous flow in a cylindrical tube containing a line of spherical particles. *J. Fluid Mech.* **38**, 75–96.

Whittembury, G., and Carpi-Medina, P. (1988). Renal absorption of water: Are there pores in proximal tubule cells? *NIPS* **3**, 61–65.

Zeidel, M. L., Ambudkar, S. V., Smith, B. L., and Agre, P. (1992). Reconstitution of functional water channels in liposomes containing purified red cell CHIP28 protein. *Biochemistry* **31**, 7436–7440.

Zhang, R., Alper, S., Thorens, B., and Verkman, A. (1991). Evidence form oocyte expression that the erythrocyte water channel is distinct from band 3 and the glucose transporter. *J. Clin. Invest.* **88**, 1553–1558.

The Use of Computers in Understanding How Animal Cells Crawl

David R. Soll

Department of Biological Sciences, University of Iowa, Iowa City, Iowa 52242

Amoeboid cell motility is a complex three-dimensional process which involves pseudopod expansion, cellular translocation, and, in some cases, pseudopod retraction and complex interactions between the ventral surface of the pseudopod and substratum. In order to quantify the basic behavior of amoeboid cells and the dynamics of pseudopod extension and retraction, sophisticated two-dimensional and three-dimensional computer-assisted motion analysis systems have been developed which reconstruct digitized images and compute motility and dynamics morphology parameters. These systems provide a wealth of information of how amoeboid cells crawl and they have begun to be utilized (1) to elucidate the basic rules of amoeboid movement, (2) to identify the behavioral defects of cytoskeletal mutants, and (3) to elucidate the mechanism of chemotaxis. In addition, these systems represent powerful tools for analyzing the effects of drugs on cell behavior, most notably that of white blood cells and neoplastic cells. Since computer-assisted motion analysis is a relatively young field, the technologies are still evolving and have been underutilized in most areas involving cell motility. This review, which includes a description of these technologies and examples of their application, will hopefully serve as an impetus for expanded use.

KEY WORDS: Computer-assisted motion analysis, Amoeboid cell motility, Behavior defects of cytoskeletal mutants, *Dictyostelium* motility, Three-dimensional reconstruction of motile cells, Cell motility and chemotaxis.

I. Introduction

Single-cell motility plays a fundamental role in the life cycle of cells ranging in complexity from bacteria to man. In bacteria, motility can be essential for dispersal, feeding, and multicellular morphogenesis, and involves a number of distinct processes, including flagellar rotation (Berg, 1991). In eukaryotes, single-cell motility is involved in an even more diverse list of processes, including feeding, dispersal, gamete fusion, aggregation, embryonic development, tissue and organ genesis, maintenance of multicellular form, and the cellular immune response. Individual eukaryotic cells can move by two general mechanisms, flagellar or cilliary beat (Smith and Sale, 1994) and amoeboid locomotion or "crawling" (Allen, 1981; Stossel, 1993; Trinkaus, 1984).

It is probably safe to say that every method of organized cellular movement is complex, and that anecdotal, nonquantitative descriptions are no longer sufficient, especially in cases of amoeboid locomotion, for answering the questions now being posed concerning mechanism. This point has become particularly poignant in recent years in the analysis of cytoskeletal mutants (Condeelis, 1993a). Many distinct cytoskeletal proteins which interact with actin originally did not appear to be essential for the basic process of cellular translocation since specific mutations which resulted in the absence of functional proteins resulted in mutant strains which could still form pseudopods and translocate (Brink et al., 1990; Knecht and Loomis, 1987; DeLozanne and Spudich, 1987; Jung and Hammer, 1990). It was originally assumed that the absence of a mutant phenotype was due to functional redundancy of actin associated proteins. However, when computer-assisted methods were used to elucidate the behavioral phenotypes of several of these mutants, it was discovered that although mutant cells were able to extend a pseudopod and to translocate along a substratum, particular aspects of these dynamic processes were aberrant, and the defects were specific to the mutated protein (Wessels et al., 1988, 1991; Cox et al., 1992; de Hostos et al., 1993). Therefore, many actin-associated proteins fine tune the process of pseudopod extension and cellular translocation, and the behavior of mutants lacking these components in many cases cannot be assessed by casual inspection (i.e., nonquantitative analysis). Behavioral phenotypes of such mutants first require a refined, quantitative description of wild-type or parent strain behavior, which can then be compared with that of a mutant. Such quantitative comparisons are facilitated by computer-assisted motion analysis systems.

In recent years, the capacity to videorecord, digitize, and then computer-analyze the motile behavior of cells (Soll, 1988; Soll et al., 1988a), pseudopods (Wessels et al., 1994), intracellular particles (Wessels and Soll, 1990), and tagged surface elements (Sheetz et al., 1989; Kucik et al., 1991) has been instrumental in uncovering some of the rules governing normal behavior, and, in turn, has provided a context for defining specific abnormalities in the motile behavior of select mutants, neoplastic cells, and cells treated with inhibitors. Recording methods, image processing systems, and motion analysis software have rapidly evolved and, over the past several years, have become standard technologies in a variety of cell biology laboratories. However, computer-assisted methods for analyzing cellular and subcellular motility remain underutilized technologies. Therefore, the objectives of this review are to describe several of the computer-assisted technologies and quantitative methods which have recently evolved for analyzing amoeboid cell movement, to formulate some of the questions which can now be answered with these technologies, and to describe some of the new technologies which are emerging, especially those involving computer-assisted three-dimensional reconstruction and analysis of moving cells.

A. Quantitative Approach

Why can't we simply describe qualitatively what we see through a microscope in order to understand how an amoeboid cell moves? We can, and that is the first step in an analysis of amoeboid cell behavior. However, there are dynamics in the process of translocation which cannot be interpreted by casual observation. For instance, if a cell is moving too quickly, any complex or redundant behavior will be missed due to the slow processing time of the human mind in real-time observation, and if a cell is moving too slowly, our incapacity to compare still frames separated by minutes usually results in an interpretation of inactivity. The complexity of amoeboid translocation, therefore, requires a system for image storage and high-resolution motion analysis. Consider what we now know about the general characteristics of amoeboid motion. Amoeboid cells translocate along a substratum by the extension of a pseudopod or by the anterior expansion of a lamellipod. Except for nematode sperm (Nelson et al., 1982), these cellular extensions, which we will refer to collectively as pseudopods, either are filled with F-actin or contain an enriched F-actin cortex (Stossel, 1993; Condeelis, 1993b; Clarke et al., 1975). Extension of pseudopods, therefore, appears to be driven, or stabilized, by F-actin polymerization, but the forces involved in pseudopod extension, stabilization, and retraction, the positional information defining the cell axis, and the mechanism which moves the rest of the cell body forward during or after anterior pseudopod extension, have not been resolved. Many simple models have been proposed to explain how a cell moves forward (Condeelis, 1993b; Condeelis et al., 1990; Allen and Taylor, 1975; Jay et al., 1993; Hay, 1989), and although many of them explain one or more aspects of locomotion, none of them adequately explain the complex three-dimensional and temporal dynamics of cellular locomotion, especially for cells which form lateral pseudopods. We now know that cellular translocation is cyclic in a broad range of amoeboid cells (Wessels et al., 1994; Murray et al., 1992; Royal et al., 1995; Sylwester et al., 1995), and that in some cells the cycle involves a three-dimensional z-axis component

(Wessels *et al.,* 1994; Murray *et al.,* 1992). Cells move in surges, extending a pseudopod during a rapid translocation phase in the x,y axes. Cells move in a straight line by anterior pseudopod extension, turn in a curved path by biased expansion of the pseudopod to the left or right of the cell axis (Royal *et al.,* 1995; Sylwester *et al.,* 1995), or turn sharply by forming a lateral pseudopod which assumes the role of leading edge (Wessels *et al.,* 1994). Some amoebae do not make sharp turns due to apparent interference by very large, dominating nuclei (Sylwester *et al.,* 1993). In the case of chemotactically responsive amoebae like *Dictyostelium,* which appear to employ lateral pseudopod formation as a means for assessing the direction of a chemotactic gradient, not all pseudopods are formed on the substratum, and the interaction between the ventral surface of the pseudopod and the substratum controls turning (Wessels *et al.,* 1994). Pseudopods formed off the substratum have a propensity for retraction, and the mechanism of pseudopod retraction may be quite complex. Select cytoskeletal mutants of *Dictyostelium* have been found to behave differently from wild-type cells in the rate, frequency, and three-dimensionality of pseudopod extension (Cox *et al.,* 1992; Wessels *et al.,* 1988, 1991; Wessels and Soll, 1990; Titus *et al.,* 1992), while other select mutants have been found to behave differently from wild-type cells due to differences in shape (Alexander *et al.,* 1992) or changes in the position of pseudopods due to the loss of pseudopod anchoring (Shutt *et al.,* 1995a). Amoeboid cell behavior is, therefore, complex and the complexity can be attributed, at least in part, to a large number of actin-associated proteins (Stossel, 1993; Condeelis, 1993b; Luna and Condeelis, 1990) as well as the idiosyncrasies and functional role of the particular cell. How, then, does this description of the complexity of amoeboid behavior relate to the original question posed? Why can't we simply look at a moving cell through a microscope and immediately determine what is happening? Why do we need computer-assisted systems?

Abercrombie described fibroblast behavior in his pioneering work using time lapse recordings, direct observation, and electron microscopy (Abercrombie, 1961; Abercrombie *et al.,* 1970a,b,c), and provided us with a behavioral phenotype which, to this day, is the basis for describing how fibroblasts crawl. His description, although quantitative, was not quantitative enough to provide the kind of precise motility and dynamic morphology parameters which are now deemed necessary for describing how amoeboid cells extend pseudopods, retract pseudopods, interact with the substratum, and coordinate pseudopod extension and cellular translocation, prerequisites for assessing the behavioral defects of cytoskeletal mutations, the role of pseudopod extension in chemotaxis, and the effects of modulators, such as chemoattractant and extracellular matrix molecules, on translocation. Abercrombie did not provide the necessary time plots of parameters which would have exposed rhythms or distinguished subtle behavioral defects in mutant cells. He did not quantitatively assess the relationship of pseudopod extension, interaction with the substratum, and cellular translocation. A computer-assisted system would have provided Abercrombie and his associates with the computational power to perform such measurements, and, given Abercrombie's obvious penchant for using different techniques to describe the behavior of fibroblasts, there is little doubt that he would have vigorously applied computer-assisted motion analysis systems if they had been available at the time of his classic studies.

B. A Brief History of the Use of Computers in Analyzing Amoeboid Movement

Computers began to be used in the analysis of individual amoeboid cell movement in the 1970s, but the development of sophisticated motion analysis systems was hampered throughout that decade by the absence of inexpensive video cameras and recorders for easy storage and replay, the absence of methods for rapid or automatic digitization, and the absence of analytical software. Perhaps the most intense use of computers in the 1970s and early 1980s was in the area of sperm motility rather than amoeboid motion. There was and still is interest in correlating motility parameters of sperm with infertility in humans, and in assessing the vigor of sperm samples for use in artificial insemination for both humans and animals. The most common system for analyzing both sperm movement and amoeboid cell movement in the 1970s and early 1980s included digitizing tablets inferfaced to microcomputers (Schmassmann et al., 1979; Makler, 1980; Futrelle et al., 1982; Varnum and Soll, 1984). Since inexpensive video systems were not readily available until the early to mid 1980s, a number of less efficient methods were used in the 1970s to obtain tracks of cell positions prior to digitization, including serial still frame photographs, movies, multiple exposures on the same film frame, and live, reflected images. Plastic sheets could be overlaid on the still framed or reflected image, and the cell position could be recorded, generating a track of positions. The plastic overlay could then be placed on a digitizing tablet, and a cursor used to digitize the position of a cell over time into the microcomputer datafile. Simple programs were written for microcomputers or personal computers for measuring parameters based upon the dynamics of estimated cell position. Such parameters included speed, direction, net distance, and the proportion of motile cells, but were relatively imprecise at short time intervals because they rarely involved a consistent mathematical method for quantitating the position of the cell centroid. Rather, analyses were performed at low magnification and cell position was interpreted by the researcher. It is fair to say that the first application of computers in the 1970s and early

1980s in the analysis of amoeboid cell motion using digitizer tablets relieved the researcher of manually taking measurements with a ruler, manually computing basic parameters, and manually recording these parameters in order to present the final data. However, these systems were relatively slow and did not provide information on the changes in cell shape, especially those related to pseudopod extension, which we now know are fundamental in interpreting amoeboid cell movement. Interestingly, even with the cumbersome and time-consuming recording methods, manually operated digitizer boards and the very slow computers which were available in the 1970s, the potential did exist for developing a sophisticated motion analysis system. Manually operated digitizer boards could be used to record the x,y coordinates of a cell perimeter by clicking at points around the perimeter, and on rare occasions, this type of method was employed (e.g., Schmassmann *et al.,* 1979), but rarely exploited for amoeboid cell motility.

C. Evolution of Sophisticated Motion Analysis Systems in the 1980s

In the 1980s, several advances were made which together resulted in the development of advanced computer-assisted systems for analyzing amoeboid motion. The first was the availability of inexpensive video systems. Video provided three things, a signal which could be readily accessed for digitization, a very rapid method for recording without processing film, and the capacity for inexpensive, continuous recording rather than the inordinate dependence upon time lapse recording forced upon the user by conventional cinematography. The second advance in motion analysis systems was the development of real-time integrated digitizers, hardware which allowed real-time digitization of high-contrast contours. Although the availability of this technology (the VP110 from Motion Analysis Corp., Santa Rosa, CA) had an immediate impact in the areas of sperm analysis, it did not have a similar impact on the analysis of amoebic motility, since the necessary high-contrast images tended to decrease the detail of the cell perimeter and the first systems had no software for the analysis of changes in cell shape, a fundamental characteristic of amoebic locomotion. In the late 1980s, frame grabber boards became available. These boards allowed immediate digitization of a limited number of video frames, which could then be analyzed with edge-detection software for the digitization of cell perimeters. These boards soon became less expensive then hard-wired real-time contour digitizers, and were more accessible to software manipulation. They did not, however, provide the capacity for immediate and continuous digitization which is provided by real-time contour digitizers. The third advance which accelerated the development of sophisticated motion analysis systems was the evolution of powerful microcomputers with rapid processing capabilities and increased storage.

In the 1980s, two relatively different approaches to the analysis of cellular motility began to emerge. One method was based on processing (Fischer *et al.*, 1989) and identified movement by pixel displacement. If a cell moved a short distance, the number of pixel changes was roughly proportional to translocation (Fig. 1A). This differencing method has been effectively used to compute the general displacement activity of a cell population since a large number of cells can be included in a single computation, and morphological parameters can be gleaned from the pixel patterns of a single cell image.

The more popular approach to motion analysis, and the one which has ultimately led to the most sophisticated motion analysis systems, is based upon methods which identify and digitize the positions of pixels at the outline of a moving object, thus reducing the amount of digitized information (Fig. 1B). The first commercial system available in 1985 was the Expertvision system from Motion Analysis Corporation (Santa Rosa, CA). This

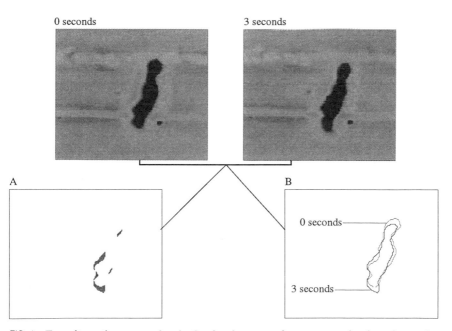

FIG. 1   Two alternative approaches in the development of computer-assisted motion analysis systems: (A) a processing method in which pixel displacement is used for analysis, and (B) an edge detection method in which pixels at the cell perimeter are used for analysis. High-contrast images of an amoeboid cell are presented at 0 and 3 s. These images are used in the examples of processing (A) and edge detection (B).

system evolved from a research tool for studying the photobiology of micro-organisms (Hand and Davenport, 1970), and included a real-time contour digitizer and a proprietary motion analysis software package based in a SUN computer. For the analysis of amoeboid cell motion, a videocamera was mounted on the ocular of a microscope, and the cell image was either videorecorded or ported directly into the real-time contour digitizer. The digitizer was tuned for thresholding and contrast in order to obtain the most noise-free outline. The system then fed the x,y coordinates of the pixels at the cell periphery into an Expertvision data file. The digitized perimeters were used by the Expertvision system to compute the position of the centroid (center of mass) of each image. The centroid paths were then used to compute velocity, directionality, and a number of other centroid-based parameters. Although the original Expertvision system, and a subsequent system, Celltrack, based in an IBM PC (AT), were highly effective in analyzing high-contrast images such as sperm, bacteria, or para-mecia (Lieberman *et al.*, 1988), they had two drawbacks for cell biologists studying amoeboid motion. First, because the software was dedicated to a real-time contour digitizer signal which required high-contrast edges, one could not analyze low-contrast images or use such microscopic configura-tions as Nomarski (differential interference contrast) or phase. Second, since the system was conceived for organisms which did not change shape, it lacked programs for analyzing the contour of an amoeboid cell perimeter even though the information, in the form of perimeter pixel positions, was stored in the data file. This latter limitation was removed by the addition to the SUN-based Expertvision system and the IBM-based Celltrack system of DMS software (Soll, 1988; Soll *et al.*, 1988a,b). The DMS software pro-gram was developed to monitor with time not only parameters based on the position of the cell centroid, but also parameters based on cell shape, and provided several unique animated presentations of the digitized image which have proved quite useful in interpreting amoeboid cell locomotion.

II. Developing a Sophisticated Two-Dimensional Motion Analysis System

A. Imaging and Recording

An effective motion analysis system should have multiple methods for digitization, since different cell types will provide different edge contrast in the same microscopic format, and different experimental formats will require different types of microscopy. For instance, cells like the soil amoeba *Dictyostelium discoideum* or human polymorphonuclear leukocytes have significant height when translocating at velocities of 7 to 20 μm per minute on glass or plastic surfaces (Wessels *et al.,* 1994; Murray *et al.,* 1992). Therefore, they produce contrasted images under standard Kohler or dark field illumination. However, other cells like fibroblasts can be extremely flat, and their edges therefore may not be highly contrasted by standard lighting. Without a significant level of contrast between the cell-free substrate and the cell perimeter, automated digitizing systems have a difficult time performing real-time digitization. For many flat cell images, there is little difference in the intensity of the cytoplasm and surrounding cell-free medium. In such cases, the low level of edge contrast may allow the viewer to interpret the cell circumference, but an automatically digitized image of the same cell may contain perimeter gaps and extracellular noise levels too close to thresholds for automatic discrimination. In such cases, manual digitization is more effective.

Since automatic digitization relies upon high-contrast edges, the cell images recorded and analyzed rarely contain intracellular details. In studies in which the dynamics of intracellular particles have been analyzed in conjunction with cellular motility (e.g., Wessels and Soll, 1990), Nomarski optics have been required, and manual rather than automatic edge detection has been employed. In situations in which high-contrast cell images have been recorded for automatic digitization, distinctions could not be made between the pseudopodial zone containing particulate-free cytoplasm and the main cell body, which contains a nucleus and particulate-rich cytoplasm. This limits the capacity of the experimenter to verify that expansion areas identified by the software program reflect the dynamics of cytologically identified pseudopods. Therefore, the type of imaging one employs depends upon both the contour and height of the cell and the experimental protocol. A generalized scheme for motion analysis of an amoeboid cell is presented in Fig. 2. It includes a number of components important in image acquisition.

1. A Platform for Cell Migration

This can vary in complexity from a simple glass slide supporting the sample and coverslip, to a temperature-controlled perfusion chamber, like a Sykes-Moore (Bellco Glass, Inc., Vineyard, NJ) or Dvorak-Stotler chamber (Nicholson Precision Instruments, Gaithersburg, MD), to a gradient chamber like the one designed by Zigmond for analyzing chemotaxis (Zigmond, 1977). The medium may consist of simple buffer solution, growth medium, or a solution containing chemoattractant or other signal molecules. A solution of chemoattractant may be introduced into a perfusion chamber in pulses or as temporal gradients (Varnum *et al.,* 1985), or it may be presented as a spatial gradient in a gradient chamber (Zigmond, 1977; Varnum and Soll, 1984). The surface of the chamber may be coated with extracellular

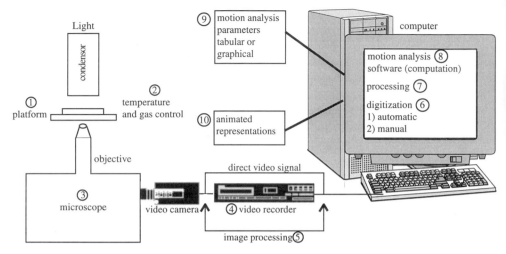

FIG. 2   A general scheme for a two-dimensional computer-assisted motion analysis system.

matrix molecules or a cell monolayer (Sylwester *et al.*, 1995). Any experimental protocol can be achieved as long as the cell image can be visualized
and recorded.

2. Temperature Control Equipment

This may be in the form of a thermostatically controlled air curtain, a
temperature-controlled microscope stage, or an environmental chamber
encapsulating the stage, which can also be used to provide gas control for
tissue culture cells.

3. A Microscope with Objectives and Condenser with the Correct Configuration for the Desired Image

In cases in which the cell is separated from the objective by a distance
greater than the thickness of a coverslip, long-distance working objectives
and condenser may be necessary. For plastic platforms, Hoffman objectives
rather than Nomraski objectives may be necessary for obtaining differential
interference microscopy. As noted, the optical configuration will depend
upon the cell contour, the platform, and the parameters of interest. The
experimenter must carefully test the optical configuration and lighting with
the desired method for digitization, as previously discussed.

4. Camera and Recorder

Videocameras should provide the highest resolution compatible with the
recording method. Although videodiscs provide extremely high-quality im-

ages, they are presently very expensive for extensive recording. For most experimental purposes, a reasonably good quality CCD camera and 3/4 inch recorders are sufficient. Although direct digitization through an automatic digitizer or immediate storage in a frame grabber are alternatives to recording on videotape, the latter has a number of advantages. First, videorecording provides an immediate, inexpensive, and accessible record which can be reviewed before digitization. Second, it provides a back-up in case the digitized image is lost. Third, if a system employs a frame grabber board, there are limits to the length of video which can be accessed in real time and, depending upon the digitization format, it may not be able to handle a live image for extended periods.

5. Image Processor

For automatic digitization and, in some cases, for manual digitization, a high-contrast image facilitates the process. Standard image processors can be used to enhance or contrast the cell image between camera and recorder or between recorder and digitizer (Fig. 2). Systems which use frame grabber boards and digitizing software for automatic digitization readily incorporate processing into the program. In most situations, trial and error will dictate your need for processing. The most important theme of imaging is to obtain the best recorded image for the experimental objectives and methods of digitization.

B. Digitizing

Whether using a real-time contour digitizer which is hard-wired, a frame grabber board, or manual digitization, the objective is the same, to digitize the perimeter of an image into the computer data file in the form of the x, y coordinates of perimeter pixels (Fig. 1B). This reduces the memory requirement for a videoframe in the file and provides information for computing the centroid location and contour parameters. To describe the method and inherent problems of automatic digitization, a system based on a frame grabber will be considered. To begin with, the frame frequency must be determined. This will depend upon the rate of cellular translocation and the information desired, and will also affect the velocity measurement, a point discussed in a later section. Common frame grabbers are based upon a continuum of intensity levels which may range, for example, from 0, which is black, to 255, which is white. Digitizer software should provide a method for thresholding all pixels above a decided level as white, and all pixels below it as black. The outline will then be generated from black pixels which have a white neighbor in any direction. This threshold method works best when the image is of even intensity and the intensity of the background is significantly different from that of the cell. However, when illumination is uneven, a gradient method is superior to a threshold method. In the gradient method, the differential between intensities rather than an absolute threshold is determined by the user. The system then searches for pixels which have a neighbor 2 or 3 pixels in any direction which differs by more than the gradient level, rather than by a fixed threshold. The gradient method can be developed for digitizing phase contrast images in which one edge converts from white to black while the other edge converts from black to white centrapedally. The gaps which would form at the points of reversal would have to be filled through logic built into the system specifically for this purpose. Although gradient methods would seem to be the most versatile approach to original complex images, they also generate the most complex image since the method will find all edges, including those inside the cell perimeter; additional size and shape thresholding is often required to rid the digitized image of noise.

With manual digitization, a digitized movie is first made, and the digitized image is then still framed at any desired frame interval. The mouse can then be used to generate a continuous trace, or it can be serially clicked to generate a series of perimeter points which are then connected by straight or curved lines. As noted previously, fairly transparent images such as those obtained by Nomarski optics, images of very flat tissue culture cells with little contrast, or phase contrast images are normally digitized more accurately by a manual mode. In addition, edges with complex contours such as those possessing spikes or filopodia are far more amenable to manual digitization. Manual digitization can be quite fast for an experienced user, and one can waste more time trying to accomodate an automatic digitization mode than applying a manual mode.

C. Editing Raw Data of Cell Contours Prior to Analysis

Although manual digitization usually provides uninterrupted perimeter data, automatic digitization may not, and therefore must be edited. It can be performed manually if one is dealing with a limited number of frames, but if there is a significant number, it must be performed by the program. The first major problem is connectivity. A digitized perimeter image is 1-connected if every point in the cell outline is immediately next to another pixel in the cell outline, in a diagonal, horizontal, or vertical direction. An object is 2-connected if every point in the cell outline is, at most, 1 pixel away from another point in the cell outline, and so on. In other words, the gap allowed is 1 less than the connectivity level. After a connectivity limit is set, automatic connections are made to fill gaps.

A second major problem in automatic digitization involves smaller or larger noncellular objects which exhibit the same contrast threshold or intensity gradient at their edge as the cell perimeter. In addition, contrasted objects or regions within a cell may also be digitized and in some cases they may be contiguous with the real cell perimeter within the connectivity limit. These artifacts are effectively removed by setting upper and lower thresholds for the number of pixels in the perimeter of an object of interest. In the same way, unwanted digitized images can automatically be eliminated from the raw data file by placing thresholds on velocity, path length, roundness, etc. A sophisticated system must have such capabilities.

A third problem involves major gaps in the outline of a cell, and one solution involves dilation and erosion algorithms. In the upper left corner of Fig. 3A, an example is presented of a cell image with a central portion out of focus. The out-of-focus portion probably represents a z-axis gap (i.e., a portion of the cell with a change in the z axis which removes it from the focal plane used for analysis). The digitized image in this case is composed of two outlined perimeters (Fig. 3A). In the upper left corner of Fig. 3B, the dilation phase of the method is applied. In this case, a dilation of 1 adds four horizontal and vertical neighbor pixels to each original pixel. The dilated image is then outlined to produce the dilated result (lower left corner, Fig. 3B). Since the dilation result is now bigger by the pixels added outwardly to the original image, it must be eroded (upper left corner, Fig. 3C). If a digitized image is dilated n times, it must be eroded n times to approximate the original image size. In the upper right corner of Fig. 3A, an example is presented of a cell image which is out of focus at two ends, resulting in a digitized image with two perimeter gaps. Rather than use a connectivity limit, one can use a dilation–erosion program to connect the in-focus perimeter sections (right upper and lower corners, Figs. 3B, 3C). Dilation–erosion methods can lead to artifactual

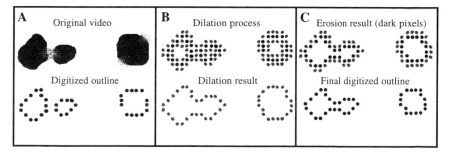

FIG. 3 Dilation and erosion techniques in generating a final reconstructed image of a digitized cell.

shapes if the image has to be dilated too many times in order to connect parts. The final solution, therefore, must always be compared with the videotape image, and the final reconstructed image must be edited, when necessary.

Once all images in a time sequence are digitized and edited for perimeter continuity, there are still a number of necessary steps prior to the genesis and review of a perimeter path. A perimeter may really represent a series of short straight lines joined to approximate a curve as a result of automatic digitization and editing methods, or as a natural consequence of manual digitization. Therefore, replacement of the original outline with a beta spline image (see next section for procedure) will generate curved surfaces which more accurately approximate the original contour of the cell. The effect is similar to that obtained with a "French Curve." Three parameters can be applied in this form of smoothing; "bias," "tension," and "resolution" (for a detailed description of this method, see Barsky, 1988). Bias determines how a tangent to a point at the perimeter affects the curve on either side, tension determines the tightness of the curve, and resolution determines the detail of the curve segments. If one compares first the initially digitized cell image and then the smoothed cell image to the original video image of the cell, it is usually the smoothed cell image which is representative of the original video image.

D. Computing the Cell Centroid

An accurately digitized contour serves three purposes. First, it provides a digitized representation of the cell which can be used in a number of animations. Second, it provides the digital information necessary for measuring morphological parameters. Third, it provides the information for computing the x,y coordinates of the center, or centroid, of the cell in each digitized frame. The cell centroid is extremely important since it is the basis for several of the most basic parameters of cell motion, including speed and directionality, and, therefore, deserves careful consideration by everyone analyzing cell behavior.

The shape of a cell at any time point will affect the position of the centroid, and it, therefore, must be considered in the computation of centroid position. The "initial shape" of a cell represents the connected circuit of pixels at the perimeter. If one uses this initial shape, which has not been refined with beta splines, and fills in the intervals between real perimeter pixels, one obtains a boxy image. However, it is most effective if the centroid is determined from the "final shape" of a cell. The final shape is the mathematically precise beta–spline replacement image and can be determined by the following procedure. A preliminary centroid is computed by averaging the x and y coordinates, respectively, of the pixels of the initial shape. From this centroid, one scans in a single direction until a pixel at the perimeter is intercepted. This pixel is considered the "base pixel," and all other pixels are sequenced in a counterclockwise direction. One can use every pixel (an increment of 1), or any increment desired. By then determining the mathematical equations of the beta splines, one obtains a mathematically defined curve connecting each pair of pixels in the perimeter. Evenly spaced vertices along the curves at increments determined by the level of resolution are then connected to generate the final shape.

The centroid of a cell can then be computed by two obvious methods. In the first method, the average of the x and y coordinates, respectively, of the evenly spaced vertices of the final shape is computed. This is referred to as the boundary method. In the second method, the area is used to compute the center of mass. This is referred to as the area method. The area method is probably more effective for amoeboid cells forming lateral pseudopods because it is less artifact prone. For instance, if a cell is crescent-shaped, the boundary method would be more prone to place the centroid outside the cell perimeter while the area method would be more prone to place it within the perimeter.

E. Parameters Based upon the Cell Centroid

Once the cell perimeter has been digitized at desired intervals into the data file, the digitized image edited and reconstructed with beta splines to generate final cell shapes, and the centroids computed, sophisticated motion analysis software should be capable of generating any type of two-dimensional computer animations and should be able to compute any parameter one can formulate which is based upon either centroid position or cell contour. More importantly, a sophisticated system should be able to generate this information at intervals as short as every thirtieth of a second, the limit of conventional video, or as long as one desires. This will provide the researcher with time plots which can be smoothed and interpreted. In most motility studies, the researcher is first interested in velocity, and, to a lesser extent, the additional parameters one can compute from the centroid tracks. In Figs. 4A and 4B, tracks have been generated from the digitized perimeters and centroids of two wild-type *Dictyostelium* amoeba, and in Figs. 4C and 4D, tracks have been generated from the digitized perimeters and centroids of two mutant *Dictyostelium* amoebae lacking myosin IA (Titus *et al.*, 1992). Such tracks are extremely informative not only in visualizing the path of translocation, but also in obtaining an impression of persistence, directionality, turning and dynamic morphology (e.g., the

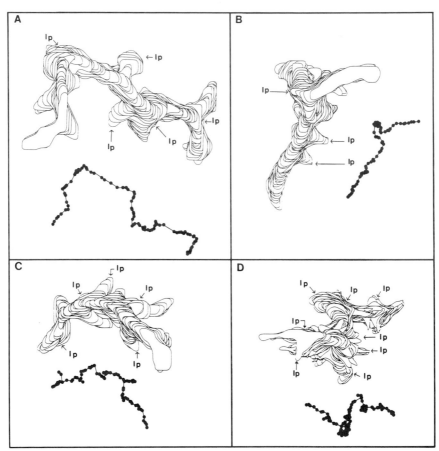

FIG. 4 Computer-generated perimeter and centroid tracks of two control wild-type amoeboid (A,B) and two myoA-minus amoeboid (C,D) amoebae of *Dictyostelium discoideum*. Note the more zig-zag behavior of the mutant cells lacking myosin A. lp, lateral pseudopod. See *Titus et al.* (1992) for details.

formation of lateral pseudopods), parameters which can also be quantified from the centroid and perimeter tracks.

1. Speed or Velocity

Velocity is a measure of how fast a cell is moving, and, in its purest form, should be computed from the centroid position. However, several things should be kept in mind. First, cells can have significant centroid velocity but actually not go anywhere, while other cells may have very little centroid velocity and still go somewhere. In the former case, an anchored cell may continue to extend and retract pseudopods, but not translocate. The centroid of such a cell would consequently move back and forth, and could generate significant velocity even though the cell body makes no net progress in any one direction. In the latter case, a cell may spread in all directions to twice its diameter in response to a stimulant, but since spreading is symmetrical, there will be no significant movement of the centroid. Such a cell will have no measured velocity, if the velocity parameter is based on centroid position. In most cases of amoeboid crawling, however, centroid translocation is a good indicator of motility.

Velocity is simply distance over time. However, one must determine what distance over what time. For the centroid track in Fig. 5A, one can compute average velocity as total distance over time, or net distance in a single direction over time. The difference in values can be fairly dramatic. For instance, if the position of the cell in Fig. 5A were measured at the first and last point in the track, the net distance traveled would be 5.56 μm and the rate would be 5.56 μm divided by the time interval between the first and last position. However, if the distances between all centroids were summed over the same time period, the total distance traveled would be 9.97 μm, and the rate would be 9.97 μm divided by the time interval. The latter velocity is 1.8 times the former. For a centroid track generated from digitized perimeter images in a computer-assisted motion analysis system, velocity can be computed as follows for each interval,

$$\text{Velocity } (f) = \text{scale} \times \sqrt{([x(f) - x(f - I)]^2 + ([y(f) - y(f-I)]/I)^2},$$

when $1 \leq f - I$, f is the current frame, $(x(f), y(f))$ are the x,y coordinates of the centroid of an object in frame f when $1 \leq f \leq F$ and F is the total number of frames, I is the frame increment, and scale is the scale units in

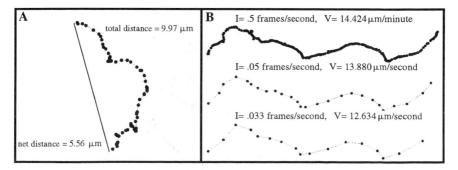

FIG. 5 The effect of interval time on the computation of average velocity over a translocation path.

distance units per pixel. As noted, velocity can vary significantly as a function of the interval time. In the example presented in Fig. 5B, the velocity computed for the interval $I = 0.5$ s is 14.4 μm per minute, the velocity for $I = 20$ s is 13.9 μm per minute, and the velocity for $I = 30$ s is 12.6 μm per minute. These differences are not insignificant (e.g., the difference between the velocity for the shortest and longest intervals in the example in Fig. 5B is 14%) and demonstrate that no published velocity is absolute. A reported velocity should always be considered in terms of interval time. Comparisons of velocity performed by different laboratories should, therefore, be scrutinized with the preceding reservations in mind.

In the computation of the forward velocity of a cell based on centroid position, lateral variation of the centroid track resulting from shape changes can cause jumps in short interval velocities and generate a form of noise in an otherwise smooth translocation path. Computing instantaneous velocity using the central difference method (Maron, 1982) dampens this effect. While the previous computation of velocity provides a measure of distance for a particular interval between centroids, instantaneous velocity provides a measure of the velocity average for two consecutive intervals which is prescribed to the intervening frame. The computation of instantaneous velocity (IV) is computed as follows for each centroid,

$$IV(f) = (scale \times frate) \times \overline{([x(f+1) - x(f)]/I)^2 + ([y(f+1) - y(f)]/I)^2},$$

when $1 \leq f - I$ and $f + I \leq F$, and

$$IV(f) = (scale \times frate) \times \overline{([x(f+1) - x(f)]/I)^2 + ([y(f+1) - y(f)]/I)^2},$$

when $f - I < 1$ and $f + I \leq F$, in the first frame. A similarly specific equation is computed for the last frame. Once velocity has been computed for each interval, or instantaneous velocity has been computed for each frame, one must average them to obtain a single average value with standard deviation for that cell. If several cells have been analyzed, one can obtain the mean value for the population. However, average and mean velocities can be quite deceiving since amoeboid cells move in surges and exhibit relatively constant periods between surges (Wessels *et al.*, 1994; Murray *et al.*, 1992; Royal *et al.*, 1995; Sylwester *et al.*, 1995). Therefore, time plots of the velocity of individual cells (Fig. 6) are usually a requisite for a complete description of the motile behavior of a translocating cell. In addition, the time plot of instantaneous velocity can provide more than 20 additional parameters if the plot exhibits cyclic dynamics. The additional parameters can be computed from the intervals and slopes of the peaks and troughs.

2. Other Parameters Based upon Centroid Paths

A centroid path provides the basic information for computing a number of parameters in addition to velocity. These include the proportion of

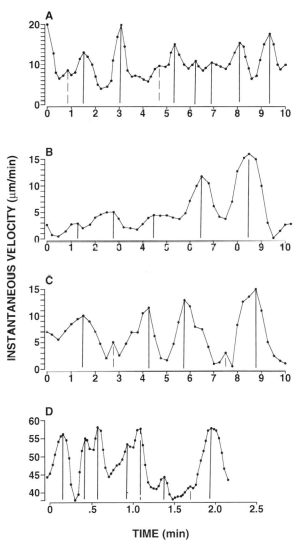

FIG. 6   Time plots of instantaneous velocity generated by computer-assisted motion analysis systems for different cell types. Note that in every case there is a velocity cycle (i.e., the cell moves in surges.) A, *Dictyostelium* discoideum; B, Sup T1 (T cell line); C, HIV-induced T cell Syncytium; D, *Ascaris* sperm.

translocating cells, the proportion of time each cell is translocating, direction, directional change, acceleration, persistence, and axis tilt. Although these parameters could be computed without a computer, the difference in the amount of time it would take is staggering and, in the end, prohibitive if calculations are made at short time intervals for significant numbers of cells.

There are a number of ways to measure direction, directional change, persistence, and acceleration. One solution based on the central difference method (Maron, 1982) is provided for each, but other, sometimes more extravagant, formulas may more effectively suit ones needs.

a. Direction (Dir)

$$\text{Dir}\,(f) = \text{angle}\,([x(f+I) - x(f-I)], [y(f+I) - y(f-I)])$$

when

$$1 \leq f - I \text{ and } f + I \leq F,$$

$$\text{Dir}\,(f) = \text{angle}\,([x(f+I) - x(f-I)], [y(f+I) - y(f)])$$

when

$$f - I < \text{ and } f + I \leq F,$$

$$\text{Dir}\,(f) = \text{angle}\,([x(f) - x(f-I)], [y(f) - y(f-I)])$$

when

$$1 \leq f - I \text{ and } f + I > F,$$

where angle (x,y) is the angle of the vector $\overrightarrow{x,y}$. In generating a graph, multiples of $\pm$ 360 degrees are applied in order to keep the plot continuous. Therefore, an object moving in a spiral would have sequential directions of 0,45,90,135,180,225,270,315,360,405,etc. Direction can be an important parameter in migration studies.

b. Directional Change (Dir Ch)

$$\text{Dir Ch}\,(f) = 0$$

when $f - I < 1$.

$$\text{Dir Ch}\,(f) = |(\text{Dir}[f] - \text{Dir}[f - I])|$$

for all other cases.

If the directional change is greater than 180°, it is subtracted from 360°. This produces values which are always between 0 and 180°. Directional change reflects the directional persistence of a translocating cell.

c. Acceleration (Acc)

$$\text{Acc}(f) = (\text{speed}[f + I] - \text{speed}[f - I]/2$$

when $1 \le f - I$ and $f + I \le F$.

$$\text{Acc}(f) = \text{speed}(f + I) - \text{speed}(f)$$

when $f - I < 1$ and $f + I \le F$.

$$\text{Acc}(f) = \text{speed}(f) - \text{speed}(f - I)$$

when $1 \le f - I$ and $f + I > F$.

Although acceleration has rarely been used in motion analysis studies, it will definitely prove useful in analyzing the chemotaxis of single cells, the dynamics of embryonic cells *in situ*, and the behavioral changes in motility cycles.

d. Persistence (Pers)

$$\text{Pers}(f) = \text{speed}(f)/(1 + [100/360) \times \text{Dir Ch}[f]).$$

It should be noted that this computation of persistence is basically speed divided by the direction of change, and is the same as speed if the cell is not turning. In other words, persistence is proportional to speed and inversely proportional to directional change. Persistence has been computed in many studies of directed cell movement and responsiveness to extracellular signals.

e. Axis Tilt (Tilt)
Axis tile is simply the angle (in degrees and always less than 180°) that the major axis (see following section for computation) drawn through the centroid makes with the horizontal line drawn through the centroid. Multiples of ± 180° are added to give continuity for graphing. Thus if an object is spinning at a constant rate, the axis tilt graph will be a corrected line with a constant slope. Axis tilt is useful in assessing directed movement to a target site or source of a chemotactic signal.

F. Parameters Based upon the Cell Contour

Although most of the quantitation of cell locomotion which is published is usually related to velocity and, therefore, centroid translocation, amoeboid cells move by extending a pseudopod and, in many cases, undergo cyclic shape changes (Satoh *et al.*, 1985; Wessels *et al.*, 1994; Murray *et al.*, 1992; Sylwester *et al.*, 1995). Changes in cell shape during cellular translocation must, therefore, be resolved and quantitated with precision, and computer-assisted systems have evolved for this specific purpose (Soll, 1988; Soll *et al.*, 1988a,b). Using the x,y coordinates of perimeter pixels, rather than the position of the centroid, a number of parameters and unique animations have been formulated to quantitate the dynamic morphology of the whole cell perimeter and portions of the perimeter like the pseudopod.

1. Size Measurements

Size measurements computed from the entire cell contour provide information which, when related to velocity and directionality, is useful in understanding how cell shape and changes in the z axis are involved in translocation. Indeed, plots of cell area, not velocity, were instrumental in first suggesting that *Physarum* amoebae undergo cyclic behavior (Satoh *et al.*, 1985). Size measurements include mean width, maximum width, central width, mean length, maximum length, mean radial length, radial deviation, area, and perimeter. One can also compute predicted volume and predicted surface area from the two-dimensional contour. The formulas or descriptions of several of these parameters depend upon definition of the major axis of the cell. To define the major axis, the pixel on the final shape outline farthest from the centroid is found. This pixel becomes one end of the major axis. The pixel in the contour farthest from that initial point is then found, and the straight line connecting the two points is considered the major axis.

a. Mean Width (Mean Wid)

$$\text{Mean Wid}(f) = \text{Area}(f)/\text{Max Len}(f),$$

where Max Len is maximum length (see below).

b. Maximum Width
The maximum width represents the length of the longest chord perpendicular to the major axis.

c. Central Width
The central width represents the length of the chord drawn through the centroid and perpendicular to the major axis.

d. Mean Radial Length (Rad Len)
Rad Len(f) is the average distance from boundary pixels of the cell to the centroid. Let n be the number of vertices of the final shape, indexed from 0 to $n - 1$. Let $L(i)$ be the distance from vertex i to the centroid. Rad Len(f) can then be computed from the formula

$$\text{Rad Len}(f) = (L[0] + \ldots + L[n - 1])/n.$$

e. Radial Deviation (Rad Dev)
Rad Dev is the ratio of the standard

deviation (SD) of Rad Len to Rad Len, written as a percentage. Rad Dev(f) is then computed by the formula

$$\text{Rad Dev}(f) = 100 \times \text{SD/Rad Len}(f).$$

f. Maximum Length   Maximum length is defined as the length of the major axis.

g. Lengths and Widths Computed from Smallest Enclosing Rectangle   In some geometric studies, it may be useful to compute parameters from the smallest encapsulating rectangle. One can then compute the x bound width and the y-bound width of the rectangle.

h. Area   Area(f) is defined as the area of the final shape minus the area of any holes, which is rarely a characteristic of translocating cells. Let ($x[i]$, $y[i]$) for $i = 0, \ldots, n$ be the vertices of the final shape, such that $x[0] = x[n]$ and $y[0] = y[n]$ (so that it ends where it starts). Let $dx[i] = x[i + 1] - x[i]$ and $dy[i] = y[i + 1] - y[i]$. Then area can be computed from Green's Theorem (Rudin, 1972) by the formula

$$\text{Area}(f) = 0.5 \times \text{abs}\{(x[0] \times dy[0] - y[0] \times dx[0] + \ldots +$$
$$(x[n - 1] \times dy[n - 1] - y[n - 1] \times dx[n - 1])]\}.$$

i. Perimeter (Perim)   Perim(f) is defined as the perimeter of the final shape plus the perimeter of any holes, which is again rarely a characteristic of translocating cells. Based on the definitions for the area formula, the perimeter is computed by the formula

$$\text{Perim}(f) = \sqrt{dx(n - 1)2 + dy(n - 1)2}.$$

Using size thresholding, any perimeters of cytoplasmic particles or holes within the encapsolating cell perimeter can be subtracted before computation.

j. Pedicted Volume (P Vol)   By assigning a three-dimensional geometric configuration which can be computed by a simple formula and which best represents the shape of the cell of interest, on average, one can obtain a rough estimate of volume. For instance, if the 3D shape approximates an ellipsoid, the predicted volume can be computed by the formula

$$\text{P Vol}(f) = (4\pi/3) \times (\text{Max Len}[f]/2) \times (\text{Mean Wid }[f]/2)^2.$$

If the shape more closely approximates half of an ellipsoid because it is supported by a flat substratum, one can divide P Vol (f) by 2. Of course, the result of such a computation is far from accurate, which is one reason why three-dimensional dynamic image analysis systems have been developed (Wessels *et al.*, 1994; Murray *et al.*, 1992; Felder and Kam, 1994).

k. Predicted Surface (P Sur)   Using the same shape assigned for predicted volume, in this case an ellipsoid, one can compute the predicted surface by the formula

$$P \, Sur(f) = CF \times \pi \times Max \, Len(f) \times Mean \, Wid(f),$$

where CF is the ellipsoidal surface correction factor defined by

$$CF = \sin(x) \times \sqrt{\sin 2(x) + r2 \cos 2(x) dx,}$$

where r = Mean Wid(f)/Max Len(f).
Using Simpson's Rule (Gerald, 1980) with $N = 1000$, the computer can approximate the CF by the polynomial

$$CF = 0.15r^2 + 0.065v + 0.785.$$

l. Positive Flow (Pos Fl) and Negative Flow (Neg Fl)   Positive flow is the new area of a cell image (i.e., all two-dimensional expansion zones) formed in a time increment and negative flows is the old area of the same cell image (i.e., all two-dimensional contraction zones) lost in the same time increment. Both can be expressed as percentage total area. If a cell has no z axis change and no volume change during the time increment, Pos Fl and neg Fl values should be similar. To compute these parameters precisely, let f be the current frame, let FI be the flow increment, let A be the interior of the shape at frame f-FI, let B be the interior of the shape at frame f, let P be the area of B that is not in A, and let N be the area in A that is not in B. For the increment FI,

$$Pos \, Fl \, (f) = 100 \times Area \, (P)/Area \, (A) \text{ and}$$
$$Neg \, Fl \, (f) = 100 \times Area \, (N)/Area \, (A).$$

Positive flow and negative flow represent quantitative measurements of the expansion and contraction zones, respectively, of difference pictures (see following sections).

m. Net Flow (Net Fl)   If a cell changes its area as a result of a contour, height, or volume change, Pos Fl and Neg Fl will not be equal at all time points. Therefore, if one subtracts Neg Fl(f) from Pos Fl(f), one obtains Net Flow, which is positive when there is more expansion than contraction, and negative when there is more contraction than expansion. This parameter is useful for analyzing the behavior of cells which increase in height in a cyclic fashion during their behavior cycle. This parameter was originally developed to monitor the area changes associated with systole and diastole in ultrasound images of the left ventrical (E. Voss and D. R. Soll, unpublished observation).

2. Contour Measurements

Measuring contour changes has only begun to be employed in descriptions of cellular behavior, and has proven useful in evaluating the roles of particular cytoskeletal elements through behavioral analyses of cytoskeletal mutants (Wessels *et al.,* 1988). Contour changes of the whole cell envelope have been found to correlate with the velocity cycle of *Dictyostelium* amoebae (Wessels *et al.,* 1994) and human polymorphonuclear leukocytes (Murray *et al.,* 1992), but will eventually be most effective when applied to the dynamics of such protrusions as pseudopods and other cellular zones like uropods (see following section on windowing). Several contour parameters which are especially useful include roundness, convexity, and concavity.

a. Roundness (Rnd)   Roundness is basically a measure of how efficiently a perimeter encloses an area. A circle is most efficient, providing the most area for perimeter. A straight line encloses no area. One simple way to compute roundness is by the formula

$$\text{Rnd}(f) = 100 \times 4\pi \times (\text{Area}[f]/\text{Perim}[f]^2).$$

With this formula, a circle has a Rnd of 100% while a straight line has a Rnd of 0%. Translocating *Dictyostelium* amoebae exhibit an average roundness parameter of approximately 50%, while translocating human T cells exhibit an average roundness parameter of approximately 75%.

b. Convexity (Convex) and Concavity (Concav)   Although Rnd provides a measure of the efficiency of cellular encapsulation, it really provides no information on the complexity of the shape of a cell. Convexity and concavity measurements provide such information. To compute convexity and concavity, line segments connecting the vertices of the final shape of a cell are drawn and the angles of turning from one segment to the next measured. Counterclockwise turns are positive and clockwise turns are negative. Since the cell contour is closed, these always add up to 360°. Convex(f) is the absolute value of the sum of positive turn angles, in degrees, and Concav(f) is the absolute value of the sum of negative turn angles, in degrees. The following equation holds for a cell perimeter with no projections:

$$\text{Convex}(f) - \text{Concav}(f)t = 360°.$$

A circle has a Convex (f) of 360° and a Concav (f) of 0°. A cell with numerous projections and indentations along its perimeter will have a

Convex(f) of n x 180 and a Concav(f) of $n \times 180-360 = (n - 2) \times 180$, where n is the number of projections.

G. Sectoring and Windowing

Since translocating cells are asymmetric and polar, since expansion zones or anterior pseudopod may be shaped quite differently from the posterior uropod or lateral side of a translocating cell, and since different portions of a cell change shape in unique ways during the behavior cycle (Wessels et al., 1994), it is essential that a sophisticated motion analysis program have the capacity to separate a cell into sectors and quantitatively analyze the different sectors, or window a protrusion, like a pseudopod, and quantify the dynamics of the windowed section of the cell. In both sectoring and windowing, certain parameters such as roundness cannot be assessed since they depend upon a complete perimeter. However, parameters such as area, perimeter, positive flow, negative flow, and net flow can be computed. In addition, velocity measurements can be made with precision by using the boxed sides of the sectored image as edge pixels continuous with the true perimeter to compute the centroid position of the sector. Alternatively, one can monitor the position of the central pixel of the true perimeter of the sectored or windowed image to obtain a measure of edge velocity. This is especially useful in quantitating the speed of pseudopod expansion or retraction.

H. Unique Methods for Visualizing Cell Shape and Expansion–Contraction Dynamics Associated with Cellular Translocation

Cell perimeter and centroid tracks, such as the ones in Fig. 4, represent standard methods for reviewing cellular translocation. Both can also be displayed as animated computer movies by a sophisticated motion analysis system. The animated display can include overlapping images, in which each new perimeter image is displayed without erasing previous images, or only the newest frame image, erasing all old perimeters. However, once an image is digitized, there are a number of additional and, in some cases, unconventional methods of display (Soll, 1988; Soll et al., 1988a,b). For instance, the perimeter of the final cell shape can be unwrapped either by angle or by boundary length and stacked in a time series with time increasing downward along the vertical axis (Fig. 7B). This "unwrapped, stacked" display provides a new way of looking at cell contour, and a dot along the opened contour marks the anterior end of the cell (Fig. 7B). Such a display

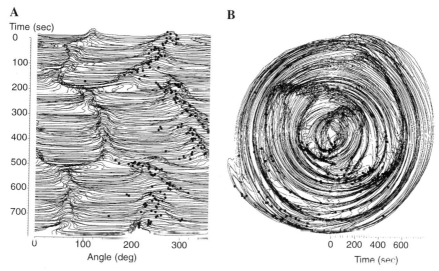

FIG. 7 Unconventional methods of displaying the cell perimeter with time. (A) Unwrapped and stacked, with time increasing downward; (B) circular and stacked radially, with time increasing centrapedally. In this case, a constant increment is added to the cell radius at each time interval. Dots represent anterior end of cell in each case.

can be informative in assessing curvature changes and the dynamics of lateral protrusions. In a second unique display, a time series of the shapes of a single cell can be drawn in a concentric fashion (Fig. 7B). This "circular, stacked" display is generated by fixing centroids of the sequential images at a single, central point in the window, and expanding the image at a constant size increment (a constant increase in radius), frame by frame. Therefore, time moves outwardly from the fixed centroid. Such a display can be informative in identifying rhythms as well as activity at particular locations along the cell perimeter, since inactive portions of the perimeter will stack tightly, while rapidly expanding portions will separate, producing space. Again, the front end of the translocating cell, or another landmark, can be noted by a dot at the perimeter. One can discern a rhythm in the circular, stacked display of the cell in Fig. 7B, which probably represents the behavior cycle of this cell (Wessels *et al.*, 1994).

Perhaps the most useful but underutilized method for displaying the area changes of a translocating cell is the "difference picture" (Soll, 1988; Soll *et al.*, 1988a,b). By superimposing the current frame (Fig. 8B) on a previous frame (Fig. 8A) separated by a selected time increment (16 s in the example in Fig. 9) and by coding "expansion zones" (regions of the later cell image not overlapping the earlier cell image; in this case filled areas), "contraction zones" (regions of the earlier cell image not overlapping the later cell

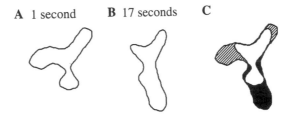

A 1 second   **B** 17 seconds   **C**

FIG. 8   Generating a difference picture. Later image (17 s) is superimposed on earlier image (1 s). The difference picture (C) is coded: filled areas represent expansion zones, hatched areas represent contraction zones, and unfilled areas represent common zones.

image; in this case hatched areas), and "common zones" (areas of the later cell image overlapping the earlier cell image; in this case unfilled areas), one generates a difference picture (Fig. 8C). The total area included in the expansion zones of a difference picture over the selected time interval is referred to as positive flow (Pos Fl(f)) at frame f, and the total area included in the contraction zones is referred to as negative flow (Neg Fl(f)) at frame f, both defined mathematically in a previous section. Frame f represents the later image in each difference picture. Difference pictures can be presented as a time series of static images, as in Fig. 9, or they can be viewed as an animated computer movie, or "dynamic difference picture," which

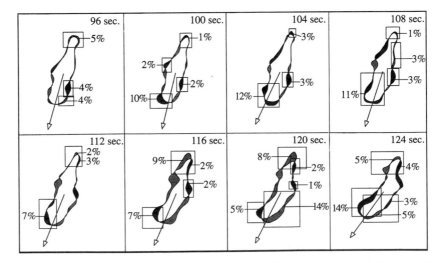

FIG. 9   A sequence of difference pictures of a *Dictyostelium* amoeba translocating in buffer over a 28-s period, during which it extends an anterior pseudopod (assessed from the original DIC images recorded on video). In this series, particular expansion and contraction zones are windowed and quantitated in % total cell area.

cannot be demonstrated in this review. The dynamic difference picture is an extraordinary visualization of cell behavior since it performs a function the human brain cannot. It provides a way of visualizing where a cell is contracting and where it is expanding as it translocates. The human brain cannot generate such a double, coded image because it does not possess geographically precise incremental memory.

An expansion zone or contraction zone in a difference picture is easily windowed and quantified as Pos Fl or Neg Fl in the percent of total cell area (Fig. 9) or square micrometers. PosFl and NegFl are rate measurements which are wholly dependent upon the time increment and which, alone, do not define a morphological entity such as a pseudopod or uropod. As the increment is increased, the distance between the overlapping images increases. If the increment is very large for a translocating cell, the two cell images which are superimposed to generate a difference picture will separate. Therefore, one cannot define a pseudopod or uropod solely from difference pictures. However, difference pictures represent a powerful method for analyzing pseudopod dynamics when coupled with direct observations of the original video.

III. Motion Analysis in Three-Dimensions

Although two-dimensional motion analysis systems are adequate for analyzing many problems related to cell motility, they do not provide a complete description because they do not quantitate cellular dynamics in three dimensions. Cells are not only three-dimensional entities which move along three dimensional paths *in vivo,* but they also perform three-dimensional functions which appear to be related to the regulation of velocity and directionality, like forming pseudopods on and off the substratum (Wessels *et al.,* 1994, 1995; Fukui *et al.,* 1991). In addition, it has been demonstrated that some (Wessels *et al.,* 1994; Murray *et al.,* 1992), but not all (Royal *et al.,* 1995; Sylwester *et al.,* 1995), amoeboid cells undergo cyclic increases and decreases in their z axis during the translocation cycle, resulting in changes in area. Therefore, to completely describe the morphological changes which accompany translocation, to adequately define the aberrant behavioral aspects of mutants defective in specific cytoskeletal elements involved in pseudopod extension, anchoring, and retraction, and to adequately assess the role of pseudopod extension in chemotaxis, a three-dimensional dynamic image analysis system is required. The components of such a system are reasonably straightforward, and several groups have developed similar front ends and dynamic reconstruction software with different levels of sophistication (Wessels *et al.,* 1994; Murray *et al.,* 1992; Felder and Kam,

1994). An advanced version of such systems (Wessels *et al.*, 1994) will be described.

A. Imaging and Recording

The first step in the three-dimensional analysis of a moving cell is obtaining optical sections within a time period short enough for reconstructing a meaningful morphology, then repeating this process at short enough time intervals to develop a meaningful sequence of morphologies which exhibits the behaviors of interest. For cells moving at average centroid velocities of 7 to 20 μm per minute and pseudopods growing at average velocities of approximately 20 μm^3 per 5 s, collecting optical sections within a 2-s interval for a single three-dimensional reconstruction and repeating the process every 5 s provides a reasonable rate for observing pseudopod dynamics, shape changes, and z axis dynamics (Wessels *et al.*, 1994; Murray *et al.*, 1992; Felder and Kam, 1994). A 2-s sectioning period represents, on average, roughly 10% of the time it takes for maximum pseudopod extension and 3% of the time it takes for a cell centroid to translocate 15 μm in the x,y axis, a distance roughly equal to the average length of a *Dictyostelium* amoeba or polymorphonuclear leukocyte. For a white- blood cell or *Dictyostelium* amoeba, which achieve heights as great as 7 to 10 μm (Wessels *et al.*, 1994; Murray *et al.*, 1992), 7 to 15 optical sections of 1.0-μm intervals usually provide the depth necessary for reconstructing cell contour and pseudopods. Although DIC optics have been used in the most recently reported methods (Wessels *et al.*, 1994; Fender and Kam, 1994), any imaging system with a short depth of field should provide reasonably good optical sections at 0.5 to 1.0-μm intervals. The increased speed of sectioning achieved by confocal microscope systems and their application to living cells (e.g., Jester *et al.*, 1991) provides a new alternative imaging system for three-dimensional motion analysis systems.

Imaging can be accomplished in a number of ways. In the simplest form with DIC optics, a moving cell is videorecorded as the microscope stage is ratcheted in the z axis at 0.5- to 1.0-μm intervals. Both stepper motors and manual ratcheting have been successfully employed, and constant movement of the stage can also be employed to obtain sections. Whether stepping or continuously moving a stage, calibration of height as a function of time is essential, and can readily be achieved with beads of known mesh (0.5- to 20-μm diameters). Optical sections are most effectively and efficiently recorded on high-quality video tape. An effective magnification should provide 50 pixels in the cell image on 1/2-inch tape and 100 pixels on 3/4-inch tape. A sequence of optical sections recorded on video is presented in Fig. 10A.

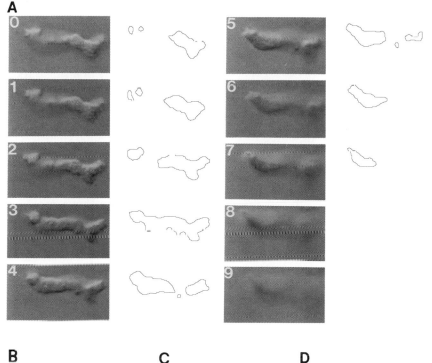

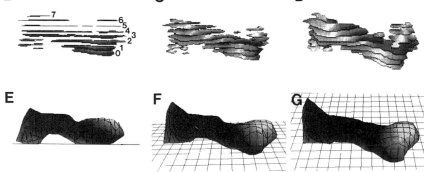

FIG. 10 The method for optically sectioning and then reconstructing a 3D image of a live translocating amoeba employing a computer-assisted three-dimension motion analysis system. (A) A series of video images of the single cell representing the optical sections of the cells beginning at 0 μm (on glass surface) and ending at 9 μm (above the glass surface). Sectioning was complete in less than 2 s. The digitized perimeters of the in focus image are presented to the right of each video frame. (B,C,D) Plate images generated from the smoothed digitized images of the optical sections, viewed at different angles. (E,F,G) Wrapped and smoothed reconstructions viewed at different angles.

B. Digitization and Reconstruction

Digitization of each perimeter is accomplished in a manner similar to that for two-dimensional analysis (Fig. 10A). However, dilation and erosion techniques cannot be used to connect multiple perimeters in a single focal plane when the multiple profiles represent separate z axis bulges in the three-dimensional contour (e.g., Fig. 10A, 0, 1, 2, 4, and 5-μm sections). The original shapes of the independent contours in each optical section, however, should be smoothed by beta splines to generate final shapes. The final shapes in the optical sections can then be stacked to generate three-dimensional plate images like the ones in Fig. 10B. With the proper stereo workstation hardware or software, one can view the plate images at any angle (e.g., Figs. 10C, 10D). Although such plate images have provided important information on the 3D dynamics of cells (e.g., Felder and Kam, 1994), they do not provide the most accurate sense of cell contour to a viewer and they are not amenable to quantitative three-dimensional contour analysis. To generate a 3D image which provides a more accurate sense of contour and which is amenable to quantitative analysis, a method has been devised for wrapping the plated image. The method involves connecting the perimeter pixels of the top, bottom, and sides to generate an encapsulated 3D form. If indentations are not a problem, edge pixels can be connected by a relatively simple process in which the stacked contours are placed on a 2D grid in which the spacing is 1 or more pixels, and at the points at which grid lines cross, vertical distances are measured to the highest and lowest periphery points to interpret top and bottom surfaces. The image is finally encapsulated by rectangular and triangular facets, generating a caged or faceted image. If a cell has lateral indentations, this technique is inadequate since the indentations will be camouflaged. In cases in which indentations must be described, the above technique is modified so that 2 plates at a time are encapsulated, beginning with the bottom most plates (1 and 2). This generates a ribbon image. Next, plates 2 and 3 are encapsulated. This overlapping wrapping procedure is continued until a faceted 3D image is complete. The caged, or faceted, image which is obtained by this method is transparent and can be viewed in dynamic 3D animations. However, in some cases a completely wrapped, nontransparent image provides the most effective means for assessing cellular behavior. To generate such an image, smoothing procedures are applied in which all nonzero grid points surrounding each surface grid point are added and then divided by the number of values used and the result replaces the original grid point. This smoothing process can be performed several times (usually six times) until a 3D cell image is obtained which is smooth and rounded (Fig. 10E) rather than plated or faceted, and which more accurately reflects the original cell contour. Such a reconstruction can again be viewed from any angle with proper hardware or software (e.g., Figs. 10F, 10G).

C. Three-Dimensional Motility Parameters

Three-dimensional reconstruction of a translocating cell at 5-s intervals can be viewed as a series of static images or, with a stereo workstation, can be viewed as a 3D animation from any vantage point. While 2D dynamic image analysis systems provide between 30 and 40 basic parameters of motion and dynamic morphology based upon the 2D dynamics of the centroid and the changes in the perimeter contour, a 3D system has the potential for more than 100 basic parameters. Since 2D parameters have been considered in detail, only a few parameters relevant to a 3D image will be briefly considered.

1. Volume

The simplest method for computing the volume of a reconstructed 3D image is by summing the volumes of blocks generated between the top and bottom grids of the faceted image. A block image is generated by sectioning the object from top to bottom and left to right, and drawing lines between square, cube, or triangular prisms, which are referred to as "blocks." The sum of all block volumes is then multiplied by the scale factor to convert to real volume.

2. Three-Dimensional Centroid and Centroid Parameters

A simple method for computing the position of the cell centroid in three dimensions first involves computing the centroids of the blocks contained in a block image. The center, or centroid of each block is computed as the average of the grid points bounding that block, and, as a weighting device, the centroid of each block is multiplied by the respective block volume. These values are then summed and divided by total volume to obtain the position of the 3D centroid of the cell. Instantaneous velocity can then be computed, as in the case of a 2D centroid track, by the central difference method (Maron, 1982). From the 3D centroid path one can then compute parameters analogous to those computed from a 2D path, such as angle change, directional change, acceleration, and persistence, but the complexity of the formulas is, in some cases, greater due to the addition of the z-axis component; the list of centroid-based parameters can also be expanded because of the third dimension.

3. Parameters Based upon the Cell Contour

The addition of z-axis information increases the complexity and number of contour-based parameters. Not only can the numerous two-dimensional width and length parameters be computed for every optical section in the x,y plane, but analogous z-axis measurements can also be computed, including mean height, maximum height, mean radial height, and radial deviation, for any plane drawn in the z axis through the cell. The same parameters can be computed for any plane drawn at an oblique angle. In addition, one can compute the surface area of the 3D capsule, positive flow in volume units, negative flow in volume units, and net flow in volume units. Contour measurements are also more complex as a result of z-axis information. Roundness, convexity, and concavity were previously defined as two-dimensional parameters, and they can be computed for any plane of the 3D image drawn at any angle. However, new measurements based on the three-dimensional shape of indentations and conical protrusions demand relatively sophisticated formulas and models, beyond the scope of this review.

D. Tracking a Cell's Location in Three-Dimensions

As researchers become more aware of the fact that many cells move along circuitous, three-dimensional routes which, in many cases, involve trajectories through tissue or extracellular matrices, they will need to follow the route of a cell in three dimensions. By staining cells and mixing them with unlabeled cells, the behavior of single cells in populations has been effectively analyzed in two dimensions, and in some cases, stained cells have been identified and analyzed at high resolution for both motility and morphology parameters (Chandrasekhar *et al.*, 1994; Sheldon and Knecht, 1995). Vital staining has, therefore, become a useful tool for identifying a single cell in a population or gel, and has been recently employed to follow the 3D path of a cell translocating in a cell aggregate (Doolittle *et al.*, 1995). Using a microstepper motor and flourescent microscopy, optical sections at approximately 1-μm intervals can be obtained through a 50- to 100-μm z-axis depth, and the fluorescent images of stained cells processed to identify cell positions at each time point (Doolittle *et al.*, 1995). Although this method provides three-dimensional paths of a cell in a matrix, mound, or tissue, the resolution of the fluorescent image is usually too low to resolve the cell perimeter. However, three-dimensional systems can be developed in which cells are recorded at high enough magnification, optical sections are obtained at short-enough distances, and the sectioning range is altered in the z axis as the cell translocates out of the original z-axis region. Such systems, now under development, self-regulate by resolving the cell position in the z axis and self-adjusting the z-axis range by controlling the stepper motor (J. Trepka, E. Voss, and D. R. Soll, personal observations). These systems will ultimately provide a three-dimensional description not only of the cell trajectory but also of the changes in cell shape and the dynamics of pseudopod extension and retraction in three dimensions.

IV. Application: What Have Computer-Assisted Studies Added to Our Understanding of How Cells Crawl?

Because computer-assisted methods for quantitating cell behavior have been underutilized, their full impact has only been fractionally realized; however, the studies so far performed in which these methods have been used have provided us with some surprising insights into how cells crawl, and have begun to impact on molecular studies into the mechanism and regulation of cellular translocation. There are four major aspects of amoeboid cell motility in which computer-assisted studies have begun to play a fundamental role: (1) the basic behavioral process of amoeboid cell translocation; (2) the behavioral consequences of mutations in select cytoskeletal or sensory transduction elements; (3) the effects of extracellular signals on cellular translocation, especially chemotactic signals; and (4) single-cell behavior in cell aggregates or tissues.

V. Basic Behavioral Process of Amoeboid Translocation

A. Instantaneous Velocity

Early quantification of amoeboid cell motility employing computer systems was restricted primarily to velocity measurements and, to a lesser extent, directionality measurements; in many cases, these were performed in order to assess the effects of environmental perturbations or extracellular signal molecules, such as chemoattractants, on cell migration (Futrelle *et al.,* 1982; Varnum and Soll, 1984; Burton *et al.,* 1986; Cheung *et al.,* 1985). Many of these studies were useful in interpreting how fast cells move or respond to external agents, but they provided little insight into the dynamic changes in shape, particularly pseudopod extension, which are basic to cellular translocation (Wessels *et al.,* 1994). What was needed was a complete quantitative description of cellular behavior, including both centroid dynamics and morphological changes, on a substrate which supports cellular translocation and in a neutral medium (i.e., a medium lacking chemotactic or chemokinetic signals). Presumably, such a description would represent the basic behavioral phenotype of a cell. In developing computer-assisted descriptions of this type, several assumptions which affect how we now think about cell motility have been challenged. For instance, although many models describing cell motility and pseudopod extension incorporate extracellular signals and signal translocation pathways in the regulation of the spatial and temporal dynamics of actin polymerization basic to pseudopod extension and cellular translocation, many animal cells will translocate at relatively high velocities on quite neutral substrates like glass and plastic, and will do so at very low cell density in chambers perfused with buffer. In some of these systems, the surfaces have not been conditioned with matrix molecules, perfusion precludes the accumulation of soluble signalling molecules, and low density precludes cell–cell interactions. This is true for *Dictyostelium* amoebae (Varnum and Soll, 1984) and human polymorphonuclear leukocytes (Murray *et al.*, 1992). In the case of T cells, it has been demonstrated that instantaneous velocity is similar on plastic, glass, dehydrated collagen, and hydrated collagen, but slightly reduced on bovine aortic endothelium (BAE) (Sylwester *et al.*, 1995). However, even though the speed and directionality of translocating cells are less sensitive to the quality of the substrate than one would expect, the proportion of translocating cells and the periods of persistent translocation can be affected significantly by the nature of the substrate, in many cases as a result of anchoring. For instance, on BAE, the proportion of translocating T cells is dramatically reduced due to penetration of the cell's pseudopods into the endothelial sheet (Sylwester *et al.*, 1995). In many cases, glass or plastic does not provide a sufficiently adhesive surface for a cell to exert the forces necessary for two-dimensional translocation, and, in such cases, cells may be released from the substratum or actually move their cell bodies backward when they extend a pseudopod because the cell body is not sufficiently anchored. In direct contrast, if a substratum provides too great an adhesive force (e.g., a glass or plastic surface coated with too high a concentration of polylysine), the proportion of anchored cells increases. In some amoeboid cell populations, like nematode sperm on glass, the majority of cells can form pseudopods, but only a small fraction of the population translocates (Royal *et al.*, 1995; Sepsenwol and Taft, 1990). In the case of anchored nematode sperm, the pseudopod, which does not retract or significantly change shape (Sepsenwol and Taft, 1990), still exhibits MSP Treadmilling, demonstrating that the machinery for translocation is active even though the cell is anchored (Royal *et al.*, 1995), and in the case of fibroblasts, the lamellapodium may exhibit ruffling behavior in the absence of translocation. Therefore, in many cases the adhesive properties of the substratum exert more of an effect on the proportion of cells which translocate than on the actual speed of translocation.

How fast do amoeboid cells translocate? As noted, the actual velocity of an amoeboid cell will depend to some degree upon the interval between measurements, but even with this caveat in mind, there are some reasonable estimates of how fast cells move on average. For instance, the average rate of translocation of *Dictyostelium* amoebae has been reported to be between 7 and 14 μm per minute (Varnum and Soll, 1984; Fischer *et al.*, 1989; Segall, 1992), that of polymorphonuclear leukocytes between 5 and 15 μm per minute (Murray *et al.*, 1992; Parrot and Wilkinson, 1981; Burton *et al.*, 1986), that of keratinocytes between 5 and 40 μm per minute (Cooper and Schliwa, 1986), and that of nematode sperm between 5 and 50 μm per minute or even higher (Royal *et al.*, 1995; Sepsenwol and Taft, 1990). However, as noted in discussing the method for computing velocity, any average velocity must be interpreted not only within the context of the interval time used in measurements and computation, but also in relation to whether all cells in a population or only those exceeding a minimum velocity threshold were used in the computation.

B. Velocity and Behavior Cycles

Detailed time plots of centroid velocity provided by computer-assisted systems have revealed that all cells so far tested move in surges and that the resulting rhythmicity represents a complex behavior cycle which, in some cells, includes z-axis dynamics (Wessels *et al.*, 1994, Murray *et al.*, 1992). Velocity cycles have been demonstrated for *Dictyostelium* amoebae (Wessels *et al.*, 1994), human polymorphonuclear leukocytes (Murray *et al.*, 1992), human peripheral blood T cells (Sylwester *et al.*, 1995), HIV-induced, giant syncytia (Sylwester *et al.*, 1995), and *Ascaris* sperm (Royal *et al.*, 1995), and they have been suggested in rhythmic shape changes of individual *Physarum* amoebae (Satoh *et al.*, 1985). The average period of these cycles have been estimated to be between 1.0 and 1.7 min in actin-based systems (Wessels *et al.*, 1994; Murray *et al.*, 1992) and less than 0.5 min in MSP-based, nematode sperm (Royal *et al.*, 1995). These velocity cycles are characterized by low standard deviations for the average period computed for single cells. Examples of velocity plots exhibiting cycling are presented for a *Dictyostelium* amoeba, a SupT1 cell (a human T cell line), an HIV- induced T cell syncytium more than 20 times the size of a single cell, and an *Ascaris* sperm (shown in Figs. 6A, 6B, 6C, and 6D, respectively). Amoeboid velocity cycles, therefore, appear to span the animal kingdom, from soil amoebae to human white blood cells, and the periodicity appears to be independent of average cell velocity. Slow and fast cells of the same strain exhibit the same period, and a single cell may greatly alter velocity between surges and still exhibit a constant interval between velocity peaks (Sylwester *et al.*, 1995). Since the amoeboid nematode sperm has replaced actin with MSP as the major polymerizing protein in the pseudopod and still exhibits a cycle (Fig. 6D), it has been tentatively concluded that motility cycles are not an exclusive characteristic of actin polymerization. Perhaps the most startling characteristic of the cycle is its size-independence. Motile HIV-induced T cell syncytia, which can achieve volumes up to 1000 times that of individual T cells (Sylwester *et al.*, 1993), exhibit approximately the same average velocity cycle as individual T cells (Fig. 6C) (Sylwester *et al.*, 1995).

In the only two studies in which the basic behavior cycle of amoeboid cells was analyzed in three dimensions (Wessels *et al.*, 1994; Murray *et al.*, 1992), cyclic increases and decreases in the z axis were observed, and usually correlated with velocity troughs and peaks, respectively. During the rapid phases of translocation along the substratum, cells, on average, underwent a decrease in maximum height, and during the interspersed phases of depressed translocation along the substratum (i.e., velocity troughs), the leading edge of the pseudopod usually lifted off of the substratum, the center of mass of the cell body moved forward along the cell axis, the maximum z axis increased, and new pseudopods, in many cases, extended from the lateral surface of the cell body (Wessels *et al.*, 1994; Murray *et al.*, 1992). The rhythmicity of *Dictyostelium* amoebae and polymorphonuclear leukocytes translocating in buffer, therefore, correlates with a three-dimensional behavior cycle. However, T cells, which have a round cell body resulting from the dominant, lobular round nucleus that fills roughly 70% of cell body volume, exhibit a 1.6-min velocity cycle without an apparent z-axis component suggested by constant area (Sylwester *et al.*, 1995). Nematode sperm exhibit no significant change in morphology, but still exhibit a velocity cycle (Royal *et al.*, 1995).These latter observations suggest that velocity cycles need not involve height changes, shape changes, or pseudopod retraction.

C. How Amoeboid Cells Turn

Computer-assisted analyses have also provided the first insights into how cells surge and how cells make sharp turns by providing information on the spatial and temporal dynamics of pseudopod extension in relation to whole cell motility. Because computer-assisted two- and three-dimensional systems are both based upon the digitization of cell contours, expansion zones formed from both the anterior end and the lateral sides of a translocating cell can be quantitated in time, and the spatial dynamics of these zones, especially in relation to the substratum, can be assessed in the turning process (Wessels *et al.*, 1994).

In such a detailed 3D analysis of pseudopod formation by *Dictyostelium* amoebae translocating in buffer (Wessels *et al.*, 1994), pseudopods were defined according to shape and minimum growth dynamics, and anterior and lateral positions were determined according to the translocation vector defining the long axis of the cell in the preceding translocation step. Of 31 anteriorly extended pseudopods, 13 were initially formed off of the substratum (i.e., their ventral surface did not contact the substratum), but 7 of these descended to the substratum and retained the role of leading edge, resulting in anterior cellular translocation (Wessels *et al.*, 1994). The 6 remaining anterior pseudopods which formed off of the substratum were retracted. Eighteen of the 31 analyzed anterior pseudopods formed on the substratum and in all cases retained the role of leading edge, resulting in anterior cellular translocation (Wessels *et al.*, 1994). Therefore, 19% of anterior pseudopods were retracted and in all cases were initially formed off of the substratum. An example of retraction of an anterior pseudopod is presented in the wrapped three-dimensional reconstructions in Fig. 11. In this example, the original anterior pseudopod (a) is retracted into the main cell body as the cell forms a lateral pseudopod (l) which assumes the role of leading edge.

Of 60 pseudopods formed laterally, 37 formed off of the substratum (Wessels *et al.*, 1994). Of these, 25 were retracted. The 12 remaining pseudopods which initially formed off of the substratum fell to the substratum, became the new leading edge, and initiated a sharp turn. Twenty-three of the 60 analyzed lateral pseudopods formed on the substratum. Nineteen of these became the new leading edge and initiated a sharp turn. Four were retracted. An example of a lateral pseudopod assuming the role of leading edge and initiating a sharp turn is presented in the wrapped three-dimensional reconstruction in Fig. 11. An analysis of pseudopod formation in additional strains of *D. discoideum* revealed similar proportions and dynamics (Shutt *et al.*, 1995a). These results demonstrate that the majority of lateral pseudopods initially formed off of the substratum by *Dictyostelium* amoebae are retracted, while the majority of lateral pseudopods formed on the substratum assume the role of leading edge and result in a sharp turn. These results suggest that contact of the ventral surface of a lateral pseudopod with the substratum usually leads to a sharp turn.

However, turning is not always effected by the extension of a lateral pseudopod. *Dictyostelium* amoebae can also make gradual turns through extension of anterior pseudopods at an angle. Such biased expansion appears to be the major mechanism of turning either in cells which exhibit difficulties in redirecting motility into a new lateral pseudopod (Sylwester *et al.*, 1993; Shutt *et al.*, 1995b), or in cells which do not form lateral

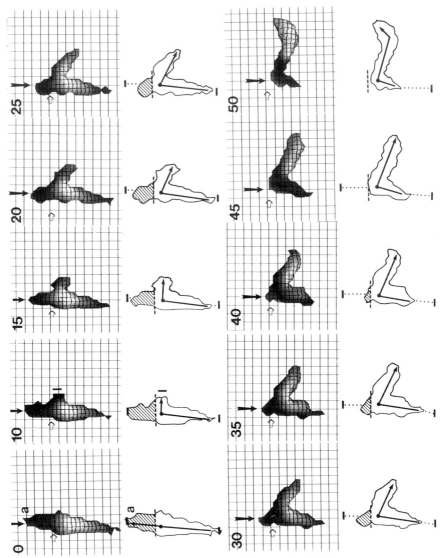

FIG. 11   Three-dimensional reconstructions viewed dorsally from the top and diagrams of expansion of the lateral pseudopod and retraction of the anterior pseudopod are presented for a *Dictyostelium* amoeba translocating in buffer. A wrapped reconstruction viewed from the top is presented in the upper portion of each panel. Time is given in seconds. In the wrapped images, the blunt unfilled arrow denotes the constricted interphase between the main cell body and the anterior pseudopod. The filled long arrow follows pseudopod retraction. In the perimeter image in the lower portion of each panel, the interface between main cell body and the prior anterior pseudopod is presented as a dashed line, the prior anterior pseudopod is hatched, the most anterior point of the prior pseudopod and the most posterior point of the uropod are noted as short horizontal bars, and retraction of the respective entities are monitored as dotted lines. The cell axis is noted by a line arrow within the cell perimeter. A small filled circle notes the pivot in the axis. The distance between lines in the grid is 1.1 μm.

pseudopods, as is the case for *Ascaris* sperm (Royal *et al.,* 1995). *Ascaris* sperm do not form lateral pseudopods, do not retract their anterior pseudopod during persistent translocation (Sepsenwol and Taft, 1990; Nelson *et al.,* 1982), and exhibit very little change in either cell or pseudopod morphology during a behavior cycle which can include velocity peaks of 60 μm per minute and an average period of 0.35 min (Royal *et al.,* 1995). Although these crawling sperm do not alter the crescent shape of their pseudopod, they can move at an angle by biased expansion of the pseudopod to one side, maintaining pseudopod shape by localized contraction kittycorner to localized expansion.

D. Pseudopod Formation during Translocation

The capacity to quantify the dynamics of cellular translocation through such parameters as instantaneous velocity and directionality, to monitor changes in shape through such parameters as roundness and convexity, and simultaneously to quantify the frequency, position, rates of expansion and retraction, and shape of every pseudopod-like protrusion emanating from the main cell body, provides a means for interpreting the rules governing amoebic translocation. This type of analysis has been performed primarily on *Dictyostelium* amoebae translocating in buffer (Wessels *et al.,* 1994, 1995). The results of these studies suggest an ordered process of pseudopod expansion and retraction, with the following characteristics, or rules. (1) Only one pseudopod expands at any one time, on average. Therefore, if the rate of expansion is plotted as a function of time for sequential pseudopods of a crawling cell, there is little overlap in the periods of expansion (Fig. 12). (2) Expansion of a pseudopod usually occurs within a period averaging 37 ± 8 s, with few disruptions (Fig. 12); when there is a disruption, it is usually accompanied by expansion of an alternative pseudopod. (3) A cell is almost always in the process of expanding a pseudopod, which is evident from the expansion periods diagrammed at the top of Fig. 12. (4) A cell translocating in buffer extends anterior and lateral pseudopods in an interspersed fashion, with rarely more than two lateral or two anterior pseudopods formed in sequence (Fig. 12); for example, the sequences of anterior and lateral pseudopods formed by the two *D. discoideum* amoebae described in Fig. 12 were a-l-a-l-l-a-l-l-a-a, and a-l-a-a-l-a-l-a-l-a, where "a" denotes an anterior pseudopod and "l" denotes a lateral pseudopod.

E. Pseudopod Retraction During Translocation

Computer-assisted, dynamic three-dimensional reconstructions of translocating *D. discoideum* amoebae (Wessels *et al.,* 1994) and human polymor-

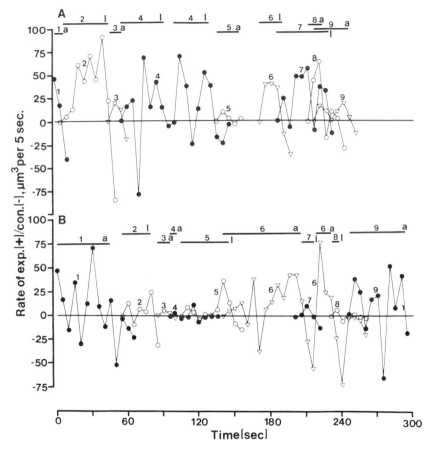

FIG. 12 The sequential extension of pseudopods analyzed by a computer-assisted three-dimensional motion analysis system. Velocity plots are presented of anterior (a) and lateral (l) pseudopods formed in sequence by two representative wild-type *Dictyostelium* amoebae (A,B). The rate of volume change for each pseudopod (numbered sequentially from 1 to 9 for both cells) is plotted as a function of time during its expansion phase. The periods of expansion are diagrammed at the top of the velocity plots. Positive rate values represent expansion and negative values contraction. The dynamics of pseudopod retraction are not present.

phonuclear leukocytes (Murray *et al.*, 1992) have revealed that pseudopod retraction represents a precise process, highly regulated both spatially and temporally. As noted in a preceding section, chemotactically responsive amoebae translocating in buffer form many lateral pseudopods which do not become the new leading edge of the cell and are retracted (Wessels *et al.*, 1994), and it has been suggested that these pseudopods may serve as a sensing device in chemotaxis (Wessels *et al.*, 1994). This suggestion is supported by the observation that lateral pseudopod formation is suppessed when cells are moving toward the source of a spatial gradient of chemoattractant (Varnum-Finney *et al.*, 1987a), and when cells are translocating toward the aggregation center in the front of a natural wave of chemoattractant in an aggregation territory (Wessels *et al.*, 1992). In both cases, a high chemotactic index suppresses the putative lateral sensing system, since the cell is on track. Therefore, amoebae in the process of assessing a chemotactic gradient extend and retract pseudopods off of the substratum, and the decision-making process as well as the retractive process very likely involves specialized actin-associated proteins. As previously noted, anterior and lateral pseudopods formed off the substratum have a much higher propensity for retraction than pseudopods formed on the substratum (Wessels *et al.*, 1994). Retraction follows a common scenario in which the pseudopod, which is usually extended from the anterior half of the main cell body, is retracted back into the main cell body. After forming a pseudopod off the substratum, the cell usually moves anteriorly by extending an anterior pseudopod on the substratum. The pseudopod formed off the substratum is retracted to a position which remains fixed in relation to the substratum. The rigidly fixed position of the base of a retracting pseudopod is impervious to the forward motion of the cell body. The fixed position of the base of a retracting pseudopod is demonstrated in Fig. 11. In effect, the cell body walks past the retracting pseudopod. Since the ventral surface of a retracting pseudopod is, in most cases, not in contact with the substratum, an anchoring structure at the base of the pseudopod must contact and remain fixed to the substratum, and the translocating cell body must somehow walk through the pseudopod base. The three-dimensional description of pseudopod extension, pseudopod retraction, and cellular body dynamics during the act of translocation in buffer has, therefore, revealed behavioral details which must involve specific cytoskeleton elements, and provides new hypotheses for the roles played by the many actin-binding elements in search of a function (see the following section on ponticulin).

VI. Aberrant Phenotypes of Cytoskeletal Mutants

A. Use of Computer-Assisted Motion Analysis Systems for Describing Aberrant Phenotypes

Perhaps the most profound impact that computer-assisted motion analysis systems have had in cellular motility studies is in the elucidation of motility defects exhibited by specific cytoskeletal mutants. The *Dictyostelium*

amoeba translocates in a fashion very similar to polymorphonuclear leukocytes, and has become a model for investigations into the role of actin-
accessory proteins in motility because *Dictyostelium* is haploid and, therefore, accessible to targeted gene disruption (Spudich, 1989; Knecht *et al.,*
1993). When actin-accessory proteins, including the myosin II heavy chain
gene (Knecht and Loomis, 1987; DeLozanne and Spudich, 1987), were first
disrupted in *Dictyostelium,* the surprise was that they were viable, at least
in vitro, and that most of them could change position on a substrate (a
very loose definition of crawling). The response to these initial qualitative
observations was that each mutated accessory protein had a functional
counterpart (Gerisch *et al.,* 1991, 1992). However, subsequent computer-
assisted analyses of several of these cytoskeletal mutants demonstrated that
they indeed had aberrant behavioral phenotypes, and that several of these
cytoskeletal elements played roles in the efficiency, order, frequency, rate,
extent, and location of pseudopod extension and retraction (Condeelis,
1993a; Cox *et al.,* 1992; de Hostos *et al.,* 1993; Wessels *et al.,* 1988, 1991,
1995; Wessels and Soll, 1990; Titus *et al.,* 1992; Alexander *et al.,* 1992; Soll
et al., 1990). Select examples of computer-assisted analyses of cytoskeletal
mutants are reviewed below, with special emphasis given to the phenotypic
traits uncovered by computer analysis and the suggested roles played by
the mutated gene products in cell motility.

B. Myosin II Heavy Chain Mutants

In 1987, two laboratories, working in concert, simutaneously generated a
myosin II disruptant referred to as hmm (DeLozane and Spudich, 1987),
and a transformant with an antisense construct referred to as *mhc*A (Knecht
and Loomis, 1987) which suppressed myosin II levels to less than 5% of
the normal level. Both mutant strains did not develop past the aggregate
stage. However, since aggregation is dependent on cellular motility, it was
concluded that cells were motile in the absence of myosin II, and casual
observations indicated that cells were indeed capable of changing location
on a substratum. However, subsequent computer-assisted analysis of both
mutant strains demonstrated that although the individual cells could sometimes change position on a substratum, motility was highly aberrant. It was
first demonstrated that individual hmm and *mhc*A amoebae were much
rounder on average and translocated at a markedly reduced instantaneous
velocity compared to wild-type cells (Wessels *et al.,* 1988; Wessels and Soll,
1990). Although *myo*II−cells could extend pseudopods, the dynamics of
pseudopod expansion were defective. First, the initial area, rate of expansion, and final area of the pseudopods were drastically reduced (Wessels
et al., 1988). Second, pseudopods were formed in random directions, demon-

strating that in *myo*II− or *myo*II−diminished cells, there was a break-down in cellular polarity (Wessels *et al.,* 1988; Wessels and Soll, 1990). In Fig. 13, a comparison is presented of computer-generated centroid tracks of a parental wild type cell (Fig. 13A) and a myosin heavy-chain null mutant cell (Fig. 13B) before and after the addition of 10^{-6} *M* cAMP (the time of cAMP addition is noted in both cases by an arrow at 0 s) (Wessels *et al.,* 1988). It is clear from the centroid track that the wild-type cell translocated in a directionally persistant fashion prior to inhibition by cAMP, while motility of the mutant cell was not persistent and the direction of the centroid was random. In a subsequent computer-assisted analysis of a myosin heavy-chain null mutant, HS2215 (Manstein *et al.,* 1989), it was demonstrated that intracellular particle velocities were depressed and there was no anterior bias in the tracks of individual particles, demonstrating

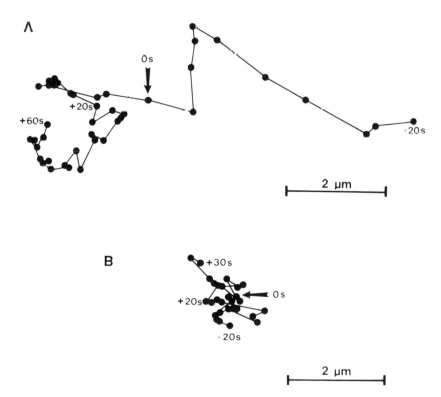

FIG. 13 Centroid plots of a representative myoII⁺ wild-type cell (A) and a representative myoII⁻ mutant cell (B) prior to and after the rapid addition of 10^{-6} *M* cAMP. Negative values represent time in seconds (s) prior to addition and positive values represent time in seconds after addition of cAMP. The arrow represents the time of addition of cAMP. A, W.T.; B, HS2215. Reproduced from The Journal of Cell Biology, 1990 **111,** 1137–1148 by copyright permission of The Rockefeller Press.

again the loss of cell polarity in myosin II mutants (Wessels and Soll, 1990). Therefore, it was concluded that intact myosin II was necessary for (1) rapid, persistent, and directional translocation; (2) the elongate morphology of a translocating cell; (3) cellular polarity (i.e., definition of an anterior end and a uropod; (4) normal pseudopod extension; and (5) normal intracellular particle motility (Wessels *et al.*, 1994; Wessels and Soll, 1990). The consequences of this dramatic behavioral phenotype on aggregation and morphogenesis are profound. *Dictyostelium* depends on efficient single-cell motility in the early stages of aggregation, when single cells chemotax toward aggregation centers (Konijn and Raper, 1961; Devreotes, 1989; Gerisch *et al.*, 1975; Wessels *et al.*, 1992), in the later stages of aggregation, when single cells enter streams and move as intimate cellular cables into the aggregate (Bonner, 1947), and in the subsequent stages of morphogenesis, when there are complex movements of groups of cells in the construction of the fruiting body (Bonner, 1944). The immediate effect of the absence of myosin II is a dramatically reduced capacity to move in a directed fashion in a spatial gradient of the chemoattractant cAMP (Wessels *et al.*, 1988). This was demonstrated by measuring the average chemotactic index (CI) of mutant and wild-type cells in a spatial gradient of chemoattractant. A CI of +1.0 reflects direct movement toward the source of attractant and a CI of −1.0 reflects direct movement away from the source. A CI of 0.0 reflects randomness. The average CI of parental wild-type cells was computed to be +0.50 ± 0.40 (N = 17 cells), which reflects a strong positive chemotactic response, while those of hmm and *mhc*A cells were 0.1 ± 0.3 and +0.07 ± 0.33, respectively (Wessels *et al.*, 1988), both latter values barely above randomness. Therefore, although *myo*II⁻ cells acquire chemotactic machinery (Peters *et al.*, 1988), they are unable to use it effectively to chemotax, presumably because of the loss of cellular polarity and the defective extension of pseudopods. In a subsequent analysis of *myo*II⁻ cells in aggregation mounds, it was demonstrated that the spiral flow of cells in mounds was substantially reduced in comparison to wild-type cells (Eliot *et al.*, 1993); in a similar analysis in which single, stained mutant cells were monitored in three dimensions in mounds, it was demonstrated that the three dimensional paths of *myo*II⁻ cells were dramatically truncated (Doolittle *et al.*, 1995), in a manner quite similar to the paths of individual *myo*II⁻ cells in buffer (Wessels *et al.*, 1988). Finally, in an analysis in which fluorescently labeled *myo*II⁻ cells were mixed with a majority of unlabeled wild-type cells and their behavior in streams was analyzed by computer-assisted methods, it was demonstrated that *myo*II⁻ cells were displaced to the lateral edges of predominantly wild-type mixed streams; although mutant cells moved faster in streams than as individual cells, their shape was significantly distorted and they did not assume the behavior cycle exhibited by wild-type cells (Sheldon and Knecht, 1995). These results suggest that *myo*II⁻ cells not only are far less motile, but also are cortically flacid and prone to distortion. Previous biophysical measurements demonstrated that myoII$^-$ cells exhibit a significant reduction in cortical tension (Pasternak $et\ al.$, 1989). In addition, it was demonstrated that ATP did not stimulate contraction of cortical cytoskeleton models of myoII$^-$ cells as it did with wild-type cells (Kuczmarski $et\ al.$, 1991). Together, these results demonstrate that myosinII, which is an actin motor (Endow and Titus, 1994; Spudich, 1989) and is localized in the cortex of the posterior two-thirds of a cell (Yumura and Fukui, 1985; Nachmias $et\ al.$, 1989), plays a key role in cellular morphology and motility, and that the defects in polarity, velocity, and pseudopod extension exhibited by myoII$^-$ cells nearly obliterate the capacity of these mutant cells to chemotax efficiently (Wessels $et\ al.$, 1988), stream (Sheldon and Knecht, 1995), and undergo multicellular morphogenesis (Knecht and Loomis, 1987; DeLozanne and Spudich, 1987; Eliot $et\ al.$, 1993).

C. Myosin I Mutants

Although computer-assisted methods played an important role in elucidating the behavioral defects of myoII$^-$ cells, the behavioral phenotype of these mutants was very strong; therefore, a good portion of it could have been interpreted qualitatively (i.e., in the absence of computer-assisted systems). This was not, however, the case in interpreting the behavioral defects of mutants of the group I myosins. The group I myosins represent a family of single-headed myosins, which remain monomeric under all tested $in\ vitro$ conditions (Korn and Hammer, 1990). Although the group I myosins differ primarily in their C terminal domains, their sequence similarities suggest that there may be functional redundancy among members of the group, and that a mutation in one group I myosin gene would have no behavioral consequence since it would be compensated for by another group I myosin. The first disruption of a member of the group I myosins was in the myoB gene, and the major developmental defect was a reproducible delay in the aggregation process and subsequent program of morphogenesis (Jung and Hammer, 1990). However, the general aspects of growth, streaming, and morphogenesis seemed relatively normal, and when wild-type and mutant cells were compared by casual observation, no differences were obvious in cellular morphology or the capacity to translocate. However, when the behavioral phenotype of myoB$^-$ cells was scrutinized with computer-assisted methods, a surprising defect was observed. MyoB$^-$ cells formed lateral pseudopods at least two times as frequently as parental wild-type cells and turned sharply more frequently (Wessels $et\ al.$, 1991). Perimeter and centroid tracks of myoB$^-$ cells were not as persistent as those of myoB$^+$ cells, difference pictures of myoB$^-$

cells exhibited more lateral expansion zones, on average, than wild-type cells, and the average instantaneous velocity of $myoB^-$ cells was depressed. Therefore, a mutation in the $myoB$ gene was not compensated for, or, at least, not completely compensated for, by one or more of the other group I myosins. These results demonstrate that cells lacking $myoB$ are not behaviorally less active than cells possessing $myoB$; rather, $myoB^-$ cells are more active in forming lateral pseudopods, which apparently results in increased turning and a depression in instantaneous velocity. Therefore, $myoB$ appears to be involved in suppressing lateral pseudopod formation rather than in the basic mechanism of pseudopod extension.

The second group I myosin gene selectively disrupted was $myoA$ (Titus *et al.,* 1992). Surprisingly, $myoA^-$ cells translocating in buffer exhibited a behavioral phenotype which was similar to that of $myoB^-$ cells. Again $myoA^-$ cells formed pseudopods at least twice as frequently as $myoA^+$ cells and turned sharply into these lateral pseudopods at least twice as frequently as wild-type cells. This behavior was clearly articulated in perimeter plots of cells translocating in buffer. While wild-type cells moved in a relatively persistent fashion (Figs. 4A,B), mutant cells zig-zagged, generating shorter and less persistent perimeter and centroid tracks (Figs. 4C,D). Since $myoA^-$ cells formed more lateral pseudopods and turned more frequently, it was assumed that suppression of lateral pseudopod formation by a spatial gradient of cAMP (Varnum-Finney *et al.,* 1987b) would normalize the path of $myoA^-$ cells. $MyoA^-$ cells did exhibit a reduction in the frequency of lateral pseudopod formation in a spatial gradient of cAMP to a level close to that of wild-type cells, but $myoA^-$ cells still turned 3 times as frequently as wild-type cells in a spatial gradient (Titus *et al.,* 1992).

Since it was demonstrated that lateral pseudopods which formed on the substratum had a higher propensity to initiate a sharp turn, it was proposed that the increase in the frequency of turning by $myoA^-$ cells was due to an increase in the number of pseudopods formed on the substratum (Wessels *et al.,* 1994). This possibility has been tested using a computer-assisted three-dimensional analysis system. The three-dimensional results support the prediction that $myoA^-$ cells form more pseudopds on the substratum, but they have also uncovered several additional behavioral defects due to the absence of myosin A. First, while $myoA^+$ cells extend one pseudopod at a time (e.g., Fig. 12), $myoA^-$ cells extend multiple pseudopods at the same time. Second, while $myoA^+$ cells extend pseudopods in a relatively continuous fashion, $myoA^-$ cells extend pseudopods in a disrupted fashion, leading to a longer average period of extension. Third, $myoA^-$ cells form 3 to 4 times as many pseudopods on the substratum as wild-type cells, resulting in the potential for up to 3 to 4 times as many sharp turns. It is clear, therefore, that myosin A is not involved in the basic process of pseudopod extension. Rather, it appears to play a regulatory role in the orderly extension of pseudopods without overlap, in the continuous expansion of pseudopods and in the proportion of pseudopods formed off the substratum (Wessels *et al.*, 1995). Although the behavioral phenotype of the *myo*A⁻ mutant now seems to be a strong one, it would very likely have been missed without the quantitation afforded by both two- and three-dimensional computer-assisted motion analysis systems.

D. ABP-120 Mutants

Another example of the value of computer-assisted systems is in the elucidation of the behavioral phenotype of ABP-120 mutants. The properties of ABP-120 suggest that it is a mini-filamin which cross-links F-actin (Condeelis *et al.*, 1982, 1984; Noegel *et al.*, 1989; Bresnick *et al.*, 1990); localization of ABP-120 in the cortex of pseudopods supports the suggestion that it is involved in pseudopod extension and motility (Condeelis, 1981; Ogihara *et al.*, 1988; Cox *et al.*, 1992). Initial computer-assisted analysis of the motile properties of an ABP-120⁻ mutant cell line generated by 1-methyl-3-nitro-1-nitrosoguanidine mutagenesis revealed no differences between parental ABP-120⁺ cells and the chemically induced ABP-120⁻ cells for average velocity and chemotaxis in a spatial gradient of cAMP (Brink *et al.*, 1990). However, a subsequent computer-assisted analysis of two APB-120⁻ cell lines generated by targeted gene disruption in two separate parental strains, Ax3 and Ax4, demonstrated that these APB-120⁻ cells were, on average, rounder, did not reextend pseudopods in a normal fashion after the rapid addition of 10^{-6} M cAMP, and translocated in buffer at one-third the velocity of APB-120⁺ cells (Cox *et al.*, 1992). In addition, APB-120⁻ cells extended pseudopods at a lower frequency during translocation in buffer and the pseudopods which were formed achieved smaller average volumes (Cox *et al.*, 1992). Electron microscopic analyses of the microfilament mesh in pseudopods demonstrated a clear abnormality in ABP-120⁻ cells (Cox *et al.*, 1995a). These behavioral, morphological, and ultrastructural defects resulted in a significant decrease in the efficiency of chemotaxis reflected by a reduced average chemotactic index (Cox *et al.*, 1992). Recently, one of the ABP-120⁻ mutants was rescued either by the insertion of an intact ABP-120 gene at a nonhomologous site or at the original ABP-120 locus (Cox *et al.*, 1995b). Both rescued strains, which were referred to as ABP-120⁻/⁺, were demonstrated to be biochemically, behaviorally, morphologically, and ultrastructurally normal. Normalization of the ABP-120⁻ phenotype by the addition of the original gene verified that the behavioral defects of the original disruptants were direct results of the original ABP-120 disruption and not the result of an unidentified second mutation, and supports the conclusion that ABP-120 plays a fundamental role in stabilizing the actin filament cytoskeleton during pseudopod extension (Condeelis, 1993a,b). The original behavioral phenotype of ABP-120⁻, which included a decrease in the final volume of pseudopods, a rounder cell shape, and a decrease in instantaneous velocity, would have been difficult to resolve without computer-assisted analysis of large numbers of cells.

E. Ponticulin Mutants

Computer-assisted two-dimensional analyses of the *myo*II mutants (Wessels *et al.*, 1988; Wessels and Soll, 1990; Sheldon and Knecht, 1995; Doolittle *et al.*, 1995), the *myo*A and B mutants (Wessels *et al.*, 1991; Titus *et al.*, 1992), and the ABP-120 mutants (Cox *et al.*, 1992, 1995b) revealed defects in the behavior of single cells. However, identifying a behavioral defect for ponticulin⁻ cells using a two-dimensional system was more elusive, at first, because general motility and morphology parameters, such as mean instantaneous velocity, mean directionality, mean cell length, and mean cell shape, revealed no significant differences between wild-type and mutant cells (Shutt *et al.*, 1995a). The defective phenotype of ponticulin⁻ cells was revealed only after analyzing dynamic 3D reconstructions of ponticulin⁻ cells translocating in buffer (Shutt *et al.*, 1995a).

Ponticulin is a unique integral membrane protein which possesses both a glycosyl anchor and a membrane spanning peptide domain. Ponticulin accounts for 96% of *in vitro* actin-binding activity and most of the actin-nucleation activity of *Dictyostelium* plasma membranes (Wuestehube and Luna, 1987; Luna *et al.*, 1990; Hitt *et al.*, 1994a). However, ponticulin⁻ cells are still capable of aggregating and progressing through morphogenesis, albeit in a somewhat more asynchronous fashion in low-density cultures than parental wild-type cells (Hitt *et al.*, 1994b), suggesting that ponticulin⁻ cells are motile, and, as noted, the mean velocity, directionality, and shape of ponticulin⁻ cells computed with a motion analysis system revealed no differences between them and wild-type cells (Shutt *et al.*, 1995a). However, perimeter tracks of ponticulin⁻ cells appeared to be, on average, wider than those of wild-type cells. Since the lateral edges of perimeter tracks reflect, in part, the dynamics of lateral pseudopods formed off the substratum, an analysis of computer-generated three-dimensional reconstructions was performed (Shutt *et al.*, 1995a). Threse reconstructions demonstrated that in contrast to wild-type cells, the base of ponticulin⁻ cells was not fixed in relation to the substratum. Without ponticulin, lateral pseudopods still extended and retracted off the substratum, but the majority tended to slip posteriorly. This resulted in an inordinate proportion of pseudopod retraction at the uropod rather than the more central main cell body, and a significant decrease in the top-end chemotactic indices in spatial gradients of cAMP (Shutt *et al.,* 1995a). Since the majority of lateral pseudopods formed off the substratum by cells translocating in buffer are retracted (Wessels *et al.,* 1994), one would predict that slippage of lateral pseudopods in ponticulin⁻ cells should have little effect on mean instantaneous velocity, and that is exactly what was observed. This is why the two-dimensional motility parameters of ponticulin⁻ cells and wild type cells were approximately similar (Shutt *et al.,* 1995a). The computer-assisted measurements of pseudopod position, the dynamics of pseudopod extension, and the dynamics of pseudopod retraction have, therefore, uncovered a unique role for ponticulin in anchoring the base of lateral pseudopods to the substratum. This role most likely would not have been resolved without the application of a computer-assisted three-dimensional motion analysis system.

F. Other Mutants Behaviorally Analyzed with Computer Assisted Systems

Only select cytoskeletal mutants analyzed in detail with computer-assisted motion analysis systems have been described in this review in order to provide discrete examples of the power of computer-assisted methodologies. Several additional mutants of interest are worth noting. First, *Dictyostelium* mutants lacking the lectin discoidin have been demonstrated to translocate with at least the same instantaneous velocity and similar directionality as wild-type cells, even though their morphology is aberrant (Alexander *et al.,* 1992). Cells lacking discoidin also lack a traditional elongate morphology, apparently due to the absence of a tapered uropod. The developmental consequence of this loss is the absence of multicellular streaming, which depends upon cohesive forces at cell ends and cell sides (Loomis, 1988). The morphological defect of discoidin⁻ cells was surprising since it was originally believed that discoidin was a cell-surface lectin involved in adhesion (Barondes *et al.,* 1982), but subsequent staining experiments demonstrated that it was cytoplasmic (Erdos and Whitaker, 1983; Alexander *et al.,* 1992) and the defect in the shape of mutants suggests it may play a role in the cytoskeleton (Alexander *et al.,* 1992). The unusual abundance of discoidin in the cytoplasm rather than in association with the plasma membrane of developmentally motile cells (Rosen *et al.,* 1973) lends credibility to this suggestion. The role of discoidin in cell shape and its possible role in lateral pseudopod formation deserves further biochemical and behavioral analysis.

Mutant cells lacking coronin, an actin binding protein with sequence homologies to the β-subunit of trimeric G proteins and localized in the cell cortex, exhibit at least a twofold decrease in instantaneous velocity and a small increase in the rate of turning (de Hostos *et al.,* 1993). This preliminary behavioral phenotype suggests that coronin may play a role in pseudopod extension and/or retraction and that coronin⁻ cells warrant a more detailed computer-assisted analysis of pseudopod dynamics. Finally, double mutant cells lacking profilin, a G-actin sequestering protein, exhibit a decrease in velocity of approximately 40% in the absence and presence of cAMP (Haugwitz *et al.*, 1994). This preliminary behavioral phenotype again suggests that profilin plays a role in pseudopod extension, but it also points out that, as in the case of coronin, the demonstration that velocity is affected by a mutation is not sufficient to determine the role of the molecule in the complex process of pseudopod extension. Again, profilin⁻ cells warrant a more detailed computer-assisted analysis of pseudopod dynamics.

VII. Role of Pseudopod Extension and Retraction in Assessing Spatial and Temporal Gradients of Chemoattractant

We have demonstrated that the quantitative description of cellular translocation in space and time provided by computer-assisted motion analysis systems has been instrumental in defining the rules which govern amoeboid translocation and has been crucial in determining the defective aspects of the behavioral phenotype of several cytoskeletal mutants. The quantitative description of a cell translocating in buffer also provides the basis for understanding how extracellular matrix molecules and chemoattractants manipulate cellular behavior in developmental processes. Although the use of computers in interpreting cell–cell interactions and cell–matrix interactions is in its infancy, the use of computers in interpreting amoeboid cell chemotaxis has a long history; only some of the more recent applications will be highlighted here, especially those related to the decision-making process, which may very well be performed at the level of pseudopod extension and retraction.

Although it has been demonstrated that chemotactically responsive amoeboid cells like human polymorphonuclear leukocytes and *Dictyostelium* amoebae can "read" the direction of a spatial gradient of chemoattractant and move up the gradient (i.e., in the direction of higher concentration) (Zigmond, 1977; Varnum-Finney *et al.*, 1987a), it is still not clear how this is achieved. Bacteria read spatial gradients through a temporal mechanism. In other words, they can assess whether the concentration of chemoattractant is increasing or decreasing with time as they swim up or down a gradient of attractant by the dynamics of receptor occupancy, and this temporal information is used to regulate their behavior (Berg, 1991; Koshland, 1980). Bacteria tumble by changing the direction of flagellar rotation and, in so doing, can change direction (Larsen *et al.,* 1974). When they encounter an increase in chemoattractant, they tumble less fequently and this keeps them on track moving up a spatial gradient (Berg and Brown, 1972; Brown and Berg, 1974). Chemotactically responsive amoeboid cells also monitor extracellular chemoattractant through the occupancy of receptors on the cell surface (Devreotes, 1989), but it is not clear whether they use a temporal or a spatial mechanism to assess the direction of a spatial gradient of attractant. If cells used a temporal mechanism, they would assess the direction of the gradient either by moving or by extending a pseudopod through the gradient and concomittantly measuring the change in receptor occupancy as a function of time. If they sensed that the concentration was increasing with time (i.e., the cell or pseudopod was moving up the spatial gradient and receptor occupancy was increasing with time), they would maintain movement in that direction, but if they sensed that the concentration was decreasing with time (i.e., the cell or pseudopod was moving down the gradient and receptor occupancy was decreasing with time), they would terminate movement in that direction and reinitiate movement in a reverse direction. If cells used a spatial mechanism, they would integrate spatial differences in receptor occupancy at a single time-point along their long cell axis or along the axis of an extended pseudopod. This positional information would then be used to direct pseudopod extension and cellular translocation up the gradient. Several computer-assisted analyses of the behavior of *Dictyostelium* amoebae, in particular, have been used to make a case for both temporal and spatial mechanisms. Although the issue of mechanism has not been resolved, it seems clear that elucidating the actual mechanism of chemotaxis in amoeboid cells will be accomplished through computer-assisted two- and three-dimensional analyses of behavior.

Initial studies of cellular translocation established that individual PMNs (Zigmond, 1977) and individual *Dictyostelium* amoebae (Varnum-Finney *et al.,* 1987b) moved in a directed fashion up a concentration gradient of chemoattractant generated in a gradient chamber. These types of experiments were superior to those involving Boyden chambers (Boyden, 1962) because they allowed visualization of the crawling cell and the computation of a chemotactic index (McCutcheon, 1946). In addition, the release of chemoattractant from micropipettes positioned close to single cells demonstrated that the generated gradient stimulated extension of pseudopods and crawling in the direction of the micropipette (Gerisch and Keller, 1981). These original experiments established that *Dictyostelium* amoebae could assess the direction of a spatial gradient of cAMP, but they did not distinguish between the alternate mechanisms of assessment. In the case of *Dictyostelium* amoebae, the natural aggregation signal is in the form of a wave (Tomchik and Devreotes, 1981). Cells in the center of an aggregation territory release cAMP in a pulsatile fashion, and cells in the immediate vicinity of the aggregation center respond to this initial signal by in turn releasing a pulse of cAMP. This form of relay results in outwardly moving, nondissipating waves of the attractant cAMP (Devreotes, 1989; Tomchik and Devreotes, 1981). Cells use these waves for guidance into the aggregation center (Bonner, 1947; Konijn and Raper, 1961). The natural wave creates a paradox for a strictly spatial mechanism since it has been demonstrated that the wave is relatively symmetrical (Tomchik and Devreotes, 1981). Therefore, cells experience a spatial gradient increasing toward the aggregation center in the front of the wave and a spatial gradient increasing away from the aggregation center in the back of the wave, and they probably experience these oppositely directed spatial gradients for the same period of time (Soll *et al.*, 1992). In addition, cells experience a positive temporal gradient in the front of a wave and a negative temporal gradient in the back of the wave (Varnum *et al.*, 1985).

To test whether *Dictyostelium* amoebae can discriminate between increasing and decreasing concentrations of attractant, a simple experiment was performed. Cells were videorecorded in a round chamber through which buffer was perfused containing increasing, then decreasing concentrations of cAMP (Varnum *et al.*, 1985; Varnum-Finney *et al.*, 1987b). In this way, the temporal dynamics of sequential natural waves were roughly simulated without establishing spatial gradients. Cells responded to the increasing and decreasing phases of the simulated temporal waves by moving rapidly and persistently in the increasing phases, which roughly imitated the temporal dynamics of the front of the wave, and by moving slowly and in a nonpersistent fashion in the decreasing phases, which roughly imitated the temporal dynamics of the back of the wave. These results, which were based simply on centroid tracks and velocity measurements (Varnum *et al.*, 1985), demonstrated that amoeboid cells readily discriminated between increasing and decreasing temporal gradients and altered their behavior accordingly. In subsequent experiments using a more sophisticated computer-assisted motion analysis system, it was demonstrated that an increasing temporal gradient of chemoattractant suppressed the frequency of lateral pseudopod formation while increasing directionality, instantaneous velocity, and persistent translocation; a decreasing temporal gradient of chemoattractant stimulated the frequency of lateral pseudopod formation and turning, but disrupted cell polarity, depressed instantaneous velocity, and resulted in a dramatic decrease in persistent translocation (Varnum-Finney *et al.*, 1987b). The relatively dramatic effects on behavior by the direction of a simulated temporal gradient correlated with the cyclic changes in cell behavior observed in a natural wave (Wessels *et al.*, 1992). Individual cells in a natural aggregation territory exhibited increases and decreases in velocity consistent with the front and back of waves, and exhibited the same changes in the frequency of pseudopod formation, polarity, and persistent translocation described in simulated temporal waves. The only difference was in orientation. While cells in simulated temporal waves in perfusion chambers moved in all directions, cells in a natural wave moved in a directed fashion toward the aggregation center during the translocation phase in the deduced front of the wave (Wessels *et al.*, 1992). Interestingly, the orientation decision appears to be made anew at the beginning of the front of each new wave (Wessels *et al.*, 1992). Therefore, in spite of the observation that most of the changes in cellular behavior in a natural wave are regulated by the temporal characteristics of the wave, we still do not know whether cells use a temporal or spatial mechanism for the decision on directional orientation.

Computer-assisted studies of cell behavior in spatial gradients have shed additional light on this process. First, it was demonstrated that when cells are oriented at an oblique angle to the direction of a spatial gradient, they extend just as many lateral pseudopods toward the source as away from the source, but turn into the pseudopods directed toward the source at least twice as frequently, suggesting that discrimination of the spatial gradient is at the level of the pseudopod (Varnum Finney *et al.*, 1987b). However, there is also evidence that cells assess directionality across the cell body since, as a cell becomes more closely aligned with the source of a spatial gradient, the cell (1) moves faster toward the source, (2) exhibits longer intervals between pseudopod extensions (i.e., exhibits a decrease in the frequency of pseudopod extension), and (3) exhibits a decreased frequency of turning. Therefore, cells appear to be able to assess the direction of a spatial gradient both through extension of a lateral pseudopod and during cellular translocation, in which lateral pseudopod formation is dramatically depressed, suggesting that there is both a pseudopod and cell body mechanism (Varnum-Finney *et al.*, 1987b). The discovery that lateral pseudopods formed off the substratum by amoebae translocating in buffer are usually retracted, while lateral pseudopods formed on the substratum usually become the new leading edge (i.e., cause a turn) (Wessels *et al.*, 1994) may be the key to elucidating the mechanism of chemotactic orientation. If the same rules hold for amoebae moving in a spatial gradient of attractant, then one would predict that lateral pseudopods are formed off the substratum to sense the direction of a spatial gradient, and must fall to the substratum to effect a turn when the correct direction is assessed. A computer-assisted three-dimensional analysis of lateral pseudopod dynamics of cells in temporal gradients of chemoattractant, in spatial gradients of chemoattractant, and in natural waves must be performed to test this prediction.

VIII. Concluding Remarks

It should be clear from this general review that relatively sophisticated computer-assisted two- and three-dimensional motion analysis systems have been developed to characterize the spatial and temporal dynamics of amoeboid cell crawling, but they have been underutilized. Their application is threefold. First, they can be used to characterize normal behavior in a quantitative fashion in order to obtain a list of the apparent rules which govern the basic process of amoeboid translocation. Second, they can be used to assess aberrant behavior in mutants which selectively lack signal transduction and cytoskeletal components in order to interpret the role of these components in the basic process of amoeboid translocation. Third, they can be used to assess the effects of chemotactic agents, other extracellular signal molecules, and extracellular matrix molecules on the basic process of amoeboid translocation. In addition, computer-assisted motion analysis systems can be used to compare normal and neoplastic cells (Brady-Kalnay *et al.,* 1991a,b) and they can be used to screen the effects of drugs on the normal behavior of developing and migratory cells, most notably metasticizing cancer cells and migratory and phagocytic white blood cells.

From this review, it should also be clear that the refined details of behavior should not be overlooked, since they reflect the complexity of crawling and they provide us with the context for interpreting mutant and chemotactic behavior. There is a prejudice that behavioral studies are "phenomenological" and therefore have no relevance in their own right. In other words, behavioral studies are only relevant when describing mutants or a behavioral response to an extracellular signal or drug. It should be underscored that molecular components must combine to generate the behaviors which can be described by computer-assisted systems; it may be no accident that the correct extension of a pseudopod involves a cytoskeletal machine composed of hundreds of molecular components, many of which fine-tune the basic process of actin polymerization and the majority of which are in search of an *in vivo* function. Therefore, the more we know about the behavioral details of pseudopod extension, pseudopod retraction, and cellular translocation, the easier it will be to elucidate the role of cytoskeletal and sensory transduction elements through mutation, and the easier it will be to understand how extracellular signals direct cell behavior.

Although two-dimensional motion analysis systems are commercially available for analyzing amoeboid cell behavior, there is no three-dimensional motion analysis system commercially available. In addition, there are no systems which can simultaneously motion-analyze both intracellular particles and the cell envelope in three dimensions. However, as more sophisticated front ends evolve for obtaining optical sections for 3D reconstructions of translocating cells, especially confocal microscopy, sophisticated software packages will follow. It will not be too long before the three-dimensional reorganization of tagged cytoskeletal elements can be visualized and quantitated in space and time, and correlated with the three-dimensional changes in the morphology and motility of a cell. This review, therefore, represents only the beginning of the story.

Acknowledgments

The author is indebted to J. Condeelis, D. Cox, J. Hartwig, E. Sheldon, D. Knecht, K. W. Doolittle, I. Reddy, J. G. McNally, D. Shutt, B. Luna, A. Hitt, A. Chandrasekhar, T. Roberts, D. Royal, M. Royal, J. Italiano, and M. Titus for sharing unpublished observations, to D. Wessel and E. Voss for reviewing the manuscript, and to Z. Sunleaf for help with figures. The research from the Soll Lab was supported in part by National Institutes of Health Grant HD18577 and technical development by a grant from the Iowa Economic Development Commission.

References

Abercrombie, M. (1961). The basis of the locomotory behavior of fibroblasts. *Exp. Cell Res., Suppl.* **8,** 188–198.

Abercrombie, M., Heaysman, J. E. M., and Pegrum, S. M. (1970a). The locomotion of fibroblasts in culture. I. Movements of the leading edge. *Exp. Cell Res.* **59,** 393–398.

Abercrombie, M., Heaysman, J. E. M., and Pegrum, S. M. (1970b). The locomotion of fibroblasts in culture. II. Ruffling. *Exp. Cell Res.* **60,** 437–444.

Abercrombie, M., Heaysman, J. E. M., and Pegrum, S. M. (1970c). The locomotion of fibroblasts in culture. IV. Electron mycroscopy of the leading lamella. *Exp. Cell Res.* **67,** 359–367.

Alexander, S., Sydow, L., Wessels, D., and Soll, D. R. (1992). *Dictyostelium* mutants lacking the discoidin lectins are defective in pseudopod extension and morphology, but still maintain cell polarity. *Differentiation (Berlin)* **51,** 149–161.

Allen, R. D. (1981). Motility. *J. Cell Biol.* **91,** 1485–1555.

Allen, R. D., and Taylor, D. L. (1975). The molecular basis of amoeboid movement. *In* "Molecules and Cell Movement" (S. Inoué and R. E. Stevens, eds.), pp. 239–258. Raven Press, New York.

Barondes, S. H., Springer, W. R., and Cooper, D. N. (1982). Cell adhesion. *In* "The Developmental Biology of *Dictyostelium discoideum*" (W. F. Loomis, ed.), pp. 195–231. Academic Press, New York.

Barsky, B. A. (1988). "Computer Graphics and Geometric Modeling Using Beta Splines," pp. 99–107. Spring-Verlag, Berlin.

Berg, H. C. (1991). Bacterial motility: Handedness and symmetry. *Ciba Found. Symp.* **102,** 58–52.

Berg, H. C., and Brown, D. A. (1972). Chemotaxis in *E. coli* analyzed by three dimensional tracking. *Nature (London)* **239,** 500–504.

Bonner, J. T. (1944). A descriptive study of the development of the slime mold *Dictyostelium discoideum. Am. J. Bot.* **31,** 175–182.

Bonner, J. T. (1947). Evidence for the formation of cell aggregate by chemotaxis in thedevelopment of the slime mold *Dictyostelium discoideum. J. Exp. Zool.* **106,** 1–26.

Boyden, S. (1962). Chemotactic effect of mixtures of antibody and antigen or polymorphonuclear leukocytes. *J. Exp. Med.* **115,** 453–466.

Brady-Kalnay, S. M., Soll, D. R., and Brackenbury, R. (1991a). Invasion of rous sarcoma virus-transformed retinal cells: Role of cell motility. *Int. J. Cancer* **47,** 560–568.

Brady-Kalnay, S. M., Boghaert, E. R., Zimmer, S., Soll, D. R., and Brackenbury, R. (1991b). Invasion by WC5 rat cerebellar cells in independent of RSV-induced changes in growth and adhesion. *Int. J. Cancer* **49,** 239–245.

Bresnick, A. R., Warren, V., and Condeelis, J. (1990). Identification of a short sequence essential for actin binding by *Dictyostelium* ABP-120. *J. Biol. Chem.* **265,** 9236–9240.

Brink, M., Gerisch, G., Isenberg, G., Noegel, A. A., Segall, J. E., Wallraft, E., and Schleicher, M. (1990). A Dictyostelium mutant lacking an F-actin cross-linking protein, the 120-KD gelatin factor. *J. Cell Biol.* **111,** 1477–1489.

Brown, D. A., and Berg, H. C. (1974). Temporal stimulation of chemotaxis in *Escherichia coli. Proc. Natl. Acad. Sci. U.S.A.* **71,** 1388–1392.

Burton, J. L., Law, P., and Bank, H. L. (1986). Video analysis of chemotactic locomotion of stored human polymorphonuclear leukocytes. *Cell Motil. Cytoskel.* **6,** 485–491.

Chandrasekhar, A., Wessels, D., and Soll, D. R. (1994). A mutation that depresses cGMP phosphodiesterase activity in Dictyostelium affects cell motility through an altered chemotactic signal. *Dev. Biol.* **169,** 109–122.

Cheung, A. T., Miller, M. E., Donovan, R. M., Goldstein, E., and Kimura, G. M. (1985). Reactivation of tritonated models of human polymorphonuclear leukocytes (PMNs): A computer-assisted analysis. *J. Leuk. Biol.* **38,** 203–211.

Clarke, M., Schatten, G., Mazia, D., and Spudich, J. A. (1975). Visualization of actin fibers associated with the cell membrane in amoebae of *Dictyostelium discoideum. Proc. Natl. Acad. Sci. U.S.A.* **72,** 1758–1762.

Condeelis, J. (1981). Microfilament-membrane interactions in cell shape and surface architecture. *In* "International Cell Biology 1980–1981" (H. G. Schweiger, ed.), pp. 306–320. Springer-Verlag, Berlin.

Condeelis, J. (1993a). Understanding the cortex of crawling cells: Insights from *Dictyostelium. Trends Cell Biol.* **3,** 371–376.

Condeelis, J. (1993b). Life at the leading edge: The formation of cell protrusions. *Annu. Rev. Cell Biol.* **9,** 411–444.

Condeelis, J., Geosits, S., and Vahey, M. (1982). Isolation of a new actin-binding protein from *Dictyostelium discoideum. Cell Motil.* **2,** 273–285.

Condeelis, J., Vahey, M., Carboni, M., DeMey, J., and Ogihara, S. (1984). Properties of the 120,000- and 95,000-dalton actin-binding proteins from *Dictyostelium discoideum* and their possible functions in assembling the cytoplasmic matrix. *J. Cell Biol.* **99,** 199–126.

Condeelis, J., Demma, B., Dharmawardhane, S., Eddy, R., and Hall, A. (1990). Mechanisms of amoeboid chemotaxis: An evaluation of the cortical expansion model. *Dev. Genet.* **11,** 333–340.

Cooper, M., and Schliwa, M. (1986). Motility of cultured fish epidermal cells in the presence and absence of direct current electric fields. *J. Cell Biol.* **102,** 1384–1399.

Cox, D., Condeelis, J., Wessels, D., Soll, D. R., Kern, H., and Knecht, D. A. (1992). Targetted disruption of the ABP-120 gene leads to cells with altered motility. *J. Cell Biol.* **116,** 943–955.

Cox, D., Ridsdale, J. A., Condeelis, J., and Hartwig, J. (1995). Genetic deletion of ABP-120 alters the three dimensional organization of actin filaments in Dictyostelium pseudopods. *J. Cell Biol.* **128,** 819–835.

Cox, D., Wessels, D., Soll, D. R., Hartwig, J. and Condeelis, J. (1995). Submitted.

de Hostos, E. L., Rehfuess, C., Bradtke, B., Waddell, D. R., Albrecht, R., Murphy, J., and Gerisch, G. (1993). *Dictyostelium* mutants lacking the cytoskeletal protein coronin are defective in cytokinesis and cell motility. *J. Cell Biol.* **120,** 163–173.

DeLozanne, A., and Spudich, J. A. (1987). Disruption of the *Dictyostelium* myosin heavy chain gene by homologous recombination. *Science* **236,** 1086–1091.

Devreotes, P. N. (1989). *Dictyostelium discoideum*: A model system for cell-cell interactions in development. *Science* **245,** 1054–1058.

Doolittle, K. W., Reddy, I., and McNally, J. G. (1995). 3D analysis of cell movement during normal and myosin-II-null cell morphogenesis in *Dictyostelium*. *Dev. Biol.* **167**, 118–129.

Eliot, S., Joss, G. H., Spudich, A., and Williams, K. L. (1993). Patterns in *Dictyostleium discoideum*: The role of myosin II in the transition from the unicellular to the multicellular phase. *J. Cell Sci.* **104**, 457–466.

Endow, S., and Titus, M. (1992). Genetic approaches to molecular motors. *Ann. Rev. Cell Biol.* **8**, 29–66.

Erdos, G. W., and Whitaker, D. (1983). Failure to detect immunocytochemically reactive endogenous lectin on the cell surface of *Dictyostelium discoideum*. *J. Cell Biol.* **97**, 993–1000.

Felder, S., and Kam, Z. (1994). Human neutrophil motility: Time-dependent three-dimensional shape and granule diffusion. *Cell Motil. Cytoskel.* **28**, 285–302.

Fischer, P. R., Merkl, R., and Gerisch, G. (1989). Quantitative analysis of cell motility and chemotaxis in *Dictyostelium discoideum* by using an image processing system and a novel chemotaxis chamber providing stationary chemical gradients. *J. Cell Biol.* **108**, 973–984.

Fukui, Y., Voss, E., Riddelle, K. S., and Soll, D. R. (1991). Cell behavior and actomyosin organization in *Dictyostelium* during substrate exploration. *Cell Struct. Funct.* **16**, 289–301.

Futrelle, R. P., Trant, J., and McKee, W. G. (1982). Cell behavior in *Dictyostelium discoideum*: Preaggregation response to localized cyclic AMP pulses. *J. Cell Biol.* **92**, 807–821.

Gerald, C. (1980). "Applied Numerical Analysis," 2nd ed., pp. 214–217. Addison-Wesley, Reading, MA.

Gerisch, G., and Keller, H. U. (1981). Chemotactic reorientation of granulocytes stimulated with micropyettes containing fMet-Leu-Phe. *J. Cell Sci.* **52**, 1–10.

Gerisch, G., Huelser, D., Malchow, D., and Wick, U. (1975). Cell communication by periodic cyclic AMP pulses. *Philos. Trans. R. Soc. London Ser. B* **272**, 181–192.

Gerisch, G., Noegel, A., and Schleicher, M. (1991). Genetic alteration of proteins in actin-based motility systems. *Annu. Rev. Physiol.* **53**, 607–628.

Gerisch, G., Schleicher, M., and Noegel, A. (1992). Dynamics of the cytoskeleton during chemotaxis. *New Biol.* **4**, 461–472.

Hand, W. G., and Davenport, D. (1970). The experimental analysis of phototaxis and photokinesis in flagellates. *In* "Photobiology of Microorganisms" (P. Halldal, ed.), pp. 253–282. Wiley–Interscience, New York.

Haugwitz, M., Noegel, A. A., Karakesisoglou, J., and Schleicher, M. (1994). *Dictyostelium* amoebae that lack G-actin-sequestering profilins show defects in F-actin content, cytokinesis, and development. *Cell (Cambridge, Mass.)* **79**, 303–314.

Hay, E. D. (1989). Theory for epithelial-mesenchymal transformation based on the "fixed cortex" cell motility model. *Cell Motil. Cytoskel.* **14**, 455–457.

Hitt, A. L., Lu, T. H., and Luna, E. J. (1994a). Ponticulin is an atypical membrane protein. *J. Cell Biol.* **126**, 1421–1431.

Hitt, A. L., Hartwig, J. H., and Luna, E. J. (1994b). Ponticulin is the major high affinity link between the plasma membrane and the cortical actin network in *Dictyostelium*. *J. Cell Biol.* **126**, 1433–1444.

Jay, P. Y., Pasternak, C., and Elson, E. L. (1993). Studies of mechanical aspects of amoeboid locomotion. *Blood Cells (New York)* **19**, 375–386.

Jester, J. Y., Andrews, P. M., Petroll, W. M., Lemp, M. A., and Cavanagh, H. D. (1991). *In vivo*, real-time confocal imaging. *J. Electron Microsc. Tech.* **18**, 50–60.

Jung, G., and Hammer, J. A. (1990). Generation and characterization of *Dictyostelium* cells deficient in a myosin I heavy chain isoform. *J. Cell Biol.* **110**, 1955–1964.

Knecht, D. A., and Loomis, W. F. (1987). Antisense RNA inactivation of myosin heavy chain gene expression in *Dictyostelium discoideum*. *Science* **236**, 1081–1085.

Knecht, D. A., Kern, H., and Sherezinger, C. (1993). Bidirectional transcription from actin promoters in *Dictyostelium*. *Biochim. Biophys. Acta* **1216**, 105–109.

Konijn, T. M., and Raper, K. B. (1961). Cell aggregation in *Dictyostelium discoideum*. *Dev. Biol.* **3**, 725–756.

Korn, E. D., and Hammer, J. A, III (1990). Small myosins. *Curr. Opin. Cell Biol.* **2**, 57–61.

Koshland, D. E. (1980). Biochemistry of sensing and adaptation in a simple bacterial system. *Annu. Rev. Biochem.* **50**, 765–782.

Kucik, D., Kuo, S., Elson, E., and Sheetz, M. P. (1991). Preferential attachment of membrane glycoproteins to the cytoskeleton at the leading edge of lamella. *J. Cell Biol.* **114**, 1029–1036.

Kuczmarski, E. R., Pavlios, L., Aquado, C., and Yao, Z. (1991). Stopped flow measurement showed the *Dictyostelium* myosin II is specifically required for contraction of amoeba cytoskeletons. *J. Cell Biol.* **114**, 1191–1200.

Larsen, S. H., Reader, R. W., Kort, E. N., Tso, N. W., and Adler, J. (1974). Change indirection of flagellar rotation is the basis of the chemotactic reponse in *Escherichia coli. Nature (London)* **249**, 74–77.

Lieberman, S. J., Hamasaki, T., and Satir, P. (1988). Ultrastructure and motion analysis of permeabilized *Paramecium* capable of motility and regulation of motility. *Cell Motil. Cytoskel.* **9**, 73–84.

Loomis, W. F. (1988). Cell-cell adhesion in *Dictyostelium discoideum. Dev. Genet.* **9**, Suppl., 549–560.

Luna, E. J., and Condeelis, J. S. (1990). Actin-associated proteins in *Dictyostelium discoideum. Dev. Genet.* **11**, 328–332.

Luna, E. J., Wuestehube, L. J., Chia, C. P., Shariff, A., Hitt, A. L., and Ingalls, H. M. (1990). Ponticulin, a developmentally-regulated plasma membrane glycoprotein, mediates actin binding and nucleation. *Dev. Genet.* **11**, 354–361.

Makler, A. (1980). Use of microcomputer in combination with multiple exposure photography technique for human sperm motility measurements. *J. Urol.* **124**, 372–374.

Manstein, D. J., Titus, M. A., DeLozanne, A., and Spudich, J. A. (1989). Gene replacement in *Dictyostelium:* Generation of myosin null mutants. *EMBO J.* **8**, 923–932.

Maron, M. J. (1982). "Numerical Analysis: A Practical Approach," pp. 283–284. Macmillan, New York.

McCutcheon, M. (1946). Chemotaxis in leukocytes. *Physiol. Rev.* **26**, 401–404.

Murray, J., Vawter-Hugart, H., Voss, E., and Soll, D. R. (1992). Three dimensional motility cycle in leukocytes. *Cell Motil. Cytoskel.* **22**, 211–223.

Nachmias, V. T., Fukui, Y., and Spudich, J. A. (1989). Chemoattractant-elicited translocation of myosin in motile *Dictyostelium. Cell Motil. Cytoskel.* **13**, 158–169.

Nelson, G. A., Roberts, T. M., and Ward, S. (1982). Caenorhabditis elegans spermatozoan locomotion: Amoeboid movement with almost no actin. *J. Cell Biol.* **92**, 121–131.

Noegel, A. A., Rapp, S., Lottspeich, F., Schleicher, M., and Stewart, M. (1989). The *Dictyostelium* gelation factor shares a putative actin binding site with alpha-actinins and dystrophin and also has a rod domain containing six 100-residue motifs that appear to have a cross-beta conformation. *J. Cell Biol.* **109**, 607–618.

Ogihara, S., Carboni, J., and Condeelis, J. (1988). Electron microscopic localization of myosin II and ABP-120 in the cortical actin matrix of *Dictyostelium* amoebae using IgG-gold conjugates. *Dev. Genet.* **9**, 505–520.

Parrot, D. M. V., and Wilkinson, P. C. (1981). Lymphocyte locomotion and migration. *Prog. Allergy* **28**, 193–284.

Pasternak, C., Spudich, J. A., and Elson, E. J. (1989). Capping of surface receptors and concomitant cortical tension are generated by conventional myosin. *Nature (London)* **341**, 541–549.

Peters, D. J. M., Knecht, D. A., Loomis, W. F., DeLozanne, A., Spudich, J., and Van Hasstert, P. J. M. (1988). Signal transduction, chemotaxis, and cell aggregation in *Dictyostelium discoideum* cells without myosin heavy chain. *Dev. Biol.* **128**, 158–163.

Rosen, S. D., Kafka, J. A., Simpson, D. L., and Barondes, S. H. (1973). Developmentally regulated carbohydrate-binding protein in *Dictyostelium discoideum. Proc. Natl. Acad. Sci. U.S.A* **70**, 2554–2557.

Royal, D., Royal, M., Italiano, J., Roberts, T., and Soll, D. R. (1995). *Ascaris* sperm pseudopods, MSP fibers move proximally at a constant rate regardless of the forward rate of cellular translocation. *Cell Motil. Cytoskel.* **31**, 241–253.

Rudin, W. (1972). "Principles of Mathematical Analysis," 3rd ed. McGraw-Hill, New York.

Satoh, H., Ueda, T., and Kobatake, Y. (1985). Oscillations in cell shape and size during locomotion and in contractile activities of *Physarum polyaphalum, Dictyostelium discoideum, Amoeba proteis* and macrophages. *Exp. Cell Res.* **156**, 79–90.

Schmassmann, A., Mikug, G., Bartsch, G., and Rohr, H. (1979). Quantification of human sperm morphology and motility by means of semi-automatic image analysis systems. *Microsc. Acta* **82**, 163–178.

Segall, J. E. (1992). Behavioral responses of streamer F mutants of *Dictyostelium discoideum:* Effects of cGMP on cell motility. *J. Cell Sci.* **101**, 589–597.

Sepsenwol, S., and Taft, S. J. (1990). In vitro induction of crawling in the amoeboidsperm *Ascaris suum. Cell Motil. Cyotskel.* **15**, 99–110.

Sheetz, M. P., Turney, S., Jian, H., and Elson, E. (1989). Nanometre level analysis demonstrates that lipid flow does not drive membrane glycoprotrin movements. *Nature (London)* **340**, 284–288.

Sheldon, E., and Knecht, D. A. (1995). Mutants lacking myosin II cannot resist forces generated during multicellular morphogensis. *J. Cell Sci.,* (in press).

Shutt, D., Wessels, D., Chandrasekhar, A., Luna, B., Hitt, A., and Soll, D. R. (1995a). Submitted for publication.

Shutt, D., Stapleton, J, Kennedy, R., and Soll, D. R. (1995b). HIV-induced syncytia of peripheral blood T cells mimic the organization of individual cells and translocate through extension of giant pseudopodia. *Cell. Immun.,* (in press).

Smith, E. F., and Sale, W. S. (1994). Mechanisms of flagellar movement: Functional interactions between Dynein arms and radial spoke-central apparatus complex. *In* "Microtubules" (J. S. Hyams and C. W. Lloyd, eds.), pp. 381–392. Wiley-Liss, New York.

Soll, D. R. (1988). "DMS," a computer-assisted system for quantitating motility, the dynamics of cytoplasmic flow and pseudopod formation: Its application to *Dictyostelium* chemotaxis. *Cell Motil. Cytoskel., Suppl.* **10**, 91–106.

Soll, D. R., Voss, E., Varnum-Finney, B., and Wessels, D. (1988a). The "dynamic morphology system": A method for quantitating changes in shape, pseudopod formation and motion in normal and mutant cells of *Dictyostelium discoideum. J. Cell Biochem.* **37**, 177–192.

Soll, D. R., Voss, E., and Wessels, D. (1988b). Development and application of the "Dynamic Morphology System" for the analysis of moving amebae. Proc. SPIE—*Int. Soc. Opt. Eng.* **832**, 21–30.

Soll, D. R., Wessels, D., Murray, J., Vawter, H., Voss, E., and Bublitz, A. (1990). Intracellular vesicle movement, cAMP and myosin II in *Dictyostelium. Dev. Genet.* **11**, 341–353.

Soll, D. R., Wessels, D., and Sylwester, A. (1992). The motile behavior of amoebae in the aggregation wave in *Dictyostelium discoideum. In* "Experimental and Theoretical Advances in Biological Pattern Formation" (H. G. Othmer, P. K. Maine, and J. D. Murray, eds.), pp. 325–338. Plenum, New York.

Spudich, J. A. (1989). In pursuit of myosin function. *Cell Regul.* **1**, 1–11.

Stossel, T. P. (1993). On the crawling of animal cells. *Science* **260**, 1086–1093.

Sylwester, A., Wessels, D., Anderson, S. A., Warren, R. Q., Shutt, D., Kennedy, R., and Soll, D. R. (1993). HIV-Induced syncytia of a T cell line form single giant pseudopods and are motile. *J. Cell Sci.* **106**, 941–953.

Sylwester, A., Shutt, D., Wessels, D., Stapelton, J., Stites, J., Kennedy, R., and Soll, D. R. (1995). T cells and HIV-induced T cell syncytia exhibit the same velocity cycle when translocating on plastic, collagen or endothelial monolayers. *J. Leuk. Biol,* **57**, 643–650.

Titus, M., Wessels, D., Spudich, J., and Soll, D. R. (1992). The unconventional myosinencoded by the myo A gene plays a role in *Dictyostelium* motility. *Mol. Biol. Cell* **4**, 233–246.

Tomchik, S. J., and Devreotes, P. N. (1981). cAMP waves in *Dictyostelium discoideum:* Demonstration by an isotope dilution fluorographic technique. *Science* **212**, 443–446.

Trinkhaus, J. P. (1984). "Cells into Organs." Prentice-Hall, Englewood Cliffs, NJ.

Varnum, B., and Soll, D. R. (1984). Effects of cAMP on single cell motility in *Dictyostelium*. *J. Cell Biol.* **99**, 1151–1155.

Varnum, B., Edwards, K., and Soll, D. R. (1985). *Dictyostelium* amebae alter motility differently in response to increasing versus decreasing temporal gradients of cAMP. *J. Cell Biol.* **101**, 1–5.

Varnum-Finney, B., Edwards, K., Voss, E., and Soll, D. R. (1987a). Amebae of *Dictyostelium discoideum* respond to an increasing temporal gradient of the chemoattractant cAMP with a reduced frequency of turning: Evidence for a temporal mechanism in ameboid chemotaxis. *Cell Motil. Cytoskel.* **8**, 7–17.

Varnum-Finney, B., Voss, E., and Soll, D. R. (1987b). Frequency and orientation of pseudopod formation of *Dictyostelium discoideum* amebae chemotaxing in a spatial gradient: Further evidence for a temporal mechanism. *Cell Motil. Cytoskel.* **8**, 18–26.

Wessels, D., and Soll, D. R. (1990). Myosin II heavy chain null mutant of *Dictyostelium* exhibits defective intracellular particle movement. *J. Cell Biol.* **111**, 1137–1148.

Wessels, D., Soll, D. R., Knecht, D., Loomis, W. F., DeLozanne, A., and Spudich, J. (1988). Cell motility and chemotaxis in *Dictyostelium* amoebae lacking myosin heavy chain. *Dev. Biol.* **128**, 164–177.

Wessels, D., Murray, J., Jung, G., Hammer, J. A., III, and Soll, D. R. (1991). Myosin IB null mutants of *Dictyostelium* exhibit abnormalities in movement. *Cell Motil. Cytoskel.* **20**, 301–315.

Wessels, D., Murray, J., and Soll, D. R. (1992). The complex behavior cycle of chemotaxing *Dictyostelium* amoebae is regulated primarily by the temporal dynamics of the natural wave. *Cell Motil. Cytoskel.* **23**, 145–156.

Wessels, D., Vawter-Hugart, H., Murray, J., and Soll, D. R. (1994). Three dimensional dynamics of pseudopod formation and the regulation of turning during the motility cycle of *Dictyostelium*. *Cell Motil. Cytoskel.* **27**, 1–12.

Wessels, D., Titus, M., and Soll, D. R. (1995). Submitted for publication.

Wuestehube, L. J., and Luna, E. J. (1987). F-actin binds to the cytoplasmic surface of ponticulin, a 17-kD integral glycoprotein from *Dictostelium discoideum* plasmamembranes. *J. Cell Biol.* **105**, 1741–1751.

Yumura, S., and Fukui, Y. (1985). Reversible cyclic AMP-dependent change in distribution of myosin thick filaments in *Dictyostelium*. *Nature (London)* **314**, 194–196.

Zigmond, S. H. (1977). The ability of polymorphonuclear leukocytes to orient in gradients of chemotactic factors. *J. Cell Biol.* **75**, 606–616.

Programmed Cell Death in Development

E. J. Sanders and M. A. Wride
Department of Physiology, University of Alberta, Edmonton, Alberta T6G 2M7, Canada

Although cell death has long been recognized to be a significant element in the process of embryonic morphogenesis, its relationships to differentiation and its mechanisms are only now becoming apparent. This new appreciation has come about not only through advances in the understanding of cell death in parallel immunological and pathological situations, but also through progress in developmental genetics which has revealed the roles played by death in the cell lineages of invertebrate embryos. In this review, we discuss programmed cell death as it is understood in developmental situations, and its relationship to apoptosis. We describe the morphological and biochemical features of apoptosis, and some methods for its detection in tissues. The occurrence of programmed cell death during invertebrate development is reviewed, as well as selected examples in vertebrate development. In particular, we discuss cell death in the early vertebrate embryo, in limb development, and in the nervous system.

KEY WORDS: Programmed cell death; Apoptosis; Development; Morphogenesis; Invertebrates; Vertebrates.

I. Introduction

It has become very clear in recent years that cell death in developing systems is not merely a degenerative process, or only a response to injury, but an active and controlled phenomenon. In this respect, cell death is analogous to cell proliferation and differentiation, in that it is centrally involved in morphogenesis and in the regulation of cell populations in embryos. The extent to which cell death contributes to embryogenesis, and its importance, was already apparent by the time that two early, but definitive, reviews on the subject were published by Glücksmann (1951) and by Saunders (1966). The former classified and catalogued

105

the evidence to that date and still provides an unprecedented wealth of information on the early literature, much of which was published in German (Hamburger, 1992). The latter reviewed the putative control mechanisms for cell death, as they were understood at that time, and, in light of more recent developments, has proved to be an extremely insightful review. Study of the roles and mechanisms of cell death in development has become an increasingly active area of research over the last 30 years, aspects of the literature having been reviewed on a number of previous occasions (Pexieder, 1975; Hinchliffe, 1981; Beaulaton and Lockshin, 1982; Snow, 1987; Hurlé, 1988; Clarke, 1990; Schwartz, 1991). However, cell death is not only of interest to developmental biologists; there is also an extensive literature from both pathology (Arends and Wyllie, 1991) and immunology (Schwartz and Osborne, 1993). Investigators in each of these fields have come to look on cell death from slightly different points of view (Schwartz and Osborne, 1993), with the result that matters of terminology have become an issue. The terminology relating to the various forms of cell death has been defined and extensively reviewed by others (Wyllie *et al.,* 1980; Walker *et al.,* 1988; Alles *et al.,* 1991; Lockshin and Zakeri, 1991; Gerschenson and Rotello, 1992; Sen, 1992; Schwartz and Osborne, 1993), and will not be dealt with exhaustively here, but an overview of the problems and considerations is necessary for an understanding of the later discussion.

The term *programmed cell death* was introduced in a series of papers by Lockshin and Williams (1964, 1965; Lockshin, 1969) to describe the temporally regulated degeneration of intersegmental muscles that occurs during the metamorphosis of the silkmoth. It is a term used primarily by developmental biologists, and carries with it the implication that there is an intrinsic genetic program, or biological clock, that is activated at a controlled time and location to cause the death of the cell. The classic example of programmed cell death in development is the precisely timed elimination of identified cells during the embryogenesis of *Caenorhabditis elegans* (Sternberg, 1991). This type of cell death is predictable and presumably results from the activation, by essentially unknown means, of "cell death genes" coding for "cell death proteins" (Ellis *et al.,* 1991a; Schwartz, 1991), although it may not always be as precise as that in *C. elegans.* As with other aspects of *C. elegans* development, such as lineage decisions, this organism is in a class of its own and it may represent the only case to date in which the use of the term "programmed" is truly warranted, as it applies to individual identified and invariant cell death. Programmed death of a cell population could also reflect the timed activity of adjacent or remote cell populations, as in cases where there appears to be controlled withdrawal of trophic factors, or the production of factors which activate the "suicide" program in the target cells. Clearly, there could also be a timed appearance of receptors for such factors on the target cells. This type of cell death is essential for normal morphogenesis.

The term *apoptosis* was introduced by Kerr *et al.* (1972) to describe the morphological features of cells undergoing degeneration in a variety of pathological situations. Although originally used primarily by pathologists, the term gradually became applied to a number of immunological and developmental phenomena so that apoptosis and programmed cell death tended to become synonymous. While it is true that apoptosis can be recognized both in normal development and in normal adult tissues, where it may be involved in the physiological control of growth and in immune function (Gerschenson and Rotello, 1992; Kerr, 1993), the two terms are not equivalent. There have been a number of attempts to reestablish the separate identities of these two terms (Alles *et al.*, 1991; Lockshin and Zakeri, 1991; Gerschenson and Rotello, 1992; Bowen, 1993; Schwartz and Osborne, 1993), because although apoptosis appears also to be programmed in many cases, the term "apoptosis" carries with it morphological connotations which do not apply to all instances of programmed cell death seen during development. Among the distinguishing features of apoptosis are characteristic condensation of the chromatin, cytoplasmic blebbing, and internucleosome cleavage of DNA as a result of the activation of endonucleases. The blebbing gives rise to apoptotic bodies which are subsequently phagocytosed. The cascade of apoptotic events may be induced in the target cell population, but not in surrounding cells, by a variety of agents, often hormones or growth factors (Sen, 1992), and is ultimately lethal. As with programmed cell death in development, the decision to undergo apoptosis in response to a signal is a controlled, differentiative, cell process requiring the activation of a genetic program. So the question arises: are all examples of programmed cell death occurring in development apoptotic? Again, this question has been addressed in the recent literature (Lockshin and Zakeri, 1991; Schwartz and Osborne, 1993; Schwartz *et al.*, 1993). The evidence supports the view of these authors that the two terms are not interchangeable, but that apoptosis may be a type of programmed cell death, and that the two phenomena may share certain common characteristics and pathways. In short, we do not yet know how many different types of cell death there are. Cells undergoing programmed cell death in developing systems may or may not show all of the morphological and biochemical distinguishing features of apoptosis (Schwartz *et al.*, 1993); however, since this term is appearing increasingly in the literature of developmental biology we will describe its characteristics in some detail. Further, in spite of the recommendation by Alles *et al.* (1991) that the term "cell deletion" be used, we retain the term programmed cell death because of its familiarity to developmental biologists.

One other term needs brief clarification: *necrosis* is characterized by a general cell lysis as a result of direct or indirect membrane damage (Wyllie *et al.,* 1980; Wyllie, 1981; Walker *et al.,* 1988; Kerr and Harmon, 1991; Lockshin and Zakeri, 1991; Gerschenson and Rotello, 1992; Kerr, 1993; Schwartzman and Cidlowski, 1993). The morphological characteristics are a consequence of impaired membrane transport leading to uncontrolled ion and fluid movements. This is accompanied by pycnosis (condensed, hyperchromatic nuclei), mitochondrial swelling and rupture, and lysosomal rupture and inflammation, and is often the pathological response to generally lethal chemical or physical insult. Necrosis does not require the expression of a specific set of gene products. Necrosis is nonphysiological and is not usually seen in developmental situations, even though "necrotic zones" are referred to in the course of development of the vertebrate limb (Saunders *et al.,* 1962). It should be noted that whereas necrosis, as opposed to apoptosis, is always pathological, the reverse is not always the case; i.e., apoptosis may be either physiological or pathological. Physiological cell death is almost always apoptotic.

In surveying the developmental events in which cell death plays an important role, it is clear that there are several categories of event where programmed cell death is involved: (a) in metamorphosis, best known in insect and amphibian development, where in the former the intersegmental musculature becomes unnecessary, and in the latter the tail structures are resorbed; (b) in general morphogenetic reorganization, such as the elimination of tissue between the digits of the developing amniote limb, in heart and kidney morphogenesis, and in many other situations discussed below; (c) in fusion processes, such as in the formation of the secondary palate and closure of the neural tube; (d) in the developing nervous system, where cell death appears to be intimately involved in the establishment of patterns of synaptic organization and axonal pathways (Oppenheim, 1989, 1991; Oppenheim *et al.,* 1992b; Johnson and Deckwerth, 1993). Clearly there is some overlap between some of these arbitrary categories; this is not meant to be a classification in any sense, but simply a useful *aide-mémoire.*

Here, we review the features, both morphological and biochemical, of programmed cell death and of apoptosis; we survey the occurrence of these phenomena during invertebrate and vertebrate development; and we examine some recent advances in our understanding of the molecular mechanisms that have been proposed as controlling elements for these processes, including the potential roles of growth factors and oncogenes.

II. Apoptosis

A. Morphology of Cell Death

The ultrastructural and light microscopical characteristics of cells undergoing apoptosis have been comprehensively described several times in the literature (Wyllie *et al.*, 1980; Wyllie, 1981; Allen, 1987; Kerr *et al.*, 1987; Walker *et al.*, 1988), so that a brief account will suffice here. Most of the earlier reviews are compiled with specific reference to pathological material, and apoptosis in embryonic cells may not always show all of the features described.

Apoptosis is typically seen in individual cells distributed within a cell population that is generally unaffected, so that in contrast to necrosis there is not usually a mass of dying cells to be observed. Also, since some phases of the process are relatively rapid, taking a matter of a few minutes, identification of cells in certain stages of apoptosis in tissue sections may be difficult. Events occur simultaneously in the nucleus and in the cytoplasm (Fig. 1), but the most striking changes are generally the nuclear ones. Chromatin tends to become condensed, often into crescent-shaped aggregates of compact material (pycnosis), and also accumulates on the inner surface of the nuclear membrane. The nucleus itself may become markedly indented and eventually fragments. In the cytoplasm, the ground substance also becomes condensed, with compaction of organelles into restricted regions. This is accompanied by vacuolation, swelling of endoplasmic reticulum cisternae, distention and rupture of mitochondria, and blebbing of the cell membrane (zeiosis). Mitochondria tend to remain of normal appearance, at least in the early phases of the process. The exocytosis of fluid-filled vacuoles presumably accounts for the characteristic shrinkage that accompanies apoptosis (hence the original name of this phenomenon: "shrinkage necrosis"; Kerr, 1971), since there are no gross changes in membrane permeability at this early stage in the sequence of events (Sen, 1992). The affected cells lose their intercellular connections, including desmosomes in the case of epithelial cells, they then detach and are pushed out from the tissue layer into the extracellular space or lumen. Blebbing continues and membrane-bound bodies are pinched off. These apoptotic bodies are of varying size and may contain nuclear fragments or cytoplasmic material, or both. These bodies are relatively long-lived (12–18 h; Wyllie *et al.*, 1980) and are eventually phagocytosed by macrophages or by normal adjacent cells. Apoptotic bodies that are not phagocytosed undergo degeneration similar to that seen in necrosis ("secondary degeneration"; Kerr *et al.*, 1987), and Wyllie (1981) makes the important point that apoptotic bodies exclude vital dyes until the onset of this secondary degeneration.

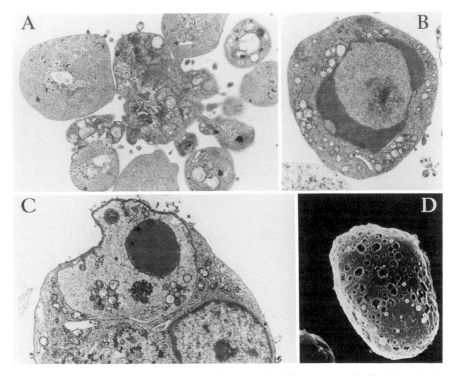

FIG. 1   The morphological characteristics of apoptosis. (A) An apoptotic fibroblast in the course of fragmentation into membrane-bounded bodies. (B) Nuclear chromatin condensation. (C) Phagocytosis of an apoptotic cell (its condensed nucleus prominent at the top of the picture) by one of its viable neighbours, a portion of whose nucleus is seen at the bottom of the picture. (D) A scanning EM of the surface of an apoptotic thymocyte; the pitted surface results from the fusion of dilated ER with the plasma membrane. Reprinted with permission of Dr. A. H. Wyllie and Kluwer Academic Publishers, from Wyllie (1992).

Although some of these morphological features, such as pycnosis, are shared with those seen during necrosis (Lockshin and Zakeri 1991), the full picture of apoptosis is distinct and, at least when all phases are present, should not be confused with necrosis (Wyllie *et al.,* 1987). A relatively common occurrence in embryonic tissue is the appearance of cells in the electron microscope with strikingly darkly staining cytoplasm. These "dark cells" appear to be moribund, and the changes indicate that they are undergoing a form of necrosis (Walker *et al.,* 1988).

In most instances of cell death in embryos, phagocytosis appears to occur while the chromatin is condensed and before the fragmentation of the nucleus, but phagocytosis can occur at any stage of apoptosis (Hurle, 1988). In some dying cells of *C. elegans,* for example, phagocytosis occurs very quickly (Ellis *et al.,* 1991a), while in some neurons phagocytosis does not occur until after extensive autolytic degeneration has occurred (Hornung *et al.,* 1989; Nixon and Cataldo, 1993). The form and extent of lysosomal involvement in the cell death process has been made the basis of a classification of types of cell death (Clarke, 1990). In the limb bud, there seems to be a combination of autophagy leading to autolysis, and phagocytosis leading to lysosomal degradation (Hinchliffe, 1981). In this case, there is both early lysosomal involvement, in the dying cells themselves, and later lysosomal involvement from the engulfing cells, but the phagocytes are not responsible for cell death (Clarke, 1990). In the limb bud, the dying cells first produce acid phosphatase-rich autophagic vacuoles, then autolyse and fragment, and are then phagocytosed by macrophages or by normal mesenchymal cells in the region which become phagocytic. This conversion of neighboring normal cells into phagocytic cells has been documented in a number of different embryonic situations (Beaulaton and Lockshin, 1982). A similar series of events occurs during the resorption of the tadpole tail at amphibian metamorphosis (Weber, 1964), and in the degeneration of the intersegmental muscles in the silkworm at metamorphosis (Beaulaton and Lockshin, 1982). It is notable that early lysosomal involvement, in the form of autophagy and autolysis, is not a feature of classical apoptosis, where there is no rise in acid phosphatase levels in the apoptotic cells themselves or in the apoptotic bodies (Wyllie *et al.,* 1980; Lockshin and Zakeri, 1991). This, then, is an example of a way in which programmed cell death in development does not conform to the general scheme of apoptosis as it is usually described.

It is understood that, in some cases at least, cell surface changes accompany apoptosis, which allow recognition by phagocytes (Savill *et al.,* 1993). These changes may involve the cleavage of terminal sialic acids, with consequent lowering of surface charge, to reveal subterminal saccharides which are not normally exposed. These sugars may then be recognized by lectins on the phagocyte cell surface, triggering engulfment. There is also evidence for the involvement of the $\alpha_v\beta_3$ integrin or vitronectin receptor and of thrombospondin and CD36 in the recognition process. In *C. elegans,* the *ced* series of genes is required for the recognition of dying cells and/or for their phagocytosis (Ellis *et al.,* 1991b; and see below).

B. Mechanisms of Cell Death

With the recognition of cell death as an important and controlled phenomenon has come the study of its underlying mechanisms. As with the study of many biological mechanisms, the initial phase of experimentation appeared to reveal some generalities, such as the importance of endonucleases and of calcium ions, which have been widely and probably too readily accepted. As research has progressed, more and more exceptions to these initial generalities have been recorded, leading to the conclusion that there must be many different pathways to death.

1. DNA Fragmentation and Endonuclease Activation

Apoptosis is frequently accompanied by a characteristic fragmentation of DNA in the affected cell. This fragmentation is the consequence of the activation of an endogenous endonuclease which causes nonrepairable internucleosomal cleavage of the DNA, resulting in the production of fragments of some 180–200 base pairs, corresponding to the length of the DNA in the nucleosomes (Arends et al., 1990; Compton, 1992; Schwartzman and Cidlowski, 1993). Preferential, though not exclusive, double-strand cleavage occurs at the linker regions between the nucleosomes, giving the familiar ladder seen by agarose gel electrophoresis, a feature clearly seen in one of the paradigms for apoptosis—the cell death experimentally induced in thymocytes by their exposure to glucocorticoids (Wyllie, 1980; Wyllie and Morris, 1982). The nucleosomes are apparently protected, and it is important to note that there is no random digestion of the chromatin by DNases or proteases; instead, the cleavage is specifically internucleosomal, and this further distinguishes apoptosis from necrosis where there is indiscriminate DNA degradation. In addition, several studies have indicated that some of the nucleases possess RNase as well as DNase activity, and more recent evidence points to the possibility that some of the enzymes have exonuclease activity (Fraser et al., 1993). It has also been claimed that the characteristic dense fibrillar nucleolar core structure seen during apoptosis may be a reflection of the DNA cleavage, and that therefore the endonucleases are responsible for the morphological changes seen in the nucleus (Arends et al., 1990). On the other hand, other investigators have failed to find such a relationship (Oberhammer et al., 1993).

DNA fragmentation is a relatively early event in the cell death process, and it clearly precedes death itself as judged by the uptake of vital dyes. In thymocytes, for example, fragmentation occurs some 2 h after glucocorticoid treatment, while cell viability does not begin to drop for 8 h (Compton and Cidlowski, 1986). DNA cleavage having occurred, it is not possible to rescue the cells, and their death is inevitable. By contrast, the treatment of thymocytes with nuclease inhibitors or with zinc ions blocks both DNA fragmentation and apoptosis (Cohen and Duke, 1984; Martin et al., 1991), providing further evidence that DNA degradation is an integral step in the cell death process. The cleavage process most probably causes rapid loss of transcription, but may not itself be lethal (Arends et al., 1990), and studies have shown that in some examples of induced apoptosis the DNA fragmentation may be blocked without an effect on cell lysis (Zheng *et al.,* 1991). However, lethality may be linked to particular stages of the cell cycle (Sen, 1992; Walker *et al.,* 1993b), or, in the case of neurons, apoptosis may be associated with a failure to reenter the cell cycle normally (Rubin *et al.,* 1993). In some cells, for example, there may be a sensitive period at the G_2/M phase transition (Sorenson *et al.,* 1990), so that affected cells may continue to progress as far as G_2 and then die instead of entering M phase.

There are a number of outstanding issues in relation to DNA fragmentation and apoptosis (discussed by Sen, 1992; Schwartzman and Cidlowski, 1993): is DNA cleavage a direct cause of apoptosis or is it a secondary effect; is there synthesis of new endonucleases (Compton and Cidlowski, 1987) or are only preexisting ones activated by changes in calcium fluxes; the specific endonucleases have not been identified (Fraser *et al.,* 1993), but may include DNase I (Peitsch *et al.,* 1993) and/or DNase II (Barry and Eastman, 1993)—is there a family of related isozymes and if so, what are its characteristics? Most recently, apoptosis has been demonstrated in the absence of DNA fragmentation, thus questioning the universality of this characteristic (Schultze-Osthoff *et al.,* 1994).

2. Endonucleases in Development

Although the presence of a DNA ladder has been considered diagnostic for apoptosis, this is not strictly correct, since apoptotic cell death can occur in the absence of DNA cleavage (Gromkowski *et al.,* 1988; Cohen *et al.,* 1992; Schnellmann *et al.,* 1993). In particular, it is still not known whether DNA fragmentation invariably accompanies programmed cell death in development, the difficulty being the relatively small numbers of scattered cells that are affected by cell death in embryos. Although there are accounts of DNA cleavage in developing tissues, for example in retinal ganglion cell death (Ilschner and Waring, 1992), there is little evidence that endonucleases of the type seen in thymocytes are active during development, and certainly laddering of DNA has only rarely been demonstrated (Lockshin and Zakeri, 1991). Zakeri *et al.* (1993) attempted to demonstrate DNA fragmentation in dying interdigital tissue as well as during the metamorphic cell death occurring in the labial glands and intersegmental muscle of *Manduca* larvae. No evidence for early DNA degradation was found in these systems, and it was concluded that endonuclease activity neither is an initiator of cell death nor is it required for cell death in these tissues. However, these authors are careful to point out that there may be confounding factors responsible for these negative results (Lockshin and Zakeri, 1991; Zakeri *et al.,* 1993), although clearly endonuclease activity cannot be a major factor in the degradation of DNA in these tissues. The absence of early DNA fragmentation in interdigital tissue was also found by Garcia-

Martinez *et al.* (1993), who reported that such degradation is relatively late, tends to be nonspecific (i.e., not solely internucleosomal), and appears at the same time that lysosomal activity is occurring.

Clearly, in a number of embryonic systems crucial changes are happening long before the DNA is fragmented, suggesting that programmed cell death in development differs significantly in this regard from the thymocyte model from which many ideas have been extrapolated. So, while the thymocyte has been fruitful in defining many characteristics of apoptosis, it must be regarded as limited as a model for programmed cell death in development.

3. The Role of Calcium Ions

The endonuclease is usually, though not always (Alnemri and Litwack, 1990; Barry and Eastman, 1993), calcium, magnesium-dependent, being activated, at least in thymocytes, by changes in the calcium ion balance (Cohen and Duke, 1984; McConkey *et al.*, 1989a,b; reviewed by Sen, 1992; Trump and Berezesky, 1992; Schwartzman and Cidlowski, 1993). An early, rapid, and sustained rise in levels of cytosolic calcium is one of the characteristics of apoptosis in thymocytes (Schwartzman and Cidlowski, 1993), and in androgen-independent prostatic cells, a three- to sixfold rise in intracellular calcium concentration, for 12 h, is enough to induce the appearance of apoptosis (Martikainen *et al.*, 1991). In thymocytes, this rise likely involves an elevation in the levels of cAMP (McConkey *et al.*, 1990), and inhibition of the increase in calcium ion concentration will prevent endonuclease activation and hence cell death (McConkey *et al.*, 1989a). Conversely, increasing cytosolic calcium ion concentrations, using calcium ionophores, induces cell death (McConkey *et al.*, 1989b). The influx of calcium ions responsible for the rise intracellularly may be due to effects on existing calcium channels (Schwartzman and Cidlowski, 1993), and calcium channel antagonists are able to suppress the activity of cell death-associated genes such as *c-fos* and testosterone-repressed prostate message-2 (*TRPM-2*; Connor *et al.*, 1988). It has been suggested that intercellular signalling of cell death in neurons is mediated by calcium ion fluxes which are able to cross gap junctions between the cells (Wolszon *et al.*, 1994).

While there has been much emphasis on the connection between the rise in cytosolic calcium and apoptosis (Orrenius *et al.*, 1991), in this, as in other features of apoptosis, there are caveats, because nerve cells growing in the absence of nerve growth factor can actually be rescued from cell death by elevated calcium ion levels (Koike *et al.*, 1989). Also, in some apoptotic systems, raised levels of cytoplasmic calcium ions are not alone sufficient to induce cell death (Zheng *et al.*, 1991; Duke *et al.*, 1994), while in others calcium chelators can induce apoptosis (Kluck *et al.*, 1994). The indications are that changes in calcium ion flux are important, but a net rise in intracellular levels may not be a universal trigger for cell death.

4. Signal Transduction

In light of the above data, it is clear that calcium ions provide some link to the activation of endonucleases in many cells; however, these ions also appear, not surprisingly, to be involved in more classical signal transduction pathways (Sen, 1992; Lee et al., 1993). Both cAMP- and protein kinase C (PKC)-mediated pathways have been implicated in apoptotic events. Elevation of cAMP levels has been shown to induce DNA fragmentation in thymocytes via protein kinase A (McConkey et al., 1990), and such a pathway may be implicated in the glucocorticoid effects alluded to earlier (Dowd and Miesfeld, 1992). Although implicated, the case for PKC is unclear; it has been reported both to inhibit DNA fragmentation and cell death (McConkey et al., 1989c) and to induce apoptosis (Jin et al., 1992). The situation with regard to the incidence of tyrosine phosphorylation is equally unclear at present, the limited data indicating the possibility of apoptosis being accompanied by either enhanced (Uckun et al., 1992) or inhibited tyrosine phosphorylation (Chen and Rosenberg, 1992). It may be significant that antibody binding to the cell surface *Fas* antigen, a member of the tumor necrosis factor receptor/nerve growth factor receptor family, induces both apoptosis (Owen-Schaub et al., 1992) and rapid tyrosine phosphorylation of several proteins (Eischen et al., 1994).

That intracellular pathways are both important and poorly understood is shown by experiments using the p53 tumor suppressor gene, the expression of which appears to be able to initiate apoptosis as one means of mediating growth control (Yonish-Rouach et al., 1991; Shaw et al., 1992). p53 activity appears to be related both to the cell cycle (Wu and Levine, 1994; Martin et al., 1994) and to intercellular contact (Bates et al., 1994), and may act through its ability to activate interleukin-1β-converting enzyme (ICE; see below) which is similar to the *C. elegans ced-3* gene product (Miura et al., 1993; Yuan et al., 1993; Vaux et al., 1994). T cells deficient in p53 die normally upon exposure to glucocorticoids, but are resistant to treatment with ionizing radiation, indicating that the pathways involved in two apparently similar cell death processes are in fact different (Lowe et al., 1993; Clarke et al., 1993).

The association of p53 activity with intercellular contact (Bates et al., 1994) brings up the question of the relationship between apoptosis and events at the cell surface. Evidence is just beginning to accumulate to show that cell death is connected not only with cell-to-cell contact, but also with cell-to-matrix contact, since apoptosis may be induced by the dislodging of epithelial cells from their substratum (Frisch and Francis, 1994; Re et al.,

1994; Ruoslahti and Reed, 1994). These contact relationships may be reflected by changes in cell surface glycosylation patterns (Hiraishi *et al.,* 1993) and in the activity of cell adhesion molecules (Fujita *et al.,* 1993).

Despite the accumulating data on the potential pathways of signal transduction during the cell death process (Sen, 1992; Lavin *et al.,* 1993; Walker *et al.,* 1993b), there is not yet a clear picture of the multiple interactions between the mediators, or of the networks that must be involved (Evans, 1993).

5. Gene Activation and Protein Synthesis

It has been known for many years that programmed cell death occurring during development can often be prevented by inhibition of protein synthesis and, in some instances, also by blocking transcription. Some examples of this phenomenon are: the regression of the tadpole tail at morphogenesis (Tata, 1966), the death of the intersegmental muscle cells of *Manduca* (Lockshin, 1969), the death of palatal epithelial cells at the time of fusion of the palatal shelves (Pratt and Greene, 1976), and the death of rat and chick embryo neurons (Martin *et al.,* 1988; Oppenheim *et al.,* 1990). Such results are corroborated by the demonstration that there is expression of new proteins concomitant with the onset of the cell death program. So, for example, in *Manduca* the intersegmental muscles of the larva degenerate at metamorphosis in response to reduced levels of ecdysteroids. The muscles are accordingly induced to persist in the presence of exogenous ecdysteroids or after treatment with actinomycin-D. Although the degeneration is accompanied by the repression of some myofibrillar proteins, it may also be shown that an array of hitherto unseen proteins appears on 2D gels (Wadewitz and Lockshin, 1988). Four genes ("cell death genes") in particular are up-regulated at this time, and are considered to be required for the death process (Schwartz *et al.,* 1990a), one of these has been identified as coding for polyubiquitin (Schwartz *et al.,* 1990b). Ubiquitin is a highly conserved protein involved in the targeting of unwanted proteins to degradation pathways in the cell (Hershko, 1988). Although ubiquitin is present in *Manduca* larval muscles (Schwartz, 1991) and in nerves (Fahrbach and Schwartz, 1994), its late function in the rapid disposal of proteins makes it unlikely to be responsible for the onset of programmed cell death itself (Schwartz, 1991). However, it may be a marker for cell death and one of an assembly of proteins, synthesized *de novo,* that are required for the execution of the program.

Ubiquitin is but one of an array of genes whose expression has been associated with cell death (tabulated by Schwartz and Osborne, 1993). Another example is the testosterone-repressed prostate message-2 (*TRPM-2*; also known as SGP-2 and clusterin), which, although the function of the gene product is unknown, is induced to high levels in a variety of apoptotic cells, including some during development of the degenerating interdigital tissue (Buttyan *et al.,* 1989), though apparently not those in the palate and nervous system (Garden *et al.,* 1991). Some genes, such as p53 (see above) and *c-myc* (Buttyan *et al.,* 1988), appear to be similar to ubiquitin in that they are required for cell death and their blockage therefore prevents the demise of cells, while others, such as *bcl-2* (Vaux, 1993), seem to protect against death and their blockage correspondingly increases the incidence of death. It will be recalled that the product of the *C. elegans* cell death gene *ced-3* is probably a protease similar to the interleukin-1β (IL-1β)-converting enzyme (ICE) which can mediate apoptosis in fibroblasts (Miura *et al.,* 1993; Yuan *et al.,* 1993; Vaux *et al.,* 1994). Although cell death is not mediated by IL-1β, indicating that there must be another substrate for ICE, overexpression of the murine ICE (mICE) gene or the *C. elegans* *ced-3* gene can induce programmed cell death in mammalian cells, indicating a possible physiological role for gene products of this type in the onset of apoptosis.

The list of new gene products that appear coincident with apoptosis is currently expanding extremely rapidly (Nakashima *et al.,* 1993; Hébert *et al.,* 1994).

The genes mentioned here, together with others, are considered elsewhere in this review, but suffice it to say here that no single gene product has been associated with all forms of programmed cell death in development. Indeed, reverse cases have been recorded in which inhibition of protein synthesis allows cell death (Goldman *et al.,* 1983). It must also be said that it is not always clear to what extent the induction of new proteins is a cause or an effect of cell death (Lockshin and Zakeri, 1991).

Similar studies using the glucocorticoid-stimulated thymocyte indicate a parallel situation. The glucocorticoids induce the appearance of new proteins and increase the rate of synthesis of others (Voris and Young, 1981; Colbert and Young, 1986), and correlations have been made between this up-regulation and the onset of apoptosis using inhibitors of protein and mRNA synthesis (Wyllie *et al.,* 1984). In these circumstances, the calmodulin gene is one that is hormonally controlled, the levels of mRNA being increased 10-fold after glucocorticoid treatment, supporting suggestions for a role for this calcium-binding protein in apoptosis (Dowd *et al.,* 1991). Despite this, the requirement for protein synthesis in pathological apoptosis does not seem to be as clear as in developmental programmed cell death, and numerous cases may be cited (reviewed by Schwartzman and Cidlowski, 1993) in which inhibitors of protein synthesis are ineffective in abrogating apoptosis (Waring, 1990).

6. Regulation by Hormones and Growth Factors

Reference has already been made to the effects of glucocorticoids in the regulation of thymocyte apoptosis, but there are several other examples of cell death occurring in hormone-dependent tissues after hormone with-drawal. These include prostate and mammary gland in vertebrates (Tennis-wood *et al.*, 1992), and a variety of ecdysteroid-dependent tissues in insects (Robinow *et al.*, 1993). During strictly developmental events, efforts have been centered on the determination of the roles of various growth factors and cytokines in the promotion either of cell death or of cell survival. The function of target-derived nerve growth factor (NGF) in the maintenance of neurons is well established (Perez-Polo *et al.*, 1990), and is discussed below, but a number of other factors have been shown to promote neuronal survival during development, in both *in vitro* and *in vivo* situations. These include: somatostatin (somatotropin release-inhibiting factor, SRIF; Weill, 1991), which may act via an effect on calcium ion fluxes, neurotrophins, insulin-like growth factor-I (IGF-I), leukemia inhibitory factor (LIF), brain-derived neurotrophic factor (BDNF), ciliary neurotrophic factor (CNTF), and members of the fibroblast growth factor (FGF) family (Hughes *et al.*, 1993; Thaler *et al.*, 1994). The evidence suggests that moto-neuron and sensory neuron survival depends on the integrated action of several groups of soluble factors (see below).

Several other factors have been implicated in modulation of the cell death process during various other phases of development. For example, transforming growth factor $\beta1$ (TGF$\beta1$) is able to initiate apoptotic path-ways during the genesis of blood cells (Selvakumaran *et al.*, 1994), while colony stimulating factors (CSFs) promote survival of hemopoietic precur-sor cells (Williams *et al.*, 1990). The interleukins also apparently play a complex role in the regulation of apoptosis during the development of thymocytes (Migliorati *et al.*, 1993). In more conventional developmental systems, both FGF-2 and epidermal growth factor (EGF) have been impli-cated in rescuing cells from apoptosis, the former in the limb bud (Fallon *et al.*, 1994), and the latter in the kidney (Koseki *et al.*, 1992).

Recently, attention has been drawn to possible roles for tumor necrosis factor-α (TNF-α) in early development (Wride and Sanders, 1995). This factor, which has been associated with apoptosis in the immune system for many years, has been found to be expressed, in high-molecular-weight forms, in precise spatio–temporal patterns during the early development of the chick and mouse (Gendron *et al.*, 1991; Wride and Sanders, 1993; Jaskoll *et al.*, 1994). Although the relationship of these immunoreactive forms of the factor to native TNFα is as yet unclear, mRNA for this factor is expressed in many different tissues (Kohchi *et al.*, 1994). Several tissues are found to express the factor during early development: neural tube, spinal nerves, myotome, kidney, lens, limb bud and notochord—all of which are undergoing morphogenetic reorganizations which include cell death. In several of these locations the expression of TNFα-like proteins is corre-lated with the occurrence of cell death, and appears in cells that are undergoing apoptosis (Wride *et al.*, 1994), leading to the speculation that these proteins are involved in developmental programmed cell death (Wride and Sanders, 1995).

7. Genetic Regulation

In a number of developing systems, the genes involved in the control of cell death are being identified. This field is most advanced in invertebrate systems such as *C. elegans* (see below), where genes of the *ced* series are established regulators of normal programmed death, and others, called "degenerins," such as *mec-4* and *deg-1*, have roles in abnormal neurodegeneration (Driscoll, 1992).

In the avian embryo, the transcription factor *c-rel* is expressed at high levels in cells undergoing programmed cell death, including the limb bud, neural crest, and CNS (Abbadie *et al*, 1993). However, overexpression of *c-rel* in cultured avian fibroblasts does not lead to an increase in the incidence of cell death, although such an effect can be demonstrated with other cell types. Also, in nervous tissue, *c-fos* has been implicated in terminal differentiation and death (Smeyne *et al.*, 1993), but there is not yet any clear and coherent hypothesis as to the roles of these genes in the life of the embryo. Greater progress has been made in the elucidation of the relationship between *c-myc* expression and apoptosis, but not yet in developing systems. In 1987, Wyllie *et al.* reported that apoptosis was a much more frequent phenomenon in tumors that express *c-myc* than in those that do not. More recent work has shown that in T cell hybridomas, *c-myc*, more usually thought to be associated with cell growth, is necessary for activation-induced apoptosis (Shi *et al.*, 1992), and that this effect is modulated by the influence of *bcl-2* (Bissonette *et al.*, 1992). The action of *bcl-2* in this system, therefore, appears to be as a survival signal that redirects the action of *c-myc* to stimulate growth rather than death (Green and Cotter, 1993). It is also of note that among its other effects, *c-myc* may increase sensitivity to TNF-induced apoptosis (Klefström *et al.*, 1994).

The *bcl-2* (B-cell lymphoma/leukemia-2) gene, associated with B-cell malignancies, encodes a 26-kDa Bcl-2 protein which localizes to the nuclear envelope, endoplasmic reticulum, and mitochondrial membranes (Givol *et al.*, 1994; Nuñez and Clarke, 1994; Reed, 1994). This gene is able to block the final common pathway for programmed cell death, thus prolonging cell survival. Although a number of *bcl-2*-independent mechanisms of blocking cell death have been identified, *bcl-2* has been shown to be active in regulating cell survival in a large number of cases (Reed, 1994), and it may act by regulating the movement of calcium ions through the membranes of the endoplasmic reticulum, thus abolishing the calcium signal for apoptosis (Lam *et al.*, 1994). Among its actions, *bcl-2* appears to be able to integrate its effects with *c-myc* to modulate the apoptosis-inducing effects of p53 (Ryan *et al.*, 1994), and suppress cell death mediated by ICE (see above; Miura *et al.*, 1993). Most recently, a number of Bcl-2-related gene products have been identified, one of which, Bax, is able to complex with Bcl-2 and regulate the activity of the *bcl-2* gene (Korsmeyer *et al.*, 1993). Two other Bcl-2 homologs, $Bcl-x_L$ and $Bcl-x_S$, also modulate the effects of *bcl-2* (Nuñez and Clarke, 1994), and it has been suggested, based on patterns of expression, that genes of the *bcl-x* series may play a significant role in regulating cell death during development (González-García *et al.*, 1994).

Although it appears that *bcl-2* expression can rescue neurons from normally occurring programmed death, or from death resulting from NGF deprivation (Martinou *et al.*, 1994; Mizuguchi *et al.*, 1994; Rubin *et al.*, 1994), its effects during normal development are not yet clear. "Knockouts" of *bcl-2* in mice show that embryos are able to develop normally to birth, only to die during postnatal development (Veis *et al.*, 1993; Nakayama *et al.*, 1994). There may therefore be some redundancy in the pathway influenced by *bcl-2*, because this gene is certainly expressed in a variety of tissues in the embryo, including kidney, heart, CNS, PNS, and muscle (Eguchi *et al.*, 1992; LeBrun *et al.*, 1993; Lu *et al.*, 1993; Merry *et al.*, 1994). In the limb bud, Bcl-2 protein occurs only in the areas of cell survival, and not, for example, in the interdigital zones (Novack and Korsmeyer, 1994).

C. Techniques for Examining Cell Death

Because cell death in developing systems involves cells scattered through a generally viable population, and because some phases of the process are fleeting, it has not always been easy to identify apoptotic cells in embryos or in pathological samples (Bowen, 1981). The classical morphological criteria are subjective and identification of such cells is both time consuming and tedious. Also, embryonic cells often do not go through all of the morphological phases of apoptosis, being phagocytosed relatively early in the process of death. Whether the recent development of a cell-free system using isolated nuclei (Lazebnik *et al.*, 1993) will be suitable for embryonic cells is not yet known.

Vital stains for dying cells, such as nile blue sulfate, have a long history of use (Saunders *et al.*, 1962; Pexieder, 1975), and continue to be employed (Jeffs *et al.*, 1992; Jeffs and Osmond, 1992). However, these stains, which also include neutral red, acridine orange (see Fig. 4), and toluidine blue (Hinchliffe and Griffiths, 1986; Martín-Partido *et al.*, 1986), lack specificity for apoptotic cells, and demonstrate cells that are likely already dead rather than cells early in the pathway to death. Cytochemical methods for hydrolytic enzymes can be capricious, and are primarily useful only after the stage of phagocytosis has been reached.

The detection of DNA fragmentation remains a useful corroborative tool (Prigent *et al.*, 1993; Walker *et al.*, 1993a), though the dispersed nature of the dying cells in embryonic systems makes it difficult to be sure that a negative result is true (Lockshin and Zakeri, 1991). Further, although separation of DNA fragments gives a temporal analysis of degradation, it gives no spatial information. The latter shortcoming has been addressed by the recent refinement of *in situ* techniques. The *in situ* visualization of DNA strand breaks in whole or sectioned cells relies on the incorporation of biotin-labeled nucleotides into DNA breaks by means of DNA polymerase (Iseki and Mori, 1985; Iseki, 1986; Ansari *et al.*, 1993; Gold *et al.*, 1993; Wijsman *et al.*, 1993) or terminal deoxynucleotidyl transferase (TdT; Gavrieli *et al.*, 1992), followed by histochemical localization of the label (Figs. 2 and 3). The technique has become known as "*in situ* end labeling"

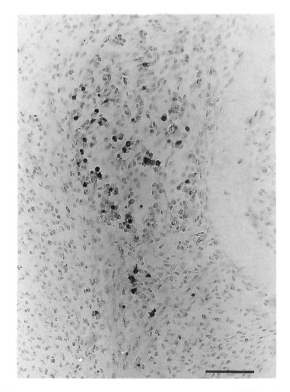

FIG. 2 The TUNEL technique demonstrates DNA fragmentation in cells destined to die. Densely labeled nuclei are seen in the dorsal root ganglion of the early chick embryo. Scale bar, 50 μm.

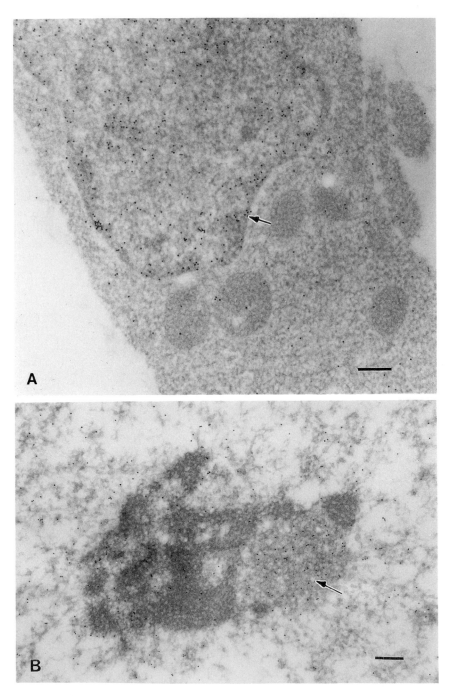

FIG. 3 At the electron microscope level, the TUNEL technique shows labeling associated with: (A) the chromatin condensed on the inner surface of the nuclear membrane (arrow), and (B) with the heterochromatin (arrow). Scale bars, 1 μm.

(ISEL), "*in situ* nick translation" (ISNT), or "TdT-mediated dUTP-biotin nick end labeling" (TUNEL), and has been successfully used for the spatial analysis of apoptotic cells in early embryonic material (Wride *et al.*, 1994).

The general applicability and reliability of antibodies generated against single-stranded DNA (Naruse *et al.*, 1994) and against surface antigens of apoptotic cells (Rotello *et al.*, 1994) remain to be determined.

III. Programmed Cell Death in Development

A. Invertebrate Development

1. *Caenorhabditis elegans*

Lineage-dependent death of cells in the nematode worm *C. elegans* was first identified by Sulston (1976). *Caenorhabditis elegans* has proven to be an ideal system in which to study programmed cell death and it has perhaps provided the greatest impetus to workers in the field. A well-characterized set of cell lineages results in the appearance of 1090 somatic cells. The nematode is transparent, and using Nomarski differential interference contrast optics the 131 cells (105 neuronal and 26 nonneuronal) that undergo programmed cell death can be identified (Driscoll and Chalfie, 1992). Genetic analysis of *C. elegans* has revealed genes that function in normal developmental cell death as well as genes that mutate to cause inappropriate cell death (reviewed by Driscoll, 1992; Driscoll and Chalfie, 1992). Molecular characterization of these "death" genes in *C. elegans* has provided insight into normal and abnormal mechanisms of cell death in this system, which are clearly applicable, in many cases, to vertebrates (Driscoll and Chalfie, 1992; Hengartner and Horvitz, 1994a,b). Cell death in *C. elegans* has many of the hallmarks of apoptosis and, as mentioned earlier, it may be one of the few systems where the term "programmed" cell death is truly warranted.

Specific genes are implicated in cell death in *C. elegans* (Ellis and Horvitz, 1986). Two of these, *ced-3* and *ced-4* (*ced = cell death abnormal*), were identified and it was shown that their wild type (*wt*) function is required for the initiation of all cell deaths in *C. elegans*. It has been shown, using genetic mosaic analysis in which some cells are genotypically mutant and others are genotypically nonmutant, that the products of the *ced-3* and *ced-4* genes act autonomously to cause death in those cells in which they are expressed (Yuan and Horvitz, 1990). A reduction or loss of *ced-3* or *ced-4* function results in a transformation of the fates of cells which normally die such that they are not subject to death (Ellis and Horvitz, 1986). The surviving cells are capable of adopting roles in the embryo normally performed by others. For example, Avery and Horvitz (1987) showed that a cell that normally dies in the *wt* can function as a pharyngeal neuron in a *ced-3* mutant. Mutant worms appear normal in both morphology and behavior, implying that programmed cell death is not an essential aspect of nematode development (Ellis and Horvitz, 1986).

In *C. elegans,* cell death mediated by *ced-3* and *ced-4,* though not essential, may be advantageous to the organism for several reasons (discussed by Ellis *et al.,* 1991a; Driscoll and Chalfie, 1992). It has been noted that mutants respond less well to chemosensory information, which is perhaps due to interference by additional neurons in neuronal circuits. In addition, mutants mature more slowly than wild types, and thus a selective advantage is conferred by cell death in wild types.

Both *ced-3* and *ced-4* have now been cloned and sequenced. *ced-4* encodes a novel protein which has no known homology to other proteins except for two regions which show homology to known calcium-binding domains (Yuan and Horvitz, 1992). *ced-3,* on the other hand, encodes a protein which is homologous to the human and murine cysteine protease ICE and to the product of the mouse *Nedd-2* gene (Yuan *et al.,* 1993) and its human homolog *Ich-1* (Wang *et al.,* 1994). *Nedd-2* is expressed in the mouse embryo brain and induces apoptosis when overexpressed in cultured fibroblast and neuroblastoma cells (Kumar *et al.,* 1994). It has been proposed that the product of *ced-3* acts as a cysteine protease during *C. elegans* cell death and that ICE would function as a cysteine protease in programmed cell death in mammalian cells (Yuan *et al.,* 1993). Indeed, in a subsequent paper, Miura *et al.* (1993) showed that ICE was able to induce apoptosis when transfected into rat fibroblasts. Furthermore, this cell death could be inhibited by overexpression of *crmA,* a cowpox virus that is a specific inhibitor of ICE, and by *bcl-2,* the mammalian homolog of *ced-9.* ICE can also promote cell death in cells which naturally undergo cell death during development of vertebrates. Gagliardini *et al.* (1994) have shown that chick embryo dorsal root ganglion neurons are rescued from death in serum-free culture by the overexpression of the *crmA* gene, implicating ICE in vertebrate neuronal cell death. Thus, there is an emerging family of genes with homology to *ced-3* including the genes for ICE, *Nedd2,* and *Ich-1.*

Ellis and Horvitz (1991) argued that because of the diversity in origin and cell type among dying cells there could be regulatory genes capable of controlling the fates of some cells, but not of others, through their actions on the *ced-3* and *ced-4* genes. *ces-1* and *ces-2* (*ces* = cell death specification) act on *ced-3* and *ced-4* to control the decisions of two cells in the pharynx to die. Mutations causing a gain in *ces-1* function or a loss in *ced-2* function prevent the death of these cells, but not of most other cell deaths.

Genes that are involved in the engulfment of dying and dead cells and the degradation of DNA in these cells have also been identified in *C. elegans*. Sulston (1976) identified a mutation, *nuc-1,* resulting in a lack of endodeoxyribonuclease. In these mutants, the DNA from dead cells was not degraded, but persisted in engulfing cells; however, despite this, cell deaths are essentially normal in this mutant. Hedgecock *et al.* (1983) identified two genes, *ced-1* and *ced-2,* in which mutations resulted in a lack of removal of DNA and nucleoplasm as well as a reduced efficiency of cell corpse engulfment. However, these genes were not essential for engulfment and it was not clear whether they were required in the dying cells or the engulfing cells. Subsequently, five other engulfment genes were identified (Ellis *et al.,* 1991b) and these were designated *ced-5, ced-6, ced-7, ced-8* and *ced-10.*

There has been some speculation that since engulfment occurs concomitant with death, engulfment might be a prerequisite for death. However, Hedgecock *et al.* (1983) showed that this is apparently incorrect for most deaths, since most deaths occurred with normal timing in the *ced-1* and *ced-2* mutants. However, *C. elegans* possesses two cells which will not die unless engulfed by neighbors (Sulston *et al.,* 1980). These are the gonadal linker cells, and if the cells that usually engulf them are microablated with a laser, they survive indefinitely. It is of interest that a similar situation exists in at least one case of vertebrate programmed cell death. Macrophages appear to be required for the programmed cell death of the pupillary membrane in the mouse eye (Lang and Bishop, 1993). In mice expressing a transgene that disrupts subsets of mature macrophages, the pupillary membrane remains intact, suggesting a direct involvement for engulfing cells, in this case macrophages, in the death of cells in this structure.

Hengartner *et al.* (1992) identified another important regulator of cell death in *C. elegans,* the *ced-9* gene. This gene acts to protect cells from death by antagonizing the effects of *ced-3* and *ced-4.* A mutation that abnormally activates *ced-9* (Hengartner *et al.,* 1992) or overexpression of wild-type *ced-9* (Hengartner and Horvitz, 1994c) results in prevention of normal cell death, while mutations that deactivate *ced-9* result in the death of cells that are not normally subject to death (Hengartner *et al.,* 1992). This latter mutation is lethal, revealing that *ced-9* function is an essential prerequisite for successful *C. elegans* development. With respect to its ability to protect cells from death, Hengartner *et al.* (1992) realized that *ced-9* showed a marked similarity to the human oncogene *bcl-2,* which is expressed in B and T cells in the immune system and protects them from apoptosis. Subsequently, an elegant experiment showed that expression of the human *bcl-2* gene in *C. elegans* resulted in a reduction in programmed cell death in *C. elegans* (Vaux *et al.,* 1992), revealing the conservation of function of *bcl-2.* Furthermore, the predicted sequence of the *CED-9* protein (Hengartner and Horvitz, 1994c) has homology with the product of mammalian *bcl-2*. It is also becoming apparent that some viral genes are also members of the *bcl-2/ced-9* family. The baculovirus p35-encoding gene (*p35*) is an inhibitor of virus-induced apoptosis in insect cells. Expression of *p35* in *C. elegans* prevents the death of cells normally destined to die and can rescue the lethality associated with a loss of function mutation in *ced-9* (Sugimoto *et al.*, 1994).

2. Insects

Programmed cell death occurs in a number of different tissues and during all stages of insect development: in the embryo and during postembryonic stages, including the larval–pupal transition (metamorphosis); and in the adult (reviewed by Fahrbach and Truman, 1987; Lockshin 1981; Schwartz, 1992; Steller *et al.*, 1994; Truman, 1984, 1990; Truman *et al.*, 1990, 1992).

Work concerned with programmed cell death during embryonic stages of insect development has mainly concentrated on grasshopper (Truman *et al.*, 1992) and, more recently, *Drosophila* embryos (Raff, 1994; Steller *et al.*, 1994). However, most work has focused on metamorphosis, since this is an ideal model for postembryonic development (Tata, 1993) and on the death of tissues in the adult following eclosion (emergence of the adult from its pupal case; Truman, 1990; Truman *et al.*, 1990, 1992; Schwartz, 1992). The large tobacco hornworm moth, *Manduca sexta*, has been highly informative because of its size and the characteristics of its tissues, which are more homogeneous than vertebrate systems (Lockshin, 1981; Fahrbach and Truman, 1987).

Programmed cell death has several functions in insect development, including the removal of obsolete neurons and muscles, the generation of segmental specializations, sexual dimorphism, and the adjustment of neuronal number (Schwartz, 1992; Truman *et al.*, 1992). Like metamorphosis itself, programmed cell death in insects is intimately linked to levels of hormones, e.g., ecdysteroids and juvenile hormone (Truman *et al.*, 1992).

The embryonic period of development is generally brief and a relatively small number of neurons are produced in the larval CNS. These neurons are important for larval behavior, but during metamorphosis dramatic changes in body form result in extensive remodeling of the musculature and the nervous system. The final period of programmed cell death occurs after the adult emerges from its pupal case and this wave removes muscles, motoneurons, and interneurons that are no longer needed by the adult. The reader is referred to the above-mentioned reviews for more comprehensive and detailed analyses, but certain aspects will be discussed in greater detail below.

a. **Drosophila** If *C. elegans* has been the organism of choice for developmental biologists interested in the phenomenon of programmed cell death, it is likely that the organism of the future is *Drosophila* (Raff, 1994). *Drosophila*, although smaller than *Manduca* and less suitable for experimental studies, is particularly amenable to genetic manipulation and it is in this area that further insights into the genetic regulation of programmed cell death are expected. Indeed, recent work has already begun to provide such information (White *et al.*, 1994; reviewed by Raff, 1994, and Steller *et al.*, 1994).

Cell death is important throughout the development of *Drosophila*, occurring during embryonic and metamorphic stages and in the adult (Steller *et al.*, 1994; Truman, 1990). This cell death is consistent with that seen in other systems (Steller *et al.*, 1994) in that it is subject to both lineage-dependent and epigenetic control, including cell–cell interactions and regulation by trophic factors and ecdysone.

It is important to emphasize that most studies of cell death in *Drosophila* have concentrated on the later stages of development, such as metamorphosis (see below), rather than on embryonic development. However, it was shown recently, (Fig. 4) that there is substantial cell death beginning at an early stage of *Drosophila* embryonic development (Abrams *et al.*, 1993). The pattern of cell death is highly reproducible at a given stage, but asymmetries in the exact numbers and positions of dying cells on either side of the midline suggest that the decision for cells to die is not strictly predetermined at these stages. Further work has now identified a gene, *reaper*, which clearly has a central regulatory role in programmed cell death in the *Drosophila* embryo (White *et al.*, 1994), and *reaper* mRNA is expressed in cells that are undergoing apoptosis.

A potential effector of programmed cell death in *Drosophila* has recently been cloned and sequenced (Brand and Bourbon, 1993). The *rox 8* gene codes for an RNA-binding protein that is similar to the human TIA-1-type nucleolysins that are involved in natural killer cell and T-lymphocyte-mediated apoptosis in the immune system. This work provides the basis for further genetic analysis for potential roles for these proteins in programmed cell death in *Drosophila*.

The *Drosophila* eye has proven to be an excellent model system in which to study how differentiation and cell death interact during development (Bonini *et al.*, 1993). Eye differentiation begins in the eye imaginal disc and is accompanied by a low level of cell death (Wolff and Ready, 1991). The *eyes-absent* gene (*eya*) has been shown to control cell survival and differentiation in *Drosophila* eye imaginal discs (Bonini *et al.*, 1993), and in *eya* mutants, progenitor cells undergo programmed cell death rather than progressing into the pathway of retinal differentiation. Thus, the *eya* gene appears to be involved in influencing the distribution of cells between

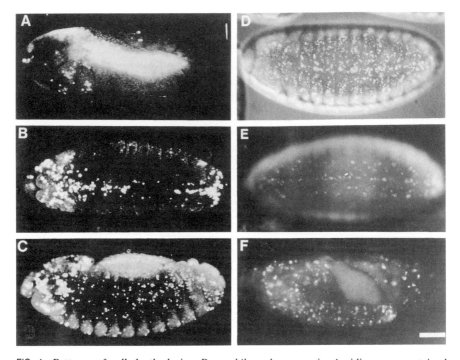

FIG. 4   Patterns of cell death during *Drosophila* embryogenesis. Acridine orange-stained embryos; anterior to the left. Yolk in the center of the embryos gives a diffuse, nonspecific staining. (A) Stage 12, lateral view. At this eary stage, dying cells are found in the gnathal segments and the clypeolabrum, and near the caudal tip of the retracting germ band. (B) Late stage 12, ventral view. Cell death spreads from both cephalic and caudal extremities along the ventral midline axis. Prominent cell death in the ventral cephalic region is also found. (C) Stage 13, lateral view. Dying cells accumulate around the cephalic ganglia and beneath the dorsal ridge. Scattered deaths occur throughout the lateral epidermis. (D) Stage 13, ventral view. Segmentally reiterated deaths are found in the ventrolateral epidermis. (E) Stage 13/14, ventral view. One central and two lateral columns of cell deaths are apparent in the ventral portions of the embryo. (F) Late stage 16, lateral view. Note that prominent numbers of cells die in the central nervous system. Scale bar, 50 μm. Reprinted with permission of Dr. H. Steller and the Company of Biologists Ltd., from Abrams *et al.* (1993).

programmed cell death and differentiation. This is analogous to the situation in *C. elegans* in which cells that survive in *ced-3* and *ced-4* mutants perform useful functions (see above).

Several other mutations that affect cell survival in *Drosophila* eye imaginal discs have been identified and these belong to the EGF-receptor family. *Ellipse* alleles are mutations to the EGF-receptor homolog that reduce the number of ommatidia in the eye imaginal disc by increasing cell death (Baker and Rubin, 1992); *torpedo* is a *Drosophila* homolog of a receptor tyrosine kinase of the EGF-receptor subfamily, mutations in which result, once again, in increased cell death at early stages (Clifford and Schüpbach, 1992). Thus, *torpedo* may transduce signals necessary for cell survival per se during early embryogenesis in *Drosophila.* These results suggest that in *Drosophila,* *eya* and EGF-receptor homologs may be functionally analogous to survival factors, such as *ced-9* and *bcl-2,* which are required to rescue cells that have an intrinsic suicide potential.

The naturally low level of cell death occurring in the *Drosophila* eye during development (Wolff and Ready, 1991) results in the removal of two to three surplus cells per ommatidium. Cells die by apoptosis as well as by less well-characterized processes. Because of the regular pattern of the *Drosophila* eye, it is a sensitive amplifier of defects in programmed cell death, and a failure of cell death to remove surplus cells results in a disorganized retinal lattice (Wolff and Ready, 1991). Two genes, *echinus* and *roughest,* have been identified that are required for elimination of cells from the retinal epithelium (Wolff and Ready, 1991). Order in the *Drosophila* retina is achieved by various cell cell interactions in which rival cells compete for existence and the loser dies (Cagan and Ready, 1989). This is analogous to the programmed death of the linker cell in *C. elegans* (see above; Sulston *et al.,* 1980).

As with *C. elegans,* the introduction of the p35 baculovirus protein into developing *Drosophila* embryos and eyes results in a dramatic reduction in natural cell death as well as susceptibility to X-ray-induced cell death throughout the embryo and in the eye (Hay *et al.,* 1994). Excess pupal eye cells that are normally eliminated by programmed cell death continue to exist in the presence of p35 and develop into pigment cells. Thus, p35 blocks a normal, highly conserved cell death pathway and the isolation and identification of molecules that interact with p35 in *Drosophila,* as well as in other systems, should result in the identification of proteins involved in this conserved cell death pathway.

Indeed, *Drosophila* has already provided some insights into proteins that may be involved in the cell death pathway. Ramos *et al.* (1993) have cloned a transmembrane protein that interacts with the *roughest* protein, affects axonal projections, and is required for programmed cell death in the developing retina, while Treier *et al.* (1992) have cloned a gene that is a highly conserved ubiquitin-conjugating enzyme that could have a role during programmed cell death in ubiquitin-dependent proteolysis.

Programmed cell death also has an important role in the sculpturing of the *Drosophila* CNS during metamorphosis and in the adult (Kimura and Truman, 1990; reviewed by Truman, 1990). Two rounds of cell death occur in the *Drosophila* ventral CNS: the first wave occurs immediately following pupariation and a second wave occurs soon after the emergence of the adult. The second wave of cell death involves the degeneration of transiently

useful structures (muscles and motoneurons) that are involved in the emergence of the fly from its puparium and in the expansion of its wings,
but which are not required thereafter. Muscle degeneration appears to be
triggered by a signal (probably eclosion hormone) that is produced from
the head region (Kimura and Truman, 1990). Two classes of mutants, *mcd-
1* and *mcd-2* (*mcd* = muscle cell death) have been isolated that delay the
death of some of these muscles (Kimura and Tanimura, 1992). Differences
in the cell death processes mediated by *mcd-1* and *mcd-2* suggest that these
are at least two of the genes acting on different steps of the process of
muscle cell death.

The death of neurons, however, is not triggered until after emergence
of the fly; it can be delayed for hours by preventing behavior that is associated with wing expansion (Kimura and Truman, 1990). The mechanisms
by which wing expansion behavior is linked to programmed cell death
remain undetermined, but electrical activity in nerves has been shown to
influence the extent of programmed cell death in vertebrates (Oppenheim,
1991; and see below).

Some evidence has recently appeared that programmed cell death of
Drosophila ventral CNS neurons is ecdysone regulated (Robinow *et al.*,
1993). A subpopulation of about 300 ventral CNS neurons expresses a 10-
fold higher level of ecdysteroid receptor A than other ventral neurons.
This higher expression begins early in metamorphosis and persists through
the pupal stages. The same population of neurons undergoes programmed
cell death when the fly emerges. It appears that a prerequisite for death is
a reduction in ecdysteroids at the end of metamorphosis, which is analogous
to the situation in *Manduca* (see below), since treatment of flies with 20-
hydroxyecdysone rescues these neurons if given at least 3 h before the time
of degeneration. Thus, variations in the patterns of ecdysteroid receptor
isoforms may be an important switch in the regulation of programmed cell
death in *Drosophila*.

b. *Manduca sexta*   The characteristics of *Manduca* (Lockshin, 1981; Fahrbach and Truman, 1987) have meant that more descriptive information and
experimental data are available on neuronal death in this system than in
any other insect.

There is a precise, highly reproducible, spatio–temporal pattern of abdominal neuron death in moths, which occurs after the emergence of the
adult moth from its old pupal skin (eclosion; Truman, 1983). Most of the
intersegmental muscles degenerate during pupation (see below), and this
muscle loss is accompanied by degeneration of interneurons and motoneurons. For example, 50% of the neurons in the ganglia of segments A3, A4,
and A5 die during the first days of adult life (Truman, 1983). Degeneration
of interneurons begins 2 h before eclosion and continues until 30 h after,

while death of motoneurons begins at 8 h after eclosion (Truman, 1983). This degeneration of neurons is dependent upon an endocrine signal—a reduction in the concentration of ecdysteroids at the end of metamorphosis (Truman and Schwartz, 1984), but the cell death program itself is flexible in that the fates of individual cells can be modified by behavior that might naturally occur in the life of the insect. For example, artificially increasing the length of time that the emerging moth spends in digging behavior (the moth undergoes metamorphosis underground, so it has to dig to the surface) delays the death of the D-IV motoneurons that are important in this digging behavior (Truman, 1983). Thus, the order of cell death ensures that a neuron will persist through the time that it is needed for a particular behavior.

It has been shown that neuronal cell death at two different stages of *Manduca* development (metamorphic and postmetamorphic) is dependent upon the active synthesis of mRNA and protein (Weeks *et al.*, 1992; Fahrbach *et al.*, 1994). During the larval–pupal transition, *Manduca* loses its abdominal prolegs. This involves the degeneration of proleg muscles, the dendritic arbors of proleg motoneurons regress, and a subset of the proleg motoneurons themselves dies. In this case, degeneration and death are triggered by the prepupal peak of ecdysteroids. Weeks *et al.* (1992) investigated the involvement of protein synthesis in these events by giving repeated injections of cycloheximide during the prepupal peak of ecdysteroid. This significantly reduced the extent of proleg motoneuron death in a dose-dependent fashion. However, dendritic regression was less susceptible to inhibition by cycloheximide, suggesting some intrinsic differences in the mechanisms of cell death between different classes of neurons. Fahrbach *et al.* (1994) have looked at the effects of both cycloheximide and actinomycin–D on the death of neurons in the abdominal ganglia of the *Manduca* CNS, both *in vivo* and *in vitro*. They found that the effectiveness of treatments in delaying or blocking neuronal death is dependent upon the time of administration, since early dying neurons could not be rescued while late dying ones could. It was shown that the ability of actinomycin-D to prevent cell death waned at the same time at which replacement of steroid hormone could no longer block death, implying that this "steroid commitment point" represents the time at which genes that mediate cell death are transcribed. Cycloheximide, on the other hand, was effective in delaying the time of neuronal death until just after the time of commitment to death, implying that ongoing protein synthesis is required for the initiation of the degeneration process.

If protein synthesis is an important prerequisite for neuronal cell death in *Manduca,* it is important to know which proteins are specifically up-regulated in dying cells. Ubiquitin is a highly conserved protein that is involved in proteolysis and it is expressed in high amounts in a number of tissues undergoing programmed cell death, including the intersegmental muscles of *Manduca*. However, other cells undergoing programmed cell death, including T-cells and *Drosophila* ommatidial cells (see above), do not express significant amounts of ubiquitin. Thus, ubiquitin accumulation may be a marker for some, but not all, types of programmed cell death. It was this rationale that Fahrbach and Schwartz (1994) have recently used in a study that localizes ubiquitin immunohistochemically in the abdominal ganglia of *Manduca*. Ubiquitin immunoreactivity was associated with the cytoplasm of dying neurons and with neurons that may have modulatory effects on dying neurons. This provides circumstantial evidence to suggest an involvement for ubiquitin in programmed cell death in the *Manduca* nervous system.

The degeneration of *Manduca* muscle is also a useful and important model for programmed cell death (Schwartz, 1992). A number of basic questions can be addressed using insect muscle as the model system. For example, how do cells know when to die? What physiological changes accompany death? What are the molecular mechanisms of death? Do all cells die by the same process?

Intersegmental muscle death in *Manduca* also occurs as a result of the reduction in ecdysteroid concentration at the end of metamorphosis, and it requires both the repression and the activation of specific sets of ecdysteroid responsive genes (Schwartz *et al.*, 1990a; Schwartz, 1992). It does not have the characteristic morphology of apoptosis (Schwartz *et al.*, 1993) and it is not accompanied by significant early DNA fragmentation by endonucleases (Zakeri *et al.*, 1993). These observations suggest that a fundamentally different mechanism of programmed cell death may exist for *Manduca* muscle, which has none of the classic hallmarks of apoptosis.

Schwartz *et al.* (1990a) showed that the gene expression pattern in *Manduca* intersegmental muscle is altered by injection of 20-hydroxyecdysone or actinomycin-D, and they isolated cDNAs for four genes that are abundantly increased concomitant with the commitment to degenerate. Furthermore, Wadewitz and Lockshin (1988) used 2D electrophoresis to demonstrate the appearance of 30 new translation products in degenerating muscle. Thus, as was implied by the studies using cycloheximide and actinomycin-D in the nervous system, programmed cell death does not appear to be due to cessation of macromolecular synthesis during intersegmental muscle cell death, but rather is due to the activation of a differentiative pathway. The exact identity of most of these up-regulated proteins remains undetermined

However, one of these up-regulated proteins has been characterized. Polyubiquitin expression is dramatically up-regulated in intersegmental muscles undergoing programmed cell death (Schwartz *et al.*, 1990b). Injection of 20-hydroxyecdysone delays the degeneration of the intersegmental muscles and prevents the increase in polyubiquitin mRNA synthesis. Thus, polyubiquitin can be considered to be a marker for programmed cell death in insect muscle, and is probably a component of the cell death pathway rather than an initiator of cell death itself (Schwartz, 1991).

c. Grasshopper Programmed cell death in the grasshopper embryo nervous system has been comprehensively reviewed by Truman *et al.* (1992). Cell death occurs in two developmental contexts in the grasshopper embryo nervous system: the removal of pioneer neurons and the establishment of segmental specializations in the CNS.

Pioneer neurons are a class of neurons that are transiently produced to provide a "scaffolding" along which growth cones of later-arising neurons migrate. They are subsequently removed by programmed cell death and therefore represent a type of neuronal programmed cell death in which the function of the obsolete neurons is developmental rather than behavioral (Truman, 1984). An example of this type of neuronal class is that of the two Ti1 pioneer neurons, which arise at the distal tip of the metathoracic leg in the grasshopper embryo and are the first neurons in the limb bud to extend to the CNS. Kutsch and Bentley (1987) showed that after these neurons have established the route for one of the major nerve trunks in the leg they undergo programmed cell death. Thus, their only function appears to be in the establishment of this nerve route.

Segmental specializations in the grasshopper CNS are established by the differential death of neurons in the various segments of the grasshopper embryo (Truman *et al.*, 1992). For example, the segmental neuroblasts are the stem cells that produce the majority of the central neurons; one of these, the median neuroblast, produces 100 neurons in T3 and 90 neurons in A1. All of the T3 neurons survive, while programmed cell death results in loss of half of the A1 cells.

Programmed cell death has also been shown to be an important component of muscle development during grasshopper development. Muscle pioneers (MPs) are an important class of large mesodermal cells that arise early in grasshopper embryo development and act as a scaffold for developing muscles and as guidance cues for motoneuron growth cones in the same manner as pioneer neurons. Ball *et al.* (1985) showed that significant amounts of programmed cell death of these MP cells are involved in the establishment of muscle pattern in the grasshopper leg.

B. Vertebrate Development

1. Early Development

The majority of studies of cell death in vertebrate development have focused on relatively late stages of morphogenesis, such as limb bud development, palate development, or the development of the nervous system (see below). Much less is known about the roles of programmed cell death during the early phases of gastrulation and neurulation, even though it has been appreciated for many years that such cell deaths occur (Jacobson, 1938; Glücksmann, 1951). In the latter work, cell death is recorded in the primitive streak, neural plate, notochord, and somites of early embryos, but it has not been until relatively recently that, with impetus derived from invertebrate work, attention has focused on these occurrences. As lineage studies have moved from invertebrate to vertebrate systems, it has become apparent that programmed cell death may play a significant and stable role in the complex lineage relationships in, for example, neural crest and muscle cell differentiation.

a. The Blastocyst and Implantation The developing mammalian blastocyst commonly shows evidence of cells in various stages of degeneration. This occurs in the inner cell mass from the very early stages and, though variable, may reach a level of 10% of the total cells present in this small population (El-Shershaby and Hinchliffe, 1974; Copp, 1978). These cells become phagocytosed by neighboring cells or by trophoblast cells, and the process persists over the entire period of blastocyst development in a variety of species (Carnegie *et al.,* 1985; Hardy *et al.,* 1989; Enders *et al.,* 1990). The total numbers of cells in the populations sustaining these deaths are relatively small, giving some hope that the phenomenon could be understood in terms of lineage dependency; however, the apparent randomness of the process has made such insights elusive. The apparent lack of patterning also makes it difficult to consider a morphogenetic significance for the phenomenon (Copp, 1978). An interesting possibility is afforded by the suggestion that at later stages the dying cells are those destined to become trophoblast cells as opposed to cells of the embryo proper (Pierce *et al.,* 1989), and that the deaths occur at the time the inner cell mass loses its trophectodermal potential. This programmed death may be precipitated by the presence of hydrogen peroxide in the blastocyst fluid which appears to be selectively cytotoxic for cells of the trophoblast lineage that would otherwise be overproduced (Gramzinski *et al.,* 1990).

Cell death in the surrounding uterine epithelium occurs simultaneously with the cell death in the blastocyst, and accompanies implantation. These cells undergo apoptotic changes and are also phagocytosed by the trophoblast cells (El-Shershaby and Hinchliffe, 1975; Parr *et al.,* 1987; Welsh and Enders, 1987).

b. Gastrulation Cell death during gastrulation is not an uncommon finding, having been recorded in most vertebrate groups; its significance, however, is enigmatic. In urodele amphibians cell death in the ectoderm is variable, accounting for between 6 and more than 25% of the total ectodermal cell population (Imoh, 1986). Its occurrence has been linked to the morphogenetic rearrangements required for the formation of the neural plate. In the chick embryo, cell death with an apoptotic morphology is widespread in all regions at the stage of gastrulation (Bellairs, 1961), but again may be associated with the presumptive neural plate (Jacobson, 1938). Similarly, early mammalian development is characterized by apparently random cell death (Daniel and Olson, 1966), which may amount to some 15–20% of the total cell number (Poelmann and Vermeij-Keers, 1976). In the CPB-S mouse, this cell degeneration appears to be intimately linked with cell proliferation in the control of cell number and embryo size, and there is some evidence for a relationship with mesoderm formation and, again, neural differentiation (Poelmann, 1980).

The primitive endoderm of the mouse is a nonpersistent layer derived from the inner cell mass. It is partially replaced by the definitive embryonic endoderm, probably of ectodermal origin, and in this process large numbers of the endodermal cells (up to 57% in some regions) are destined to die (Lawson et al., 1986). There appears to be a burst of cell death in the endoderm of the axial regions, apparent at the midprimitive streak stage of gastrulation and coincident with a phase of considerable cell movement. It is not known whether this well-defined example of early programmed cell death is lineage-related or a regional response to local conditions such as cell crowding or extracellular matrix alterations.

Clearly, in these very early stages of development the combination of extensive cell rearrangement, changes in adhesiveness and motility, and the onset of differentiation requires a high level of redundancy in the presumptive cell populations.

Recently, it has been demonstrated that in the mouse embryo disruption of the HNF-4 gene, which is a transcription factor belonging to the steroid/thyroid hormone receptor superfamily, leads to cell death and perturbed gastrulation. Although the HNF-4 gene is expressed in the visceral endoderm, it apparently normally suppresses cell death in ectoderm cells, which, in the absence of HNF-4, die in increased numbers, with the result that primitive streak formation and organogenesis are delayed (Chen et al., 1994).

c. Neural Tube Closure and the Neural Crest   Cell degeneration is extensive in the presumptive neural tissue during gastrulation (Jacobson, 1938) and in the neural plate during neurulation (Glücksmann, 1951). This cell death appears to be necessary for, or a by-product of, the successful completion of the fusion process and for the maintenance of the continuity of the ectoderm as the rolled-up neural tube pinches off. Apoptotic cells are particularly widespread at the junction of the neural plate and ectoderm, where a loosening of the epithelium and tissue segregation are occurring (Schlüter, 1973), and in the region of the anterior neuropore (Geelen and Langman, 1977). Indeed, the prevalence of teratogenic neural tube defects, including those induced by retinoic acid, in which there is a failure of closure, has been related to the possible disturbance of cell death in this area (Schlüter, 1973; Hinchliffe, 1981; Alles and Sulik, 1990). With subsequent development, the number of apoptotic cells increases, becoming centered in three regions (Homma *et al.,* 1994): the dorsal part of the spinal cord, including the neural crest (Fig. 5); a ventral zone between motoneurons and the floor plate; and the floor plate itself, where the level of cell death is very high (Wride *et al.,* 1994). On the basis of the distribution, it has been speculated that cell death at this time is associated with the early patterning of cells along the dorso–ventral axis of the neural tube (Homma *et al.,* 1994).

Although problems associated with the morphogenesis of the neural crest have been favored by developmental biologists for many years, attention has only recently been turned to the possible involvement of cell death in these processes. In the CPB-S mouse, a high frequency of degeneration has been reported in the region of the early neural crest as it separates from the neurectoderm (Vermeij-Keers and Poelmann, 1980). The authors related the high incidence of cell death in this location to the disruption of the epithelium and the deterioration of the basement membrane that occurs as the neural crest cells emigrate from the neural tube. It also appears that some populations of crest cells, specifically those giving rise to the

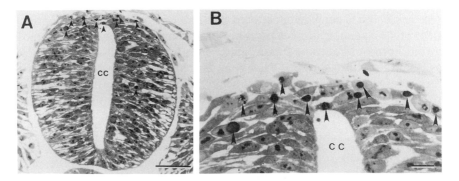

FIG. 5   Light micrographs showing pycnosis in the presumptive neural crest. (A) Transverse section of the spinal cord at 50 h of development. (B) Higher magnification of dorsal region of A. Arrowheads indicate pycnotic cells. Note that most of the pycnotic cells are found within the neural crest. CC, central canal. Scale bars: 50μm in A, 10 μm in B. Reprinted with permission of Dr. R. W. Oppenheim and Wiley-Liss, Inc. (a subsidiary of John Wiley & Sons, Inc.), from Homma *et al.* (1994).

dorsal root ganglia, depend for their survival on factors present in the adjacent neural tube, without which they die (Kalcheim and Le Douarin, 1986).

In the hindbrain of the chick embryo, domains of cell death in the dorsal midline of the rhombencephalon have been proposed as a means of patterning cranial crest cell emigration by elimination of populations, and as a cause for the failure of parts of this region to contribute to the cranial neural crest (Lumsden et al., 1991; Jeffs et al., 1992). Apoptosis in the rhombencephalon neural crest occurs in interesting patterns, being only apparent in the odd-numbered rhombomeres. Further, this cell death appears to depend on signals from the adjacent even-numbered rhombomeres that selectively effect the death of cells in the odd-numbered rhombomeres. Circumstantial evidence, based on temporo–spatial patterns of localization, correlates the onset of apoptosis in the rhombencephalic neural crest with the prior expression of members of the msx family of homeobox genes (Graham et al., 1993), thus endowing this gene family with the ability to influence tissue patterning through an effect on apoptosis.

In the trunk of the chick embryo, where migration of the neural crest seems to be governed by mechanisms different from the cranial crest, patterns of Nile blue sulfate staining suggest that neural crest cells may be undergoing extensive cell death during the course of their migrations along their various routes and that this is developmentally regulated in time (Jeffs and Osmond, 1992). The explanation for such cell death may lie simply in a response to overproduction of cells, or, more interestingly, it may be lineage-related.

d. The Tail Bud   The regression of the primitive tail in the tailless amniotes is accompanied by extensive programmed cell death (Fig. 6), consequently this highly specialized region of the embryo holds considerable interest for the understanding of morphogenesis (Griffith and Sanders, 1992). This has been studied in some detail in the tailless mutants of the mouse, such as Danforth's short tail and Brachyury (Grüneberg, 1963), in the human embryo (Fallon and Simandl, 1978), and in the chick (Schoenwolf, 1981; Sanders et al., 1986; Mills and Bellairs, 1989). The mouse mutants are characterized by profound abnormalities of the notochord, which either is absent from the start or fuses with the neural tube or tail gut. As a result of this, the tail is resorbed. In the naturally occurring short-tailed species, cell death overtakes the development of the neural tube, notochord, the mesoderm of the segmental plate and somites, the tail gut, and the ventral ectodermal ridge (Schoenwolf, 1981; Sanders et al., 1986). The greatest extent of death always occurs in, but is not restricted to, the terminal region of the tail bud. This death is integrated with the morphogenetic processes of differential growth and cell proliferation in order to remodel this region

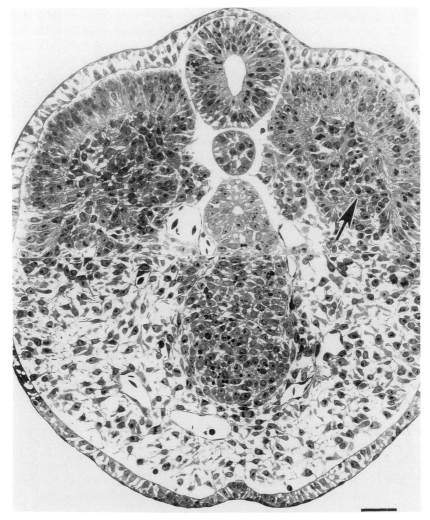

FIG. 6 Transverse section through the tail of a chick embryo (3 1/2 days of development) at the level of the most caudal somites. Pycnotic nuclei are widespread in many tissues, and particularly in the somites (arrow). Scale bar, 100 μm. Reprinted with permission of Springer-Verlag, from Sanders *et al.* (1986).

of the developing embryo (Schoenwolf, 1981; Mills and Bellairs, 1989), and it is clearly programmed, since grafting of the tail bud to ectopic sites results in the same number of somites as would have occurred *in situ,* before cell death halts segmentation of the mesoderm (Sanders *et al.*, 1986).

e. Somites and Myogenesis   Although information is sparse, cell death in the somites seems to occur primarily in the myotome. In the amphibian, the cranial myotomes degenerate in a craniocaudal sequence in the tadpole stages of development; this is apparently an autonomously programmed death, unaffected by the presence of adjacent tissues (Chung *et al.,* 1989). In the chick, signs of cell death are observed by Day 7 at the time of formation of the first myotubes (Christ *et al.,* 1983); however, muscle differentiation from the avian myotome is dependent on prior contact with the neural tube or notochord (Rong *et al.,* 1992). Prevention of this contact during critically timed windows during early development results in failure of muscle differentiation, failure of segregation of the dermatome and sclerotome, and somitic cell death. Even with the appropriate stimulus from the axial organs, continued survival of the primary and secondary myotubes depends on the establishment of a normal innervation (Fredette and Landmesser, 1991).

Although much is known about myogenic cell lineages (Stockdale, 1992), there is no information on the existence of lineage dependent cell death. The only hint that such an event might take place comes from *in vitro* studies in which the division patterns of individual colonies of myogenic cells were observed (Nameroff, 1992). In this study, the division patterns of these cells were found to be symmetrical except as a result of cell death, but whether this death is programmed or an artefact of the culture conditions is not clear.

Cell death in the subsequent fetal development of muscle has been documented (Webb, 1972), and it has been speculated that perturbation of normal death processes is related to the onset of muscular dystrophy (Webb, 1974).

f. Primordial Germ Cells The primordial germ cells (PGCs) of the mouse are first identifiable close to the primitive streak, from where they move to the endoderm of the yolk sac and hence to the hindgut (De Felici *et al.,* 1992). They then undertake an extensive and active migration to the genital ridges, where differentiation into gametes occurs. During migration their numbers increase by cell proliferation, but after arrival and before birth there is extensive elimination of cells, by a mechanism that has the morphological and biochemical characteristics of apoptosis (Coucouvanis *et al.,* 1993; Pesce *et al.,* 1993). Although the cause of this cell death is unknown, there is some evidence, at least *in vitro,* that the events are cytokine dependent, being influenced by the presence of both stem cell factor and leukemia inhibitory factor (Pesce *et al.,* 1993).

g. Heart In his monograph on this subject, Pexieder (1975) has identified no less than 31 different zones in the developing chick embryo heart, between Days 2 and 8 of incubation, in which cell death can be observed. This extensive dependence on cell death is presumably related to the very complex morphogenetic reorganizations that the organ undergoes, including the fusion of tubular primordia (Ojeda and Hurlé, 1975), the partitioning of the chambers (Manasek, 1969), and the sculpturing of the valves (Hurlé, 1979), while at the same time maintaining its vital function. During the later phases of this period of development, the cell death may also be related to the differentiation of the mesenchymal cells into fibroblasts and myocytes (Pexieder, 1975). The dying cells are clearly of apoptotic morphology (Manasek, 1969).

Because the remodeling events in the heart are diverse, there is probably no single mechanism to explain the programmed onset of cell death in this organ. However, there is experimental evidence to indicate that the characteristic hemodynamics at different places in the heart and at different times in development influence the incidence of death (Pexieder, 1975). Specifically, it appears that the cell death is directly related to the shearing stress imposed by blood flow, without which the levels of death are reduced. To date, little attention has been paid to the molecular events related to cell death in the different regions of the developing heart, although the occurrence of *bcl-2* transcripts warrants further study (Eguchi *et al.*, 1992).

h. Kidney   The development of the kidney in the amniotes (reptiles, birds, and mammals) proceeds in three stages (Torrey, 1965). The first structure to develop from the intermediate mesoderm is the pronephros, the most primitive and rudimentary kidney, which regresses entirely. This is followed by another temporary organ, the mesonephros, which also regresses. The metanephros then succeeds the mesonephros as an outgrowth from the mesonephric duct. This ureteric bud joins with the unpolarized metanephric mesenchyme and induces it to form the epithelial structures of the nephrons, which although permanent, also show cell death during their differentiation.

Because the pronephros is usually present only very fleetingly, it has not been subject to extensive study; however, an opportunity for detailed examination is offered by the lamprey whose larval stage persists for a number of years and whose pronephros degenerates over this period of time (Ellis and Youson, 1990). The regression of the pronephric tubules is complete by the time of metamorphosis, and is characterized by an apoptotic morphology accompanied by extensive infiltration by white cells. During amphibian metamorphosis there is also degeneration of the pronephros (Fox, 1971), and a similar picture is seen in the regression of the mesonephros (Salzgeber and Weber, 1966; Friebová, 1975).

In the metanephros, degenerating cells may be found around the nephric tubules after their induction from the mesenchyme (García-Porrero *et al.*, 1978); these cells show apoptotic characteristics, such as a calcium-sensitive endonuclease and DNA fragmentation (Koseki *et al.*, 1992). The degenerating cells are thought to be a fraction of the mesenchyme that is not selected for induction, and it appears that the roles of the inducer in this situation are therefore twofold, first to avert the onset of apoptosis and then to induce differentiation. Whether the rescue from apoptosis is related to the presence of the *bcl-2* gene in the kidney (Eguchi *et al.*, 1992) is unknown; however, there is some experimental evidence to indicate the *WT-1* tumor suppressor gene is involved (Kreidberg *et al.*, 1993). Mutations in *WT-1* result in urogenital malformations, the mechanisms of which include apoptosis of cells of the metanephric blastema. Further, soluble factors may be involved in the inhibition of apoptosis in the kidney, since EGF appears to be able to save otherwise condemned cells (Koseki *et al.*, 1992; Coles *et al.*, 1993). This is an interesting advance, and may indicate that this is a situation in which cell death occurs naturally because of an inadequate supply of survival factors (Coles *et al.*, 1993), but the complex and multistage nature of kidney development makes generalization difficult.

i. The Reproductive System   Before sexual differentiation, the early vertebrate embryo possesses structures characteristic of both male and female genital tracts (the Wolffian and Müllerian ducts, respectively). In the female, the absence of testosterone allows the Müllerian duct to persist, and triggers the degeneration of the Wolffian duct, while in the male, the secretion of Müllerian inhibiting substance (MIS), together with the secretion of testosterone, results in the persistence of the Wolffian duct and the apoptotic regression of the Müllerian duct. So, the Wolffian duct is destined to undergo programmed death unless saved by the presence of testosterone. By contrast, the Müllerian duct is programmed to survive, unless acted upon by MIS and testosterone (Clarke, 1982; Behringer *et al.*, 1994).

Müllerian inhibiting substance is an antiproliferative glycoprotein, produced by Sertoli cells (Tran *et al.*, 1987), which shares homology with transforming growth factor-β (Wilson *et al.*, 1993). It is apparently targeted to Müllerian duct cells via receptors on these cells and is activated by proteolytic cleavage at its site of action (Catlin *et al.*, 1993).

The degeneration of the Müllerian duct is characterized by the appearance of macrophages in the tissue, and lysosomes and condensed nuclei within the duct cells, leading to the conclusion that this event is an example of programmed cell death (Price *et al.*, 1977; Dyche, 1979). Dissolution of the basement membrane surrounding the duct results in the loss of duct integrity, and the transformation of some surviving ductal epithelial cells into mesenchyme (Trelstad *et al.*, 1982). Though less completely studied, the regression of the Wolffian duct appears to follow a similar pattern (Dyche, 1979; Jirsova and Vernerova, 1993).

j. The Lens   The cellular and molecular features of eye lens development have been reviewed by Piatigorsky (1981), while the importance of the lens as a model for chromatin degradation studies has been reviewed by Counis *et al.* (1989).

The lens is composed of two types of cell: the lens epithelial cells and the lens fiber cells. The latter are formed from the lens epithelial cells by processes of cell migration and cell elongation (Piatigorsky, 1981). Programmed cell death, by apoptosis, occurs in the lens epithelial cells during lens development (Ishizaki *et al.*, 1993), while the terminal differentiation of the lens fibers involves the coordinated loss of cell organelles and the degradation of DNA. The development of the lens is a continuous process that continues beyond embryogenesis, and thus it is also a good model for the investigation of the aging process.

It has only recently been revealed that the programmed death of lens epithelial cells is a significant component of lens development (Ishizaki *et al.*, 1993). Dying cells were found within the anterior epithelium of the rat lens and it was shown that lens epithelial cells, in serum-free culture, do not require other cell types for their survival. Lens epithelial cells were shown to undergo apoptosis only when cultured at low density. Cells could be rescued from death in low-density cultures by the addition of conditioned medium from high density cultures, suggesting that lens epithelial cells produce a soluble survival factor(s). In bovine epithelial cells, the addition of aFGF antisense primers, which inhibit endogenous aFGF expression, leads to the death of these cells (Renaud *et al.*, 1994), implicating aFGF as a survival factor for these cells.

In the lens fiber cells, death, as defined by a loss of cellular structure and function, does not occur; however, lens fiber nuclei degenerate with a strict spatio–temporal pattern (Modak and Perdue, 1970). In the chick embryo, the loss of DNA begins in the central lens fibers at Embryonic Day (ED) 6. Appleby and Modak (1977) showed that DNA fragmentation between nucleosomes occurs in differentiating lens fibers, and when DNA from these fibers is run on a gel, the pattern obtained is reminiscent of the classic DNA ladders that accompany apoptotic cell death. The degeneration of nuclei is revealed by a characteristic pattern of morphological changes (pycnosis) that, in many respects, resembles the apoptotic morphology (Modak and Perdue, 1970). Pycnotic nuclei are apparent in the central lens fibers by ED 8. As development proceeds, there is an increase in the number of pycnotic nuclei and the wave of pycnosis spreads peripherally (Fig. 7).

Endonucleases have been shown to have an important role in lens fiber DNA degradation (Counis *et al.*, 1989), and these authors have identified two molecules, of 30 and 40 kDa, which could be involved in the creation of nucleosome-size fragments during lens fiber terminal differentiation (Counis *et al.*, 1991). However, there is still no idea as to the identity of factors that may be involved in the induction of endonuclease activity in these cells. Recent work implicates a well-known mediator of apoptotic

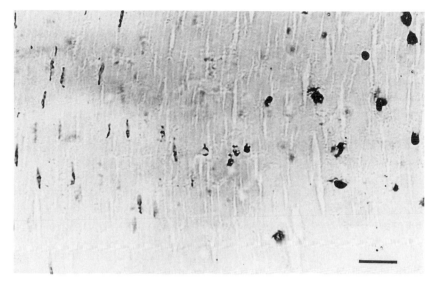

FIG. 7 A section through the lens of a Day 15 chick embryo, showing nuclei labeled with the TUNEL technique for DNA fragmentation. The least differentiated lens fibers are to the left of this illustration. As the fibers mature, the morphology of the degenerating nuclei changes from elongated to rounded and pycnotic. As degeneration progresses, the nuclei become increasingly TUNEL-positive. Scale bar, 20 μm.

cell death, tumor necrosis factor-alpha (TNFα), in this process. TNFα induces DNA fragmentation by stimulating endonuclease activity (Larrick and Wright, 1990). TNFα immunoreactivity is present in the developing chick lens (Wride and Sanders, 1993) and immunoreactivity for both types of TNF receptors is also present (M. A. Wride and E. J. Sanders, unpublished). Interestingly, TNFα immunoreactivity colocalizes with the DNA of pycnotic nuclei labeled for DNA fragmentation using TUNEL. Lens epithelial explants differentiating *in vitro* (Beebe and Feagans, 1981) undergo changes similar to those occurring in lens fibers *in vivo*.

Nuclear degeneration is accompanied by the breakdown of cell organelles in lens fibers. Bassnett and Beebe (1992) have shown that by ED 12 in the chick, the cells at the center of the lens lack both mitochondria and nuclei. This disappearance is not due to a nonspecific degradation of all intracellular structures, since the actin cytoskeleton remains intact. Disturbances in nuclear or organelle breakdown in the lens may contribute to several pathological conditions including hereditary and congenital cataracts.

2. Limb Bud Development

The role of programmed cell death in the development of the limb has been one of the models of choice in cell death studies for many years. As such, this field has been previously reviewed extensively and well (see, for example, Saunders, 1966; Hinchliffe, 1981), so that we need only to outline the key points before moving to more recent progress in this area.

The embryonic limb buds arise as paddle-shaped outgrowths which are contoured and otherwise influenced by the presence of a number of discrete zones of cell death, so that thousands of cells are programmed to die within an 8- to 10-h period. Most of the experimental work has been done on the chick wing bud (Saunders et al., 1962), and although this situation is not representative of all amniotes in the details of the "necrotic zones", the description here is nevertheless largely from that species.

The differentiation of the mesenchyme of the limb is under the control of an overlying ectodermal thickening called the apical ectodermal ridge (AER), the removal of which dramatically perturbs wing development (Summerbell, 1977; Rowe et al., 1982). A limited amount of cell death occurs in the AER itself, but its cause and function are not clear (Hinchliffe, 1981); however, this region appears to be particularly sensitive to teratogens such as retinoic acid (RA), which is able to increase the level of cell death concomitant with the production of malformation (Sulik and Dehart, 1988). In the mouse, the RA effect appears to be mediated by the nuclear receptor mRAR-beta 2 (Mendelsohn et al., 1991). The AER may exert its effect on the underlying mesenchyme through the action of FGF-2 (Fallon et al., 1994), a growth factor that is also able to rescue mesenchyme cells from programmed death in culture (MacCabe et al., 1991).

The most striking areas of death are in the mesenchyme, and first arise in three separate zones: the anterior and posterior necrotic zones (ANZ and PNZ) in the marginal regions of the bud, and the opaque patch (OP) in the core of the mesenchyme. These zones are not necessarily present to the same extent, or in the same location, in other avian species or in amniotes generally. Later, other zones of death occur between the forming digits, the interdigital necrotic zones (INZs), which are responsible for the generation of the individual digits from the solid paddle (Zakeri et al., 1994). Interdigital zones are found in all amniotes, including reptiles (Fallon and Cameron, 1977), but are reduced in the interdigital regions of webbed species (Hinchliffe, 1981). In amphibians, by contrast, INZs are absent from both webbed and free digits, and in these cases sculpting of the digits is thought to be a result of decreased cell proliferation (Cameron and Fallon, 1977). In the case of the INZ, the function of the cell death is clear, and the recent identification of a mouse mutant, *fused toes,* in which this cell death is perturbed may lead to an understanding of its mechanism (van der Hoeven et al., 1994). The roles of the ANZ, PNZ, and OP are much harder to define; they are presumably involved in patterning and in morphogenetic events such as digit formation that rely on the regulation of cell numbers. Examination of the *talpid* chick mutants supports this contention because in these animals, characterized by shortened polydactylous limbs, the limb buds do not show an ANZ or PNZ (*talpid³*: Hinchliffe and Ede, 1967; *talpid²*: Dvorak and Fallon, 1991). In these mutants there are also diminished levels of death in the OP and INZ, with associated abnormalities of development (Hinchliffe and Thorogood, 1974). Conversely, in the *wingless* mutant the ANZ in the wing bud is abnormally large and precocious in appearance, and the digits are correspondingly absent from this limb (Hinchliffe and Ede, 1973). Comparative studies suggest that the occurrence of the ANZ and PNZ may be related to the evolutionary history of limb development (Hinchliffe, 1981).

Several of the areas of cell death in chick limb buds appear to express the homeobox-containing genes GHox-7 and Ghox-8 (Coelho *et al.*, 1991, 1993), leading to the suggestion that these genes could be determinants of cell death. At periods of development during which cell death is occurring, these genes are expressed variously in the ANZ, PNZ, and interdigital mesenchyme, although expression is not limited to these areas (Krabbenhoft and Fallon, 1992), thus complicating the interpretation. In the *talpid²* mutant, which lacks some of the necrotic zones, Ghox-8 appears to be absent from the mesoderm (Krabbenhoft and Fallon, 1992; but see Nohno *et al.*, 1992). Ghox-7, however, appears to be present temporally and spatially in *talpid²* mesoderm where there are no dying cells, thus leaving its possible roles open to question (Krabbenhoft and Fallon, 1992; Coelho *et al.*, 1993). It appears to be premature to suggest a link between expression of these homeobox genes and programmed cell death in the limb.

The programmed nature of cell death in the PNZ in the chick wing was established by Saunders *et al.* (1962), who excised and grafted prospective PNZ cells from embryos at certain stages of development to ectopic sites. These cells died at the same time that they would have in their normal position, while non-PNZ cells grafted to the wing bud did not die. It was concluded that by this stage of development the cells are committed to death by their prior interactions. PNZ cells taken before the critical stage, however, did not degenerate, suggesting that a "trigger" is required to activate the death program. Further experimentation established that in organ culture, PNZ cells taken from earlier stages of development could be rescued from death by the presence of wing mesoderm, indicating the activity either of soluble factors or of direct cell–cell interaction in the expression of death as a fate (Fallon and Saunders, 1968).

The so-called "zone of polarizing activity" (ZPA), located at the posterior margin of the limb bud, appears to be fundamental for the control of the overall patterning of the developing limb, and it is possible that it may act in part by influencing the levels of cell death (Hinchliffe, 1981; Snow, 1987).

Histologically, the "necrosis" in the PNZ and OP bears many of the characteristics of apoptosis (Dawd and Hinchliffe, 1971; Hurlé and Hinchliffe, 1978), and in INZ cells internucleosomal DNA fragmentation occurs concomitant with apoptotic morphology (Garcia-Martinez *et al.*, 1993; Toné *et al.*, 1994; Zakeri *et al.*, 1994), although unspecific fragmentation may also be noted, perhaps as a result of the simultaneous early activity of phagocytes.

3. Amphibian Metamorphosis

So far, this discussion has primarily considered programmed cell death during embryonic development. There are, however, a number of opportunities to study this phenomenon during postembryonic development, including the metamorphosis of insects and amphibians where cell death is massive and may involve entire organs (Tata, 1993). The latter has long been of interest and the details have been reviewed previously (Lockshin, 1981). Work has largely been concerned with the influence of thyroxine and triiodothyronine (T_3) on the inception of the programmed death of cells of the tail tissue, with reference to the timing of appearance of macrophages and the lysosomal enzymes involved in the autolytic process (Weber, 1964). More recently, emphasis has been placed on understanding the molecular mechanisms responsible for the regional expression of thyroid hormone receptors (Kawahara *et al.*, 1991) and the interplay between thyroid hormones and other hormones such as prolactin (Tata *et al.*, 1991) and estrogen (Rabelo *et al.*, 1994). The distinct responses of tail and nontail tissues to T_3 appear to be related to separate effects of the hormone on cell proliferation and differentiation in the two types of tissue (Nishikawa *et al.*, 1989).

4. Palate Development

In mammals, the secondary palate forms by the fusion of the left and right palatal shelves which then separates the oral and nasal cavities. The fusion involves the close apposition of the epithelium covering the shelves, called the medial edge epithelium (MEE), and has attracted much attention because the event is an example of the way in which a number of biologically important phenomena are coordinated. These include the adhesion of epithelial cell surfaces, the role of the extracellular matrix, programmed cell death, and epithelial-to-mesenchymal transformation. In addition, consideration of the mechanisms of cleft palate formation has motivated much work in this area. Various aspects of the fusion process have been comprehensively reviewed previously (Shah, 1984; Ferguson, 1988), so that here, cell death is the main focus.

The elimination of the MEE at the time of fusion is accomplished by two means. The surface layer of cells undergoes apoptosis, while the basal cells undergo a transformation into mesenchymal cells which migrate away from the site of the seam (Fitchett and Hay, 1989; Schuler et al., 1991; Carette and Ferguson, 1992; Griffith and Hay, 1992). In the cells destined to die, cell division and DNA synthesis cease, followed by the appearance of lysosomes, apoptotic morphology, and autolysis. DNA fragmentation has been demonstrated using the TUNEL technique (Mori et al., 1994) among cells of the disappearing midline seam, but not in the shelves prior to contact or in the early seam. The relative contributions made by cell death and cell transformation in the disruption of the seam are a matter of debate (Ferguson, 1988), and whether distinct subpopulations of cells are involved is not known.

It is important to realize that while palate fusion and cell death are features of mammalian development, this is not the case in avians and reptiles (alligator). In the former, no fusion occurs, so that there is a natural cleft palate (Greene et al., 1983; Shah et al., 1985), while in the latter, although fusion occurs, there is no cell death but only cell migration (Ferguson and Honig, 1984).

That the cell death is programmed is shown by experiments in which palatal shelves are cultured in isolation. Under these conditions cell death occurs in the palatal epithelium (Ferguson and Honig, 1984), leading to the conclusion that medial shelf contact is not required for cell death. The use of combinations of interspecific confrontations of epithelial and mesenchymal tissues from mammalian, avian, and alligator palates has determined that it is the mesenchyme that provides signals for epithelial cell differentiation, including death (Ferguson et al., 1984). Under these conditions both chick and alligator MEE will undergo cell death in combination with mouse mesenchyme, and in the other combinations the fate of the MEE is largely determined by the origin of the mesenchyme (see Ferguson and Honig, 1984, for a complete discussion of these experiments).

The nature of the signal from the mesenchyme is uncertain. The earlier preoccupation with the demonstrable rise in cAMP in the MEE of the rat secondary palate just prior to the occurrence of cell death (Pratt and Martin, 1975) is now of somewhat uncertain significance in light of more recent knowledge of signal transduction pathways. Similarly, the observations that cell death in the MEE may be prevented by the glycosylation inhibitor diazo-oxo-norleucine (DON; Greene and Pratt, 1978) has not lead to any fundamental insights into the mechanism of death. Cleft palate can be induced in mice by inhibition of cell death with glucocorticoids, an effect which seems to be mediated by glucocorticoid receptors (Goldman, 1984), and several growth factors, including EGF, TGFβ1 and β2 (Abbott et al., 1992).

More recently, a number of growth factors have been considered as likely candidates for the mesenchymal signal (Ferguson, 1988) and these are likely to work in conjunction with extracellular matrix changes (Brinkley and

Morris-Wiman, 1984) to bring about changes in the differentiative fate of the MEE. Epidermal growth factor was the first soluble factor to be suspected of involvement in the normal fusion process (Pratt, 1980), and in culture this factor is able to suppress cell death and alter differentiation of the MEE in the presence of mesenchyme (Hassell, 1975; Tyler and Pratt, 1980). Further, EGF is able to modulate the synthesis of extracellular matrix in the developing mouse palate by stimulation of synthesis of molecules such as hyaluronate (Turley et al., 1985). It is also established that the oral and medial edge epithelia, including the cells of the seam, are positive for EGF receptor (Shiota et al., 1990). Of interest in this respect are observations on the nature of retinoic acid-induced cleft palate (Abbott et al., 1988; Abbott and Pratt, 1991). In the normal palate, expression of the EGF receptor declines prior to the onset of cell death; by contrast, MEE cells exposed to retinoic acid continue to express EGF receptor, bind EGF, proliferate, and differentiate, and fail to undergo cell death, resulting in cleft palate formation.

Other factors of potential importance that have been investigated are TGFα, TGFβ, PDGF, FGFa, and FGFb (Ferguson, 1988). Emphasis has been placed on the interactions between these molecules and the extracellular matrix, in terms both of the binding capacity of the matrix and of the stimulation or inhibition of matrix synthesis by the growth factors. The evidence is consistent with the notion that mesenchymal-derived growth factors are involved in the differentiation and death of the MEE cells, either directly or through an effect on the extracellular environment and cell adhesion. Regional heterogeneities in this interaction are visualized as occurring through regional differences in growth factor production and receptor distribution (Ferguson, 1988).

5. Nervous System Development

Cell death in the developing peripheral nervous system is one of the most celebrated naturally occurring degenerative events (Cunningham, 1982; Martin and Johnson, 1991; Oppenheim, 1991; Hamburger, 1992; Johnson and Deckwerth, 1993). It is commonly stated that, depending on the region of the nervous system, between 15 and 85% of neurons die during the course of axon outgrowth in embryogenesis (Cowan et al., 1984). This apparent overproduction of neurons presumably permits the flexibility in morphogenesis that is required for the establishment of complex interrelated neuronal networks and neuromuscular connections. During development in general, the number of cells in any given population is usually adjusted by a combination of cell proliferation and cell death. However, in the nervous system the former does not usually occur. The numbers of neurons are regulated therefore by an often massive wave of cell death which may bear the morphological features and DNA fragmentation characteristic of apoptosis (Clarke, 1990; Server and Mobley, 1991; Milligan *et al.*, 1994) or may more closely resemble necrosis (Choi, 1992). So, although neuronal cell death in development is now generally considered to be programmed (but see Oppenheim, 1991), it may not necessarily be apoptotic (Server and Mobley, 1991).

Cell death is widespread throughout the developing peripheral and central nervous systems of many species (Oppenheim, 1991). In particular, there is extensive death relatively early in development among motoneurons in the compact lateral motor columns of the spinal cord, reaching a peak in the lumbar region of the chick embryo between Days 5 and 9 of incubation (Hamburger, 1975; Chu-Wang and Oppenheim, 1978). The cell bodies of these neurons are located in the cord and will form the ventral horn. Also, peaking at 5–6 days in the chick embryo, there is considerable cell death in the cervical and thoracic dorsal root ganglia, which contain the cell bodies outside the cord (Hamburger and Levi-Montalcini, 1949). Cell degeneration peaks specifically at stages corresponding to the period during which neurons are known to be contacting their target cells, and is less apparent in ganglia at the brachial and lumbosacral levels, which are the ones innervating the limbs. In each of these cases, it seems that the extent of cell death is controlled by the peripheral fields which they innervate (Hamburger and Levi-Montalcini, 1949).

There is no firm evidence that at these stages of development cell death serves to eliminate unwanted lineages, as may be the case in invertebrates. However, such may be the case at very early stages, before neuronal outgrowth is well advanced. At this time, death is found in the neural tube in distinct zones: dorsal, including the neural crest (Fig. 6); a ventral zone lateral to the floor plate; and within the floorplate itself (Homma *et al.*, 1994; Wride *et al.*, 1994). Although it is uncertain, this death may function in the establishment of early patterning or in the deletion of certain lineages.

In the developing central nervous system, cell death is notable in the isthmo-optic nucleus (Catsicas *et al.*, 1987), the lateral geniculate nucleus (Williams and Rakic, 1988), the diencephalon (Navascues *et al.*, 1988), the hippocampus (Ferrer *et al.*, 1990), the dentate gyrus (Gould *et al.*, 1991), and the cerebral cortex (Ferrer *et al.*, 1992). In the case of death in the magnocellular nucleus of the juvenile male zebra finch, neuronal loss can be temporally correlated with the period of vocal learning (Bottjer and Sengelaub, 1989), and differential cell death may be a contributing factor to the development of sexual differences in the song patterns of the adult bird (Kirn and DeVoogd, 1989).

Cell death is a significant factor in the development of the visual system, including the early stages of optic cup formation (Schook, 1980; García-Porrero *et al.*, 1984), the differentiation of the retina, and in particular the retinal ganglion cells (Sengelaub and Finlay, 1982; Young, 1984; O'Leary *et al.,* 1986; Wong and Hughes, 1987), where death has clear apoptotic characteristics (Portera-Cailliau *et al.,* 1994), the optic nerve (Crespo *et al.,* 1985), and the optic tectum (Arees and Astrom, 1977).

In the peripheral nervous system, the primary determinants of neuronal cell death in development appear to include the size of the peripheral innervation field or target, the consequent establishment of connections allowing the action of neurotrophic supporting factors derived from the target, and the level of electrical activity at the synaptic or neuromuscular junction.

a. Availability of Synaptic Sites Although there may not necessarily be an absolute relationship between the numbers of neurons and the size of the target (see Oppenheim, 1991, for a detailed discussion), the matching of the neuron pool to the relative mass of the peripheral target tissue is apparently accomplished by competition among neuronal growth cones for a restricted number of synaptic sites. This was shown in principle by Hamburger and Levi-Montalcini (1949), who demonstrated neuronal loss when chick embryo limb buds were extirpated at 2½–3 days of incubation, and neuronal gain when additional limb buds were grafted. In the latter case, the spinal ganglia supplying the graft became enlarged due to enhanced neuronal survival. A similar set of experiments, using an amphibian, confirmed that the periphery controls the number, and size, of the neurons in the ventral horn (Prestige, 1970), although it is clear that target size alone is not necessarily a determinant of neuronal survival (Sohal, 1992).

b. Target-Derived Factors Prestige (1976) showed that some motoneurons die after having established a peripheral connection, indicating that control by the periphery does not simply depend on whether or not a stable connection is made. It is now realized that neuronal survival depends, at least partly, on an adequate retrograde supply of factors from the target, after connection is made (Brunso–Bechtold and Hamburger, 1979), and it is therefore possible to get a reduction in neuronal cell death without an increase in the number of synaptic sites, by exogenously supplying such factors *in vivo* (Hamburger *et al.,* 1981). Therefore, the effect of failed connection with the periphery is to deprive the neurons of target-derived factors, although it is possible that the primary limit on survival is the availability of synaptic sites, either in the form of limited sites at the target or of limited axonal branching, rather than the rate of synthesis of trophic factor (Pittman and Oppenheim, 1978; Oppenheim, 1989; Landmesser, 1992; Tang and Landmesser, 1993). The fundamental idea that neurons are maintained by the activity of peripheral substances again derives primarily from the work of Hamburger and Levi-Montalcini (1949; reviewed by

Hamburger, 1992), who conjectured that after limb bud extirpation, the supply of such substances became inadequate to support the corresponding spinal ganglion. The best characterized of these neurotrophic substances is nerve growth factor (NGF; Levi-Montalcini and Angeletti, 1968; Perez-Polo *et al.*, 1990), which is responsible for the maintenance of sympathetic neurons. Other candidate molecules have also been identified, such as a muscle-derived factor supporting motoneurons (Oppenheim *et al.*, 1988), brain-derived neurotrophic factor (BDNF) supporting motoneurons and neurons of dorsal root and nodose ganglia (Hofer and Barde, 1988; Oppenheim *et al.*, 1992a), and neurotrophin-3 (NT-3) supporting sensory ganglion cells (Gaese *et al.*, 1994). Ciliary neurotrophic factor (CNTF) is active during postnatal development, supporting the survival of both motoneurons (Sendtner *et al.*, 1991) and their target cells (Forger *et al.*, 1993), but is derived from glial cells rather than target tissue. Similarly, S100, a protein derived from Schwann cells, appears to be able to act as a survival factor for motoneurons (Bhattacharyya *et al.*, 1992). Furthermore, complexities are beginning to appear, such as the interdependence between factors illustrated by the sequential activity of NGF, BDNF, and NT-3 on embryonic sensory neurons (Buchman and Davies, 1993).

The characterization of the receptors for such neurotrophins has begun. In the mammalian visual system, the receptors for BDNF and NT-3 appear to be proteins belonging to the *trk* family, the structure of which changes at times that coincide with the occurrence of cell death (Allendoerfer *et al.*, 1994).

The generality of the "neurotrophic theory" of neuronal survival depends on whether as good a case can be made for the *in vivo* activity of these factors as for NGF, in which exogenous supply can be shown to enhance survival rates (Hamburger *et al.*, 1981; Oppenheim *et al.*, 1982), and blocking with NGF antiserum enhances cell death (Levi-Montalcini and Booker, 1960; Rohrer *et al.*, 1988).

It must also be noted that local deprivation of NGF in culture is able to effect the selective pruning of neurites in that location without precipitating the death of the whole cell, thereby providing a means of directing neurite outgrowth by eliminating misplaced neurites (Campenot, 1977). Conversely, the atrophy of the soma does not necessarily cause the immediate demise of the neurites (Deckwerth and Johnson, 1994), indicating some independence in the neurite degeneration process.

In line with other examples of physiological cell death, it is now considered probable that neuronal death is an active, differentiative event. In other words, the lack of neurotrophic factor permits the expression of new genetic pathways leading to death. Evidence to this effect is available. In the case of NGF-deprived sympathetic neurons, cholinergic septal neurons, and ciliary ganglia *in vitro*, inhibition of RNA or protein synthesis can prevent cell death (Martin *et al.*, 1988; Svendsen *et al.*, 1994; Villa *et al.*, 1994), and similar treatments of muscle extract-deprived motoneurons *in vitro* (Comella *et al.*, 1994; Milligan *et al.*, 1994) have similar effects. More importantly, *in vivo* treatment (Oppenheim *et al.*, 1990) can reduce the normally occurring death of motoneurons and dorsal root ganglion cells in the chick embryo. Although the interpretations of these experiments are not without difficulty (Oppenheim *et al.*, 1990), it appears most likely that neurotrophic substances such as NGF act to suppress a cell death program which normally requires active gene expression for its execution (Johnson and Deckwerth, 1993). Contrary to the situation in a number of nonneuronal cell types, the gene for SGP-2 (clusterin or *TRPM-2*) is not expressed in apoptotic neuronal cells (Garden *et al.*, 1991); however, it appears that interleukin-1β converting enzyme (ICE; see earlier) is involved since its inhibition can lead to increased neuronal survival (Gagliardini *et al.*, 1994).

To what extent such "survival signals" might be active in the suppression of cell death among nonneuronal cells is uncertain. The dependence of cells on soluble growth factors is a familiar concept, but the few clear examples relating specifically to programmed cell death so far available are as yet insufficient to establish this as a general phenomenon (Raff, 1992; Raff *et al.*, 1993). The survival of oligodendrocytes seems to be a case in point, however, since the high level of cell death among these cells in the optic nerve appears to be due to the lack of a series of factors including insulin-like growth factors (IGFs), platelet-derived growth factor (PDGF), NT-3, and CNTF (Barres *et al.*, 1992, 1993). The source of these factors may include other glial cells in the central nervous system, such as astrocytes, which produce CNTF (Louis *et al.*, 1993).

c. Synaptic Activity   There is evidence to suggest that the electrical and synaptic activity experienced by neurons during development is involved in the regulation of neuronal survival (Landmesser and Pilar, 1974; reviewed by Oppenheim, 1991). For example, there are many descriptions of reductions in the extent of neuronal cell death following pharmacological neuromuscular blockade (Pittman and Oppenheim, 1978, 1979; Creazzo and Sohal, 1979; Arens and Straznicky, 1986; Landmesser and Szente, 1986), and, conversely, there is evidence that an increased rate of death follows an increase in electrical activity (Oppenheim and Núñez, 1982). There is also reason to believe that the regulation of cell death by electrical activity may be mediated by an effect on the level of soluble factors (Lipton, 1986).

Regulation of the death of neurons appears to rely on a complex interplay between trophic support and electrical activity (Catsicas *et al.*, 1992), though the mechanism of the latter is not understood. Further, there must be a relationship between these observations and those indicating that high levels of neurotransmitters such as glutamate have "excitotoxic" effects on neurons (Mattson and Kater, 1989; Choi, 1992). Neuronal death under these circumstances appears to be mediated by elevated levels of intracellular calcium (Kater *et al.,* 1989), although the precise role of intracellular calcium in such killing is complex, being concentration-dependent (Martin and Johnson, 1991).

IV. Concluding Remarks

Studies on programmed cell death and apoptosis are progressing at an extremely rapid rate, as new insights and techniques become available. Advances in the understanding of the roles and mechanisms of cell death in invertebrate lineages will continue to provide an impetus for similar work in vertebrate systems, where cells interact with each other and with soluble and matrix-bound factors in more complex ways. Much is already known about the cellular basis of vertebrate morphogenesis, especially at the early stages of development, and this will provide a platform from which an understanding of the functions and molecular mechanisms of cell death will arise. We should also look for a clarification, in developing systems, of the relationship between what has been commonly termed "programmed cell death" and what is morphologically and biochemically recognized as "apoptosis." This field is currently experiencing a burst of activity, which can be expected to answer these questions and to uncover a multitude of unexpected facets of cell death in development.

Acknowledgments

The work from the authors' own laboratory cited here was supported by an operating grant from the Medical Research Council of Canada to E.J.S., and a studentship from the Alberta Heritage Foundation for Medical Research to M.A.W.

References

Abbadie, C., Kabrun, N., Bouali, F., Smardov, J., Stehelin, D., Vandenbunder, B., and Enrietto, P. J. (1993). High levels of c-rel expression are associated with programmed cell death in the developing avian embryo and in bone marrow cells *in vitro. Cell (Cambridge, Mass.)* **75,** 899–912.

Abbott, B. D., and Pratt, R. M. (1991). Retinoic acid alters epithelial differentiation during palatogenesis. *J. Craniofacial Genet. Dev. Biol.* **11,** 315–325.

Abbott, B. D., Adamson, E. D., and Pratt, R. M. (1988). Retinoic acid alters EGF receptor expression during palatogenesis. *Development (Cambridge, UK)* **102,** 853–867.

Abbott, B. D., Harris, M. W., and Birnbaum, L. S. (1992). Comparisons of the effects of TCDD and hydrocortisone on growth factor expression provide insight into their interaction in the embryonic mouse palate. *Teratology* **45,** 35–53.

Abrams, J. M., White, K., Fessler, L. I., and Steller, H. (1993). Programmed cell death during *Drosophila* embryogenesis. *Development (Cambridge, UK)* **117,** 29–43.

Allen, T. D. (1987). Ultrastructural aspects of cell death. *In* "Perspectives on Mammalian Cell Death" (C. S. Potten, ed.), pp. 39–65. Oxford Univ. Press, Oxford.

Allendoerfer, K. L., Cabelli, R. J., Escandon, E., Kaplan, D. R., Nikolics, K., and Shatz, C. J. (1994). Regulation of neurotrophin receptors during the maturation of the mammalian visual system. *J. Neurosci.* **14,** 1795–1811.

Alles, A. J., and Sulik, K. K. (1990). Retinoic acid-induced spina bifida, evidence for a pathological mechanism. *Development (Cambridge, UK)***108,** 73–81.

Alles, A., and 28 others. (1991). Apoptosis, a general comment. *FASEB J.* **5,** 2127–2128.

Alnemri, E. S., and Litwack, G. (1990). Activation of internucleosomal DNA cleavage in human CEM lymphocytes by glucocorticoid and novobiocin. *J. Biol. Chem.* **265,** 17323–17333.

Ansari, B., Coates, P. J., Greenstein, B. D., and Hall, P. A. (1993). *In situ* end-labelling detects DNA strand breaks in apoptosis and other physiological and pathological states. *J. Pathol.* **170,** 1–8.

Appleby, D. W., and Modak, S. P. (1977). DNA degradation in terminally differentiating lens fiber cells from chick embryos. *Proc. Natl. Acad. Sci. U.S.A.* **74,** 5579–5583.

Arees, E. A., and Astrom, K. E. (1977). Cell death in the optic tectum of the developing rat. *Anat. Embryol.* **151,** 29–34.

Arends, M. J., Morris, R. G., and Wyllie, A. H. (1990). Apoptosis: The role of the endonuclease. *Am. J. Pathol.* **136,** 593–608.

Arends, M. J., and Wyllie, A. H. (1991). Apoptosis, mechanisms and roles in pathology. *Int. Rev. Exp. Pathol.* **32,** 223–254.

Arens, M., and Straznicky, C. (1986). The development of the trigeminal (V) motor nucleus in normal and tubocurare treated chick embryos. *Anat. Embryol.* **174,** 67–72.

Avery, L., and Horvitz, H. R. (1987). A cell that dies during wild-type *C. elegans* development can function as a neuron in a *ced-3* mutant. *Cell (Cambridge, Mass.)* **51,** 1071–1078.

Baker, N. E., and Rubin, G. M. (1992). *Ellipse* mutations in the *Drosophila* homologue of the EGF receptor affect pattern formation, cell division, and cell death in eye imaginal discs. *Dev. Biol.* **150,** 381–396.

Ball, E. E., Ho, R. K., and Goodman, C. S. (1985). Muscle development in the grasshopper embryo. I. Muscles, nerves, and apodemes in the metathoracic leg. *Dev. Biol.* **111,** 383–398.

Barres, B. A., Hart, I. K., Coles, H. S., Burne, J. F., Voyvodic, J. T., Richardson, W. D., and Raff, M. C. (1992). Cell death and control of cell survival in the oligodendrocyte lineage. *Cell (Cambridge, Mass.)* **70,** 31–46.

Barres, B. A., Schmid, R., Sendnter, M., and Raff, M. C. (1993). Multiple extracellular signals are required for long-term oligodendrocyte survival. *Development (Cambridge, UK)* **118,** 283–295.

Barry, M. A., and Eastman, A. (1993). Identification of deoxyribonuclease II as an endonuclease involved in apoptosis. *Arch. Biochem. Biophys.* **300,** 440–450.

Bassnett, S., and Beebe, D. C. (1992). Coincident loss of mitochondria and nuclei during lens fiber cell differentiation. *Dev. Dyn.* **194,** 85–93.

Bates, R. C., Buret, A., van Helden, D. F., Horton, M. A., and Burns, G. F. (1994). Apoptosis induced by inhibition of intercellular contact. *J. Cell Biol.* **125,** 403–415.

Beaulaton, J., and Lockshin, R. A. (1982). The relation of programmed cell death to development and reproduction, comparative studies and an attempt at classification. *Int. Rev. Cytol.* **79,** 215–235.

Beebe, D. C., and Feagans, D. E. (1981). A tissue culture system for studying lens cell differentiation. *Vision Res.* **21**, 113–118.

Behringer, R. R., Finegold, M. J., and Cate, R. L. (1994). Müllerian–inhibiting substance function during mammalian sexual development. *Cell (Cambridge, Mass.)* **79**, 415–425.

Bellairs, R. (1961). Cell death in chick embryos as studied by electron microscopy. *J. Anat.* **95**, 54–60.

Bhattacharyya, A., Oppenheim, R. W., Prevette, D., Moore, B. W., Brackenbury, R., and Ratner, N. (1992). S100 is present in developing chicken neurons and Schwann cells and promotes motor neuron survival *in vivo*. *J. Neurobiol.* **23**, 451–466.

Bissonette, R. P., Echeverri, F., Mahboubi, A., and Green, D. R. (1992). Apoptotic cell death induced by c-myc is inhibited by bcl-2. *Nature (London)* **359**, 552–554.

Bonini, N. M., Leiserson, W. M., and Brenzer, S. (1993). The *Eyes absent* gene, genetic control of cell survival and differentiation in the developing *Drosophila* eye. *Cell (Cambridge, Mass.)* **72**, 379–395.

Bottjer, S. W., and Sengelaub, D. R. (1989). Cell death during development of a forebrain nucleus involved with vocal learning in zebra finches. *J. Neurobiol.* **20**, 609–618.

Bowen, I. D. (1981). Techniques for demonstrating cell death. *In* "Cell Death in Biology and Pathology" (I. D. Bowen and R. A. Lockshin, eds.), pp. 379–444. Chapman & Hall, London.

Bowen, I. D. (1993). Apoptosis or programmed cell death? *Cell Biol. Int.* **17**, 365–380.

Brand, S., and Bourbon, H.-M. (1993). The developmentally-regulated *Drosophila* gene rox8 encodes an RRM-type RNA binding protein structurally related to human T1A-1-type nucleolysins. *Nucleic Acids Res.* **21**, 3699–3704.

Brinkley, L. L., and Morris-Wiman, J. (1984). The role of extracellular matrices in palatal shelf closure. *Curr. Top. Dev. Biol.* **19**, 17–36.

Brunso Bechtold, J. K., and Hamburger, V. (1979). Retrograde transport of nerve growth factor in chicken embryo. *Proc. Natl. Acad. Sci. U.S.A.* **76**, 1494–1496.

Buchman, V. L., and Davies, A. M. (1993). Different neurotrophins are expressed and act in a developmental sequence to promote the survival of embryonic sensory neurons. *Development (Cambridge, UK)* **118**, 989–1001.

Buttyan, R., Zakeri, Z., Lockshin, R., and Wolgemuth, D. (1988). Cascade induction of c-fos, c-myc, and heat shock 70K transcripts during regression of the rat ventral prostate gland. *Mol. Endocrinol.* **2**, 650–657.

Buttyan, R., Olsson, C. A., Pintar, J., Chang, C., Bandyk, M., Ng, P.-Y., and Sawczuk, I. S. (1989). Induction of the TRPM-2 gene in cells undergoing programmed cell death. *Mol. Cell. Biol.* **9**, 3473–3481.

Cagan, R. L., and Ready, D. F. (1989). The emergence of order in the *Drosophila* pupal retina. *Dev. Biol.* **136**, 346–362.

Cameron, J., and Fallon, J. F. (1977). The absence of cell death during development of free digits in amphibians. *Dev. Biol.* **55**, 331–338.

Campenot, R. B. (1977). Local control of neurite development by nerve growth factor. *Proc. Natl. Acad. Sci. U.S.A.* **74**, 4516–4519.

Carette, M. J., and Ferguson, M. W. (1992). The fate of medial edge epithelial cells during palatal fusion *in vitro*, an analysis by DiI labelling and confocal microscopy. *Development (Cambridge, UK)* **114**, 379–388.

Carnegie, J. A., McCully, M. E., and Robertson, H. A. (1985). The early development of the sheep trophoblast and the involvement of cell death. *Am. J. Anat.* **174**, 471–488.

Catlin, E. A., MacLaughlin, D. T., and Donahoe, P. K. (1993). Mullerian inhibiting substance, new perspectives and future directions. *Microsc. Res. Tech.* **25**, 121–133.

Catsicas, M., Péquignot, Y., and Clarke, P. G. H. (1992). Rapid onset of neuronal death induced by blockade of either axoplasmic transport or action potentials in afferent fibres during brain development. *J. Neurosci.* **12**, 4642–4650.

Catsicas, S., Thanos, S., and Clarke, P. G. (1987). Major role for neuronal death during brain development, refinement of topographic connections. *Proc. Natl. Acad. Sci. U.S.A.* **84,** 8165–8168.

Chen, W. S., Manova, K., Weinstein, D. C., Duncan, S. A., Plump, A. S., Prezioso, V. R., Bachvarova, R. F., and Darnell, J. E. (1994). Disruption of the HNF-4 gene, expressed in visceral endoderm, leads to cell death in embryonic ectoderm and impaired gastrulation of mouse embryos. *Genes Dev.* **8,** 2466–2477.

Chen, Y.-Y., and Rosenberg, N. (1992). Lymphoid cells transformed by Abelson virus require the v-abl protein-tyrosine kinase only during early G1. *Proc. Natl. Acad. Sci. U.S.A.* **89,** 6683–6687.

Choi, D. W. (1992). Excitotoxic cell death. *J. Neurobiol.* **23,** 1261–1276.

Christ, B., Jacob, M., and Jacob, H. J. (1983). On the origin and development of the ventrolateral abdominal muscles in the avian embryo. *Anat. Embryol.* **166,** 87–101.

Chung, H.-M., Neff, A. W., and Malacinski, G. M. (1989). Autonomous cell death of amphibian (*Xenopus laevis*) cranial myotomes. *J. Exp. Zool.* **251,** 290–299.

Chu-Wang, I.-W., and Oppenheim, W. (1978). Cell death of motoneurons in the chick embryo spinal cord. I. A light and electron microscopic study of naturally occurring and induced cell loss during development. *J. Comp. Neurol.* **177,** 33–58.

Clarke, A. R., Purdie, C. A., Harrison, D. J., Morris, R. G., Bird, C. C., Hooper, M. L., and Wyllie, A. H. (1993). Thymocyte apoptosis induced by p53-dependent and independent pathways. *Nature (London)* **362,** 849–852.

Clarke, I. J. (1982). Prenatal sexual development. *In* "Reviews of Reproductive Biology" (C. A. Finn, ed.), Vol. 4, pp. 101–147. Oxford Univ. Press (Clarendon), Oxford.

Clarke, P. G. H. (1990). Developmental cell death, morphological diversity and multiple mechanisms. *Anat. Embryol.* **181,** 195–213.

Clifford, R., and Schüpbach, T. (1992). The *torpedo* (DER) receptor tyrosine kinase is required at multiple times during *Drosophila* embryogenesis. *Development (Cambridge, UK)* **445,** 853–872.

Coelho, C. N., Sumoy, L., Rodgers, B. J., Davidson, D. R., Hill, R. E., Upholt, W. B., and Kosher, R. A. (1991). Expression of the chicken homeobox-containing gene Ghox-8 during embryonic chick limb development. *Mech. Dev.* **34,** 143–154.

Coelho, C. N., Upholt, W. B., and Kosher, R. A. (1993). The expression pattern of the chicken homeobox-containing gene Ghox-7 in developing polydactylous limb buds suggests its involvement in apical ectodermal ridge-directed outgrowth of limb mesoderm and in programmed cell death. *Differentiation (Cambridge, UK)* **52,** 129–137.

Cohen, G. M., Sun, X.-M., Snowden, R. T., Dinsdale, D., and Skilleter, D. N. (1992). Key morphological features of apoptosis may occur in the absence of internucleosomal DNA fragmentation. *Biochem. J.* **286,** 331–334.

Cohen, J. J., and Duke, R. C. (1984). Glucocorticoid activation of a calcium-dependent endonuclease in thymocyte nuclei leads to cell death. *J. Immunol.* **132,** 38–42.

Colbert, R. A., and Young, D. A. (1986). Glucocorticoid-induced messenger ribonucleic acids in rat thymic lymphocytes, rapid effects specific for glucocorticoids. *Endocrinology (Baltimore)* **119,** 2598–2605.

Coles, H.S. R., Burne, J. F., and Raff, M. C. (1993). Large-scale normal cell death in the developing rat kidney and its reduction by epidermal growth factor. *Development (Cambridge, UK)* **118,** 777–784.

Comella, J. X., Sanz-Rodriguez, C., Aldea, M., and Esquerda, J. E. (1994). Skeletal muscle-derived trophic factors prevent motoneurons from entering an active cell death program *in vitro. J. Neurosci.* **14,** 2674–2686.

Compton, M. M. (1992). A biochemical hallmark of apoptosis, internucleosomal degradation of the genome. *Cancer Metastasis Rev.* **11,** 105–119.

Compton, M. M., and Cidlowski, J. A. (1986). Rapid *in vivo* effects of glucocorticoids on the integrity of rat lymphocyte genomic deoxyribonucleic acid. *Endocrinology (Baltimore)* **118**, 38–45.

Compton, M. M., and Cidlowski, J. A. (1987). Identification of a glucocorticoid–induced nuclease in thymocytes. *J. Biol. Chem.* **262**, 8288–8292.

Connor, J., Sawczuk, I. S., Benson, M. C., Tomashefsky, P., O'Toole, K. M., Olsson, C. A., and Buttyan, R. (1988). Calcium channel antagonists delay regression of androgendependent tissues and suppress gene activity associated with cell death. *Prostate* **13**, 119–130.

Copp, A. J. (1978). Interaction between inner cell mass and trophectoderm of the mouse blastocyst. I. A study of cellular proliferation. *J. Embryol. Exp. Morphol.* **48**, 109–125.

Coucouvanis, E. C., Sherwood, S. W., Carswell–Crumpton, C., Spack, E. G., and Jones, P. P. (1993). Evidence that the mechanism of prenatal germ cell death in the mouse is apoptosis. *Exp. Cell Res.* **209**, 238–247.

Counis, M. F., Chaudin, E., Allinquant, B., Muel, A. S., Sanval, M., Skidmore, C., and Courtois, Y. (1989). The lens: A model for chromatin degradation studies in terminally differentiating cells. *Int. J. Biochem.* **21**, 235–242.

Counis, M. F., Chaudin, E., Courtois, Y., and Allinquant, B. (1991). DNAase activities in embryonic chicken lens: In epithelial cells or in differentiating fibers where chromatin is progressively cleaved. *Biol. Cell* **72**, 231–238.

Cowan, W. M., Fawcett, J. W., O'Leary, D.D. M., and Stanfield, B. B. (1984). Regressive events in neurogenesis. *Science* **225**, 1258–1265.

Creazzo, T. L., and Sohal, G. S. (1979). Effects of chronic injections of α-bungarotoxin on embryonic cell death. *Exp. Neurol.* **66**, 135–145.

Crespo, D., O'Leary, D. D., and Cowan, W. M. (1985). Changes in the number of optic nerve fibers during late prenatal and postnatal development in the albino rat. *Brain Res.* **351**, 129–134.

Cunningham, T. J. (1982). Naturally occurring neuron death and its regulation by developing neural pathways. *Int. Rev. Cytol.* **74**, 163–185.

Daniel, J. C., and Olson, J. D. (1966). Cell movement, proliferation and death in the formation of the embryonic axis of the rabbit. *Anat. Rec.* **156**, 123–128.

Dawd, D. S., and Hinchliffe, J. R. (1971). Cell death in the "opaque patch" in the central mesenchyme of the developing chick limb, a cytological, cytochemical and electron microscopic analysis. *J. Embryol. Exp. Morphol.* **26**, 401–424.

Deckwerth, T. L., and Johnson, E. M. (1994). Neurites can remain viable after destruction of the neuronal soma by programmed cell death (apoptosis). *Dev. Biol.* **165**, 63–72.

De Felici, M., Dolci, S., and Pesce, M. (1992). Cellular and molecular aspects of mouse primordial germ cell migration and proliferation in culture. *Int. J. Dev. Biol.* **36**, 205–213.

Dowd, D.R, and Miesfeld, R. L. (1992). Evidence that glucocorticoid- and cyclic AMP-induced apoptotic pathways in lymphocytes share distal events. *Mol. Cell. Biol.* **12**, 3600–3608.

Dowd, D. R., MacDonald, P. N., Komm, B. S., Haussler, M. R., and Miesfeld, R. (1991). Evidence for early induction of calmodulin gene expression in lymphocytes undergoing glucocorticoid-mediated apoptosis. *J. Biol. Chem.* **266**, 18423–18426.

Driscoll. M. (1992). Molecular genetics of cell death in the nematode *Caenorhabditis elegans*. *J. Neurobiol.* **23**, 1327–1351.

Driscoll, M., and Chalfie, M. (1992). Developmental and abnormal mechanisms of cell death in *C. elegans*. *Trends Neurosci.* **15**, 15–19.

Duke, R. C., Witter, R. Z., Nash, P. B., Young, J. D.-E., and Ojcius, D. M. (1994). Cytolysis mediated by ionophores and pore-forming agents, role of intracellular calcium in apoptosis. *FASEB J.* **8**, 237–246.

Dvorak, L., and Fallon, J. F. (1991). Talpid2 mutant chick limb has anteroposterior polarity and altered patterns of programmed cell death. *Anat. Rec.* **231**, 251–260.

Dyche, W. J. (1979). A comparative study of the differentiation and involution of the Mullerian duct and Wolffian duct in the male and female fetal mouse. *J. Morphol.* **162,** 175–210.

Eguchi, Y., Ewert, D. L., and Tsujimoto, Y. (1992). Isolation and characterization of the chicken bcl-2 gene, expression in a variety of tissues including lymphoid and neuronal organs in adult and embryo. *Nucleic Acids Res.* **20,** 4187–4192.

Eischen, C. M., Dick, C. J., and Leibson, P. J. (1994). Tyrosine kinase activation provides an early and requisite signal for Fas-induced apoptosis. *J. Immunol.* **153,** 1947–1954.

Ellis, H. M., and Horvitz, H. R. (1986). Genetic control of programmed cell death in the nematode *C. elegans. Cell (Cambridge, Mass.)* **44,** 817–829.

Ellis, L. C., and Youson, J. H. (1990). Pronephric regression during larval life in the sea lamprey, *Petromyzon marinus* L. *Anat. Embryol.* **182,** 41–52.

Ellis, R. E., and Horvitz, H. R. (1991). Two *C. elegans* genes control the programmed cell deaths of specific cells in the pharynx. *Development (Cambridge, UK)* **112,** 591–603.

Ellis, R. E., Yuan, J., and Horwitz, H. R. (1991a). Mechanisms and functions of cell death. *Annu. Rev. Cell Biol.* **7,** 663–698.

Ellis, R. E., Jacobson, D. M., and Horvitz, H. R. (1991b). Genes required for the engulfment of cell corpses during programmed cell death in *Caenorhabditis elegans. Genetics* **129,** 79–94.

El-Shershaby, A. M., and Hinchliffe, J. R. (1974). Cell redundancy in the zona-intact preimplantation mouse blastocyst, a light and electron microscope study of dead cells and their fate. *J. Embryol. Exp. Morphol.* **31,** 643–654.

El-Shershaby, A. M., and Hinchliffe, J. R. (1975). Epithelial autolysis during implantation of the mouse blastocyst, an ultrastructural study. *J. Embryol. Exp. Morphol.* **33,** 1067–1080.

Enders, A. C., Lantz, K. C., and Schlafke, S. (1990). Differentiation of the inner cell mass of the baboon blastocyst. *Anat. Rec.* **226,** 237– 248.

Evans, V. G. (1993). Multiple pathways to apoptosis. *Cell Biol. Int.* **17,** 461–476.

Fahrbach, S. E., and Schwartz, L. M. (1994). Localization of immunoreactive ubiquitin in the nervous system of the *Manduca sexta* moth. *J. Comp. Neurol.* **343,** 464–482.

Fahrbach, S. E., and Truman, J. W. (1987). Mechanisms for programmed cell death in the nervous system of a moth. *Ciba Found. Symp.* **126,** 65–81.

Fahrbach, S. E., Choi, M. K., and Truman, J. W. (1994). Inhibitory effects of actinomycin D and cycloheximide on neuronal death in adult *Manduca sexta. J. Neurobiol.* **25,** 59–69.

Fallon, J. F., and Cameron, J. A. (1977). Interdigital cell death during limb development of the turtle and lizard with an interpretation of evolutionary significance. *J. Embryol. Exp. Morphol.* **40,** 285–289.

Fallon, J. F., and Saunders, J. W. (1968). *In vitro* analysis of the control of cell death in a zone of prospective necrosis from the chick wing bud. *Dev. Biol.* **18,** 553–570.

Fallon, J. F., and Simandl, B. K. (1978). Evidence of a role for cell death in the disappearance of the embryonic human tail. *Am. J. Anat.* **152,** 111–130.

Fallon, J. F., Lopez, A., Ros, M. A., Savage, M. P., Olwin, B. B., and Simandl, B. K. (1994). FGF-2, apical ectodermal ridge growth signal for chick limb development. *Science* **264,** 104–107.

Ferguson, M.W. J. (1988). Palate development. *Development (Cambridge, UK)* **103,** Suppl., 41–60.

Ferguson, M.W. J., and Honig, L. S. (1984). Epithelial-mesenchymal interactions during vertebrate palatogenesis. *Curr. Top. Dev. Biol.* **19,**, 137–164.

Ferguson, M. W. J., Honig, L. S., and Slavkin, H. C. (1984). Differentiation of cultured palatal shelves from alligator, chick and mouse embryos. *Anat. Rec.* **209,** 231–249.

Ferrer, I., Serrano, T., and Soriano, E. (1990). Naturally occurring cell death in the subicular complex and hippocampus in the rat during development. *Neurosci. Res.* **8,** 60–66.

Ferrer, I., Soriano, E., del Rio, J. A., Alcantara, S., and Auladell, C. (1992). Cell death and removal in the cerebral cortex during development. *Prog. Neurobiol.* **39,** 1–43.

Fitchett, J. E., and Hay, E. D. (1989). Medial edge epithelium transforms to mesenchyme after embryonic palatal shelves fuse. *Dev. Biol.* **131,** 455–474.

Forger, N. G., Roberts, S. L., Wong, V., and Breedlove, S. M. (1993). Ciliary neurotrophic factor maintains motoneurons and their target muscles in developing rats. *J. Neurosci.* **13,** 4720–4726.

Fox, H. (1971). Cell death, thyroxine and the development of *Rana temporaria* larvae with special reference to the pronephros. *Exp. Gerontol.* **6,** 173–177.

Fraser, M. J., Ireland, C. M., Tynan, S. J., and Papaioannou, A. (1993). Evidence for the role of an endo-exonuclease in the chromatin DNA fragmentation which accompanies apoptosis. *In* "Programmed Cell Death: The Cellular and Molecular Biology of Apoptosis" (M. Lavin and D. Watters, eds.), pp. 111–122. Harwood Academic Publishers, Chur, Switzerland.

Fredette, B. J., and Landmesser, L. T. (1991). A re-evaluation of the role of innervation in primary and secondary myogenesis in developing chick muscle. *Dev. Biol.* **143,** 19–35.

Friebová, Z. (1975). Formation of the chick mesonephros. 1. General outline of development. *Folia Morphol. (Prague)* **23,** 19–39.

Frisch, S. M., and Francis, H. (1994). Disruption of epithelial cell-matrix interactions induces apoptosis. *J. Cell Biol.* **124,** 619–626.

Fujita, N., Kataoka, S., Naito, M., Heike, Y., Boku, N., Nakajima, M., and Tsuruo, T. (1993). Suppression of T-lymphoma cell apoptosis by monoclonal antibodies raised against cell surface adhesion molecules. *Cancer Res.* **53,** 5022–5027.

Gaese, F., Kolbeck, R., and Barde, Y.-A. (1994). Sensory ganglia require neurotrophin-3 early in development. *Development (Cambridge, UK)* **120,** 1613–1619.

Gagliardini, V., Fernandez, P.-A., Lee, R. K. K., Drexler, H. C. A., Rotello, R. J., Fishman, M. C., and Yuan, J. (1994). Prevention of vertebrate neuronal death by the *crmA* gene. *Science* **263,** 826–828.

Garcia-Martinez, V., Macias, D., Gañan, Y., Garcia-Lobo, J. M., Francia, M. V., Fernandez-Teran, M. A., and Hurle, J. M. (1993). Internucleosomal DNA fragmentation and programmed cell death (apoptosis) in the interdigital tissue of the embryonic chick leg bud. *J. Cell Sci.* **106,** 201–208.

García-Porrero, J. A., Ojeda, J. L., and Hurlé, J. M. (1978). Cell death during the postnatal morphogenesis of the normal rabbit kidney and in experimental renal polycystosis. *J. Anat.* **126,** 303–318.

García-Porrero, J. A., Colvee, E., and Ojeda, J. L. (1984). Cell death in the dorsal part of the chick optic cup. Evidence for a new necrotic area. *J. Embryol. Exp. Morphol.* **80,** 241–249.

Garden, G. A., Bothwell, M., and Rubel, E. W. (1991). Lack of correspondence between MRNA expression for a putative cell death molecule (SGP-2) and neuronal cell death in the central nervous system. *J. Neurobiol.* **22,** 590–604.

Gavrieli, Y., Sherman, Y., and Ben-Sassoon, S. A. (1992). Identification of programmed cell death in situ via specific labeling of nuclear DNA fragmentation. *J. Cell Biol.* **119,** 493–501.

Geelen, J. A. G., and Langman, J. (1977). Closure of the neural tube in the cephalic region of the mouse embryo. *Anat. Rec.* **189,** 625–640.

Gendron, R. L., Nestel, F. P., Lapp, W. S., and Baines, M. G. (1991). Expression of tumor necrosis factor alpha in the developing nervous system. *Int. J. Neurosci.* **60,** 129–136.

Gerschenson, L. E., and Rotello, R. J. (1992). Apoptosis, a different type of cell death. *FASEB J.* **6,** 2450–2455.

Givol, I., Tsarfaty, I., Resau, J., Rulong, S., da Silva, P. P., Nasioulas, G., DuHadaway, J., Hughes, S. H., and Ewart, D. L. (1994). Bcl–2 expressed using a retroviral vector is localized primarily in the nuclear membrane and the endoplasmic reticulum of chicken embryo fibroblasts. *Cell Growth Differ.* **5,** 419–429.

Glücksmann, A. (1951). Cell deaths in normal vertebrate ontogeny. *Biol. Rev. Cambridge. Philos. Soc.* **26,** 59–86.

Gold, R., Schmied, M., Rothe, G., Zischler, H., Breitschopf, H., Wekerle, H., and Lassmann, H. (1993). Detection of DNA fragmentation in apoptosis, application of *in situ* nick translation to cell culture systems and tissue sections. *J. Histochem. Cytochem.* **41,** 1023–1030.

Goldman, A. S. (1984). Biochemical mechanism of glucocorticoid- and phenytoin-induced cleft palate. *Curr. Top. Dev. Biol.* **19**, 217–239.

Goldman, A. S., Baker, M. K., Piddington, R., and Herold, R. (1983). Inhibition of programmed cell death in mouse embryonic palate in vitro by cortisol and phenytoin, receptor involvement and requirement of protein synthesis. *Proc. Soc. Exp. Biol. Med.* **174**, 239–243.

González-Garcia, M., Pérez-Ballestero, R., Ding, L., Boise, L. H., Thompson, C. B., and Núñez, G. (1994). *bcl-xL* is the major *bcl-x* MRNA form expressed during murine development and its product localizes to mitochondria. *Development (Cambridge, UK)* **120**, 3033–3042.

Gould, E., Wooley, C. S., and McEwan, B. S. (1991). Naturally occurring cell death in the developing dentate gyrus of the rat. *J. Comp. Neurol.* **304**, 408–418.

Graham, A., Heyman, I., and Lumsden, A. (1993). Even-numbered rhombomeres control the apoptotic elimination of neural crest cells from odd-numbered rhombomeres in the chick hindbrain. *Development (Cambridge, UK)* **119**, 233–245.

Gramzinski, R. A., Parchment, R. E., and Pierce, G. B. (1990). Evidence linking programmed cell death in the blastocyst to polyamine oxidation. *Differentiation (Berlin)* **43**, 59–65.

Green, D. R., and Cotter, T. G. (1993). Macromolecular synthesis, *c-myc*, and apoptosis. *In* "Programmed Cell Death: The Cellular and Molecular Biology of Apoptosis" (M. Lavin and D Watters, eds.), pp. 153–166. Harwood Academic Publishers, Chur, Switzerland.

Greene, R. M., and Pratt, R. M. (1978). Inhibition of epithelial cell death in the secondary palate *in vitro* by alteration of lysosome function. *J. Histochem. Cytochem.* **26**, 1109–1114.

Greene, R. M., Shah, R. M., Lloyd, M. R., Crawford, B. J., Suen, R., Shanfeld, J. L., and Davidovitch, Z. (1983). Differentiation of the avian secondary palate. *J. Exp. Zool.* **225**, 43–52.

Griffith, C. M., and Hay, E. D. (1992). Epithelial-mesenchymal transformation during palatal fusion, carboxyfluorescein traces cells at light and electron microscopic levels. *Development (Cambridge, UK)* **116**, 1087–1099.

Griffith, C. M., and Sanders, E. J. (1992). The vertebrate tail bud, three germ layers from one tissue. *Anat. Embryol.* **185**, 101–113.

Gromkowski, S. H., Brown, T. C., Masson, D., and Tschopp, J. (1988). Lack of DNA degradation in target cells lysed by granules derived from cytolytic T lymphocytes. *J. Immunol.* **141**, 774–778.

Grüneberg, H. (1963). "The Pathology of Development. A Study of Inherited Skeletal Disorders in Animals." Blackwell, Oxford.

Hamburger, V. (1975). Cell death in the development of the lateral motor column of the chick embryo. *J. Comp. Neurol.* **160**, 535–546.

Hamburger, V. (1992). History of the discovery of neuronal cell death in embryos. *J. Neurobiol.* **23**, 1116–1123.

Hamburger, V., and Levi-Montalcini, R. (1949). Proliferation, differentiation and degeneration in the spinal ganglia of the chick embryo under normal and experimental conditions. *J. Exp. Zool.* **111**, 457–500.

Hamburger, V., Brunso-Bechtold, J. K., and Yip, J. W. (1981). Neuronal death in the spinal ganglia of the chick embryo and its reduction by nerve growth factor. *J. Neurosci.* **1**, 60–71.

Hardy, K., Handyside, A. H., and Winston, R. L. (1989). The human blastocyst, cell number, death and allocation during late pre-implantation development *in vitro*. *Development (Cambridge, UK)* **107**, 597–604.

Hassell, J. R. (1975). The development of rat palatal shelves *in vitro*. An ultrastructural analysis of the inhibition of epithelial cell death and palate fusion by the epidermal growth factor. *Dev. Biol.* **45**, 90–102.

Hay, B. A., Wolff, T., and Rubin, G. M. (1994). Expression of baculovirus P35 prevents cell death in *Drosophila*. *Development (Cambridge, UK)* **120**, 2121–2129.

Hébert, L., Pandey, S., and Wang, E. (1994). Commitment to cell death is signalled by the appearance of a terminin protein of 30kDa. *Exp. Cell Res.* **210**, 10–18.

Hedgecock, E. M., Sulston, J. E., and Thomson, J. N. (1983). Mutations affecting programmed cell deaths in the nematode *Caenorhabditis elegans*. *Science* **220**, 1277–1280.

Hengartner, M. O., and Horvitz, H. R. (1994a). Programmed cell death in *Caenorhabditis elegans*. *Curr. Opin. Genet. Dev.* **4**, 581–586.

Hengartner, M. O., and Horvitz, H. R. (1994b). The ins and outs of programmed cell death during *C. elegans* development. *Philos. Trans. R. Soc. London, Ser. B* **345**, 243–246.

Hengartner, M. O., and Horvitz, H. R. (1994c). *C. elegans* cell survival gene *ced-9* encodes a functional homolog of the mammalian proto-oncogene *bcl-2*. *Cell (Cambridge, Mass.)* **76**, 665–676.

Hengartner, M. O., Ellis, R. E., and Horvitz, H. R. (1992). *Caenorhabditis elegans* gene *ced-9* protects cells from programmed cell death. *Nature (London)* **356**, 494–499.

Hershko, A. (1988). Ubiquitin-mediated protein degradation. *J. Biol. Chem.* **263**, 15237–15240.

Hinchliffe, J. R. (1981). Cell death in embryogenesis. *In* "Cell Death in Biology and Pathology" (I. D. Bowen and R. A. Lockshin, eds.), pp. 35–78. Chapman & Hall, London.

Hinchliffe, J. R., and Ede, D. A. (1967). Limb development in the polydactylous talpid3 mutant of the fowl. *J. Embryol. Exp. Morphol.* **17**, 385–404.

Hinchliffe, J. R., and Ede, D. A. (1973). Cell death and the development of limb form and skeletal pattern in normal and wingless (ws) chick embryos. *J. Embryol. Exp. Morphol.* **30**, 753–772.

Hinchliffe, J. R., and Griffiths, P. J. (1986). Vital staining for cell death in chick limb buds: A histochemical technique in the analysis of control of limb development. *Acta Histochem. Suppl.* **32**, 159–164.

Hinchliffe, J. R., and Thorogood, P. V. (1974). Genetic inhibition of mesenchymal cell death and the development of form and skeletal pattern in the limbs of talpid3 (ta3) mutant chick embryos. *J. Embryol. Exp. Morphol.* **31**, 747–760.

Hiraishi, K., Suzuki, K., Hakomori, S., and Adachi, M. (1993). Le-y antigen expression is correlated with apoptosis (programmed cell death). *Glycobiology* **3**, 381–390.

Hofer, M. M., and Barde, Y.-A. (1988). Brain-derived neurotrophic factor prevents neuronal death *in vivo*. *Nature (London)* **331**, 261–262.

Homma, S., Yaginuma, H., and Oppenheim, R. W. (1994). Programmed cell death during the earliest stages of spinal cord development in the chick embryo, a possible means of early phenotypic selection. *J. Comp. Neurol.* **345**, 377–395.

Hornung, J. P., Koppel, H., and Clarke, P. G. H. (1989). Endocytosis and autophagy in dying neurons, an ultrastructural study in chick embryos. *J. Comp. Neurol.* **283**, 425–437.

Hughes, R. A., Sendtner, M., and Thoenen, H. (1993). Members of several gene families influence survival of rat motoneurons *in vitro* and *in vivo*. *J. Neurosci. Res.* **36**, 663–671.

Hurlé, J. M. (1979). Scanning and light microscope studies of the development of the chick embryo semilunar heart valves. *Anat. Embryol.* **157**, 69–80.

Hurlé, J. M. (1988). Cell death in developing systems. *Methods. Achiev. Exp. Pathol.* **13**, 55–86.

Hurlé, J. M., and Hinchliffe, J. R. (1978). Cell death in the posterior necrotic zone (PNZ) of the chick wing bud, a stereoscan and ultrastructural survey of autolysis and cell fragmentation. *J. Embryol. Exp. Morphol.* **43**, 123–136.

Ilschner, S. U., and Waring, P. (1992). Fragmentation of DNA in the retina of chicken embryos coincides with retinal ganglion cell death. *Biochem. Biophys. Res. Commun.* **183**, 1056–1061.

Imoh, H. (1986). Cell death during normal gastrulation in the newt, *Cynops pyrrhogaster*. *Cell Differ.* **19**, 35–42.

Iseki, S. (1986). DNA strand breaks in rat tissues as detected by *in situ* nick translation. *Exp. Cell Res.* **167**, 311–326.

Iseki, S., and Mori, T. (1985). Histochemical detection of DNA strand scissions in mammalian cells by *in situ* nick translation. *Cell Biol. Int. Rep.* **9**, 471–477.

Ishizaki, Y., Voyvodic, J. T., Burne, J. F., and Raff, M. C. (1993). Control of lens epithelial cell survival. *J. Cell Biol.* **121**, 899–908.

Jacobson, W. (1938). The early development of the avian embryo. II. Mesoderm formation and the distribution of presumptive embryonic material. *J. Morphol.* **62**, 445–501.

Jaskoll, T., Boyer, P. D., and Melnick, M. (1994) Tumor necrosis factor-α and embryonic mouse lung morphogenesis. *Dev. Dyn.* **201**, 137–150.

Jeffs, P., and Osmond, M. (1992). A segmented pattern of cell death during development of the chick embryo. *Anat. Embryol.* **185**, 589– 598.

Jeffs, P., Jaques, K., and Osmond, M. (1992). Cell death in cranial neural crest development. *Anat. Embryol.* **185**, 583–588.

Jin, L.-W., Inaba, K., and Saitoh, T. (1992). The involvement of protein kinase C in activation-induced cell death in T-cell hybridoma. *Cell. Immunol.* **144**, 217–227.

Jirsova, Z., and Vernerova, Z. (1993). Involution of the Wolffian duct in the rat. *Funct. Dev. Morphol.* **3**, 205–206.

Johnson, E. M., and Deckwerth, T. L. (1993). Molecular mechanisms of developmental neuronal death. *Annu. Rev. Neurosci.* **16**, 31–46.

Kalcheim, C., and Le Douarin, N. M. (1986). Requirement of a neural tube signal for the differentiation of neural crest cells into dorsal root ganglia. *Dev. Biol.* **116**, 451–466.

Kater, S. B., Mattson, M. P., and Guthrie, P. B. (1989). Calcium-induced neuronal degeneration, a normal growth cone regulating signal gone awry. *Ann. N. Y. Acad. Sci.* **568**, 252–261.

Kawahara, A., Baker, B. S., and Tata, J. R. (1991). Developmental and regional expression of thyroid hormone receptor genes during *Xenopus* metamorphosis. *Development (Cambridge, UK)* **112**, 933–943.

Kerr, J. F. R. (1971). Shrinkage necrosis, a distinct mode of cellular death. *J. Pathol.* **105**, 13–20.

Kerr, J. F. R. (1993). Definition of apoptosis and overview of its incidence. *In* "Programmed Cell Death: The Cellular and Molecular Biology of Apoptosis" (M. Lavin and D. Watters, eds.), pp. 1–15. Harwood Academic Publishers, Chur, Switzerland.

Kerr, J. F. R., and Harmon, B. V. (1991). Definition and incidence of apoptosis, an historical perspective. *In* "Apoptosis: The Molecular Basis of Cell Death" (L. D. Tomei and F. O. Cope, eds.), pp. 5–29. Cold Spring Harbor Lab. Press, Plainview, NY.

Kerr, J. F. R., Wyllie, A. H., and Currie, A. R. (1972). Apoptosis, a basic biological phenomenon with wide-ranging implications in tissue kinetics. *Br. J. Cancer* **26**, 239–257

Kerr, J. F. R., Searle, J., Harmon, B. V., and Bishop, C. J. (1987). Apoptosis. *In* "Perspectives on Mammalian Cell Death" (C. S. Potten, ed.), pp. 93–128. Oxford Univ. Press, Oxford.

Kimura, K.-I., and Tanimura, T. (1992). Mutants with delayed cell death of the ptilinal head muscles in *Drosophila. J. Neurogenet.* **8**, 57–69.

Kimura, K.-I., and Truman, J. W. (1990). Postmetamorphic cell death in the nervous and muscular systems of *Drosophila melanogaster. J. Neurosci.* **10**, 403–411.

Kirn, J. R., and DeVoogd, T. J. (1989). Genesis and death of vocal control neurons during sexual differentiation in the zebra finch. *J. Neurosci.* **9**, 3176–3187.

Klefström, J., Västrik, I., Saksela, E., Valle, J., Eilers, M., and Alitalo, K. (1994). c-Myc induces cellular susceptibility to the cytotoxic action of TNFα. *EMBO J.* **13**, 5442–5450.

Kluck, R. M., McDougall, C. A., Harmon, B. V., and Halliday, J. W. (1994). Calcium chelators induce apoptosis—evidence that raised intracellular ionised calcium is not essential for apoptosis. *Biochim. Biophys. Acta* **1223**, 247–254.

Kohchi, C., Noguchi, K., Tanabe, Y., Mizuno, D., and Soma, G. (1994). Constitutive expression of TNF-alpha and -beta genes in mouse embryo, roles of cytokines as regulator and effector of development. *Int. J. Biochem.* **26**, 111–119.

Koike, T., Martin, D. P., and Johnson, E. M. (1989). Role of calcium ion channels in the ability of membrane depolarization to prevent neuronal death induced by trophic-factor deprivation. *Proc. Natl. Acad. Sci. U.S.A.* **86**, 6421–6425.

Korsmeyer, S. J., Shutter, J. R., Veis, D. J., Merry, D. E., and Oltvai, Z. N. (1993). Bcl-2/Bax, a rheostat that regulates an anti-oxidant pathway and cell death. *Semin. Cancer Biol.* **4**, 327–332.

Koseki, C., Herzlinger, D., and Al-Awqati, Q. (1992). Apoptosis in metanephric development. *J. Cell Biol.* **119,** 1327–1333.

Krabbenhoft, K. M., and Fallon, J. F. (1992). Talpid2 limb bud mesoderm does not express Ghox-8 and has an altered expression pattern of Ghox-7. *Dev. Dyn.* **194,** 52–62.

Kreidberg, J. A., Sariola, H., Loring, J. M., Maeda, M., Pelletier, J., Housman, D., and Jaenisch, R. (1993). WT-1 is required for early kidney development. *Cell (Cambridge, Mass.)* **74,** 679–691.

Kumar, S., Kinoshita, M., Noda, M., Copeland, N. G., and Jenkins, N. A. (1994). Induction of apoptosis by the mouse *Nedd2* gene, which encodes a protein similar to the product of the *Caenorhabditis elegans* cell death gene *ced-3* and the mammalian interleukin-1β-converting enzyme. *Genes Dev.* **8,** 1613–1626.

Kutsch, W., and Bentley, D. (1987). Programmed death of grasshopper pioneer neurons in the grasshopper embryo. *Dev. Biol.* **123,** 517–525.

Lam, M., Dubyak, G., Chen, L., Nuñez, G., Miesfeld, R. L., and Distelhorst, C. W. (1994). Evidence that BCL-2 represses apoptosis by regulating endoplasmic reticulum-associated Ca^{2+} fluxes. *Proc. Natl. Acad. Sci. U.S.A.* **91,** 6569–6573.

Landmesser, L. T. (1992). The relationship of intramuscular nerve branching and synaptogenesis to motoneuron survival. *J. Neurobiol.* **23,** 1131–1139.

Landmesser, L. T., and Pilar, G. (1974). Synaptic transmission and cell death during normal ganglionic development. *J. Physiol. (London)* **241,** 737–749.

Landmesser, L. T., and Szente, M. (1986). Activation patterns of embryonic chick hind-limb muscles following blockade of activity and motoneurone cell death. *J. Physiol. (London)* **380,** 157–174.

Lang, R. A., and Bishop, J. M. (1993). Macrophages are required for cell death and tissue remodelling in the developing mouse eye. *Cell (Cambridge, Mass.)* **74,** 453–462.

Larrick, J. W., and Wright, S. C. (1990). Cytotoxic mechanisms of tumor necrosis factor-α. *FASEB J.* **4,** 3215–3223.

Lavin, M. F., Baxter, G. D., Song, Q., Findik, D., and Kovaks, E. (1993). Protein modification in apoptosis. *In* "Programmed Cell Death: The Cellular and Molecular Biology of Apoptosis" (M. Lavin and D. Watters, eds.), pp. 45–57. Harwood Academic Publishers, Chur, Switzerland.

Lawson, K. A., Meneses, J. J., and Pedersen, R. A. (1986). Cell fate and cell lineage in the endoderm of the presomite mouse embryo, studied with an intracellular tracer. *Dev. Biol.* **115,** 325–339.

Lazebnik, Y. A., Cole, S., Cooke, C. A., Nelson, W. G., and Earnshaw, W. C. (1993). Nuclear events of apoptosis *in vitro* in cell-free mitotic extracts, a model system for analysis of the active phase of apoptosis. *J. Cell Biol.* **123,** 7–22.

LeBrun, D. P., Warnke, R. A., and Cleary, M. L. (1993). Expression of bcl-2 in fetal tissues suggests a role in morphogenesis. *Am. J. Pathol.* **142,** 743–753.

Lee, S., Christakos, S., and Small, M. B. (1993). Apoptosis and signal transduction, clues to a molecular mechanism. *Curr. Opin. Cell Biol.* **5,** 286–291.

Levi-Montalcini, R., and Angeletti, P. U. (1968). Nerve growth factor. *Physiol. Revs.* **48,** 534–569.

Levi-Montalcini, R., and Booker, B. (1960). Destruction of the sympathetic ganglia in mammals by an antiserum to a nerve growth protein. *Proc. Natl. Acad. Sci. U.S.A.* **46,** 384–391.

Lipton, S. A. (1986). Blockade of electrical activity promotes the death of mammalian retinal ganglion cells in culture. *Proc. Natl. Acad. Sci. U.S.A.* **83,** 9774–9778.

Lockshin, R. A. (1969). Programmed cell death. Activation of lysis by a mechanism involving the synthesis of protein. *J. Insect Physiol.* **15,** 1505–1516.

Lockshin, R. A. (1981). Cell death in metamorphosis. *In* "Cell Death in Biology and Pathology" (I. D. Bowen and R. A. Lockshin, eds.), pp. 79–121. Chapman & Hall, London.

Lockshin, R. A., and Williams, C. M. (1964). Programmed cell death. II. Endocrine potentiation of the breakdown of the intersegmental muscles of silkmoths. *J. Insect Physiol.* **10,** 643–649.

Lockshin, R. A., and Williams, C. M. (1965). Programmed cell death. I. Cytology of degeneration in the intersegmental muscles of the pernyi silkmoth. *J. Insect Physiol.* **11,** 123–133.

Lockshin, R. A., and Zakeri, Z. (1991). Programmed cell death and apoptosis. *In* "Apoptosis: The Molecular Basis of Cell Death" (L. D. Tomei and F. O. Cope, eds.), pp. 47–60. Cold Spring Harbor Lab. Press, Plainview, NY.

Louis, J. C., Magel, E., Takayama, S., and Varon, S. (1993). CNTF protection of oligodendrocytes against natural and tumor necrosis factor-induced death. *Science* **259,** 689–692.

Lowe, S. W., Schmitt, E. M., Smith, S. W., Osborne, B. A., and Jacks, T. (1993). p53 is required for radiation-induced apoptosis in mouse thymocytes. *Nature (London)* **362,** 847–849.

Lu, Q. L., Poulsom, R., Wong, L., and Hanby, A. M. (1993). Bcl-2 expression in adult and embryonic non-haemopoietic tissues. *J. Pathol.* **169,** 431–437.

Lumsden, A., Sprawson, N., and Graham, A. (1991). Segmental origin and migration of neural crest cells in the hindbrain region of the chick embryo. *Development (Cambridge, Mass.)* **113,** 1281–1291.

MacCabe, J. A., Blaylock, R. L., Latimer, J. L., and Pharris, L. J. (1991). Fibroblast growth factor and culture in monolayer rescue mesoderm cells destined to die in the developing avian wing. *J. Exp. Zool.* **257,** 208–213.

Manasek, F. J. (1969). Myocardial cell death in the embryonic chick ventricle. *J. Embryol. Exp. Morphol.* **21,** 271–284.

Martikainen, P., Kyprianou, N., Tucker, R. W., and Isaacs, J. T. (1991). Programmed death of nonproliferating androgen-independent prostatic cancer cells. *Cancer Res.* **51,** 4693–4700.

Martin, D. P., and Johnson, E. M. (1991). Programmed cell death in the peripheral nervous system. *In* "Apoptosis: The Molecular Basis of Cell Death" (D. L. Tomei and F. O. Cope, eds.), pp. 247–261. Cold Spring Harbor Lab. Press, Plainview, NY.

Martin, D. P., Schmidt, R. E., DiStefano, P. S., Lowry, O. H., Carter, J. G., and Johnson, E. M. (1988). Inhibitors of protein synthesis and RNA synthesis prevent neuronal death caused by nerve growth factor deprivation. *J. Cell Biol.* **106,** 829–844.

Martin, S. J., Mazdai, G., Strain, J. J., Cotter, T. G., and Hannigan, B. M. (1991). Programmed cell death (apoptosis) in lymphoid and myeloid cell lines during zinc deficiency. *Clin. Exp. Immunol.* **83,** 338–343.

Martin, S. J., Green, D. R., and Cotter, T. G. (1994). Dicing with death, dissecting the components of the apoptosis machinery. *Trends Biochem. Sci.* **19,** 26–30.

Martinou, J.-C., Dubois-Dauphin, M., Staple, J. K., Rodriguez, I., Frankowski, H., Missotten, M., Albertini, P., Talabot, D., Catsicas, S., Pietra, C., and Huarte, J. (1994). Over-expression of BCL-2 in transgenic mice protects neurons from naturally occurring cell death and experimental ischemia. *Neuron* **13,** 1017–1030.

Martín-Partido, G., Alvarez, I. S., Rodríguez-Gallardo, L., and Navascués, J. (1986). Differential staining of dead and dying embryonic cells with a simple new technique. *J. Microsc. (Oxford)* **142,** 101–106.

Mattson, M. P., and Kater, S. B. (1989). Excitatory and inhibitory neurotransmitters in the generation and degeneration of hippocampal neuroarchitecture. *Brain Res.* **478,** 337–348.

McConkey, D. J., Nicotera, P., Hartzell, P., Bellomo, G., Wyllie, A. H., and Orrenius, S. (1989a). Glucocorticoids activate a suicide process in thymocytes through an elevation of cytosolic calcium concentration. *Arch. Biochem. Biophys.* **269,** 365–370.

McConkey, D. J., Hartzell, P., Nicotera, P., and Orrenius, S. (1989b). Calcium-activated DNA fragmentation kills immature thymocytes. *FASEB J.* **3,** 1843–1849.

McConkey, D. J., Hartzell, P., Jondal, M., and Orrenius, S. (1989c). Inhibition of DNA fragmentation in thymocytes and isolated thymocyte nuclei by agents that stimulate protein kinase C. *J. Biol. Chem.* **264,** 13399–13402.

McConkey, D. J., Orrenius, S., and Jondal, M. (1990). Agents that elevate cAMP stimulate DNA fragmentation in thymocytes. *J. Immunol.* **145**, 1227–1230.

Mendelsohn, C., Ruberte, E., LeMeur, M., Morris-Kay, G., and Chambon, P. (1991). Developmental analysis of the retinoic acid-inducible RAR-beta 2 promoter in transgenic mice. *Development (Cambridge, UK)* **113**, 723–734.

Merry, D. E., Veis, D. J., Hickey, W. F., and Korsmeyer, S. J. (1994). bcl-2 protein expression is widespread in the developing nervous system and retained in the adult PNS. *Development (Cambridge, UK)* **120**, 301–311.

Migliorati, G., Nicoletti, I., Pagliacci, M. C., D'Adamio, L., and Riccardi, C. (1993). Interleukin-2 induces apoptosis in mouse thymocytes. *Cell. Immunol.* **146**, 52–61.

Milligan, C. E., Oppenheim, R. W., and Schwartz, L. M. (1994). Motoneurons deprived of trophic support *in vitro* require new gene expression to undergo programmed cell death. *J. Neurobiol.* **25**, 1005–1016.

Mills, C. L., and Bellairs, R. (1989). Mitosis and cell death in the tail of the chick embryo. *Anat. Embryol.* **180**, 301–308.

Miura, M., Zhu, H., Rotello, R., Hartwieg, E. A., and Juan, J. (1993). Induction of apoptosis in fibroblasts by IL-1β-converting enzyme, a mammalian homolog of the *C. elegans* cell death gene ced-3. *Cell (Cambridge, Mass.)* **75**, 653–660.

Mizuguchi, M., Ikeda, K., Asada, M., Mizutani, S., and Kamoshita, S. (1994). Expression of Bcl-2 protein in murine neural cells in culture. *Brain Res.* **649**, 197–202.

Modak, S. P., and Perdue, S. W. (1970). Terminal lens cell differentiation. I. Histological and microspectrophotometric analysis of nuclear degeneration. *Exp. Cell Res.* **59**, 43–56.

Mori, C., Nakamura, N., Okamoto, Y., Osawa, M., and Shiota, K. (1994). Cytochemical identification of programmed cell death in the fusing fetal mouse palate by specific labelling of DNA fragmentation. *Anat. Embryol.* **190**, 21–28.

Nakashima, T., Sekiguchi, T., Kuraoka, A., Fukushima, K., Shibata, Y., Komiyama, S., and Nishimoto, T. (1993). Molecular cloning of a human cDNA encoding a novel protein, DAD1, whose defect causes apoptotic cell death in hamster BHK21 cells. *Mol. Cell. Biol.* **13**, 6367–6374.

Nakayama, K., Nakayama, K.-I., Negishi, I., Kuida, K., Sawa, H., and Loh, D. Y. (1994). Targeted disruption of bcl-2αβ in mice, occurrence of grey hair, polycystic kidney disease, and lymphocytopenia. *Proc. Natl. Acad. Sci. U.S.A.* **91**, 3700–3704.

Nameroff, M. (1992). A cell division counter exists in the chick embryo myogenic lineage. *Symp. Soc. Exp. Biol.* **46**, 73–78.

Naruse, I., Keino, H., and Kawarada, Y. (1994). Antibody against single-stranded DNA detects both programmed cell death and drug-induced apoptosis. *Histochemistry* **101**, 73–78.

Navascues, J., Martin-Partido, G., Alvarez, I. S., and Rodriguez-Gallardo, L. (1988). Cell death in suboptic necrotic centers of chick embryo diencephalon and their topographic relationship with the earliest optic fiber fascicles. *J. Comp. Neurol.* **278**, 34–46.

Nishikawa, A., Kaiho, M., and Yoshizato, K. (1989). Cell death in the anuran tadpole tail, thyroid hormone induces keratinization and tail-specific growth inhibition of epidermal cells. *Dev. Biol.* **131**, 337–344.

Nixon, R. A., and Cataldo, A. M. (1993). The lysosomal system in neuronal cell death, a review. *Ann. N.Y. Acad. Sci.* **679**, 87–109.

Nohno, T., Noji, S., Koyama, E., Nishikawa, K., Myokai, F., Saito, T., and Taniguchi, S. (1992). Differential expression of two msh-related homeobox genes Chox-7 and Chox-8 during chick limb development. *Biochem. Biophys. Res. Commun.* **182**, 121–128.

Novack, D. V., and Korsmeyer, S. J. (1994). Bcl-2 protein expression during murine development. *Am. J. Pathol.* **145**, 61–73.

Nuñez, G., and Clarke, M. F. (1994). The Bcl-2 family of proteins, regulators of cell death and survival. *Trends Cell Biol.* **4**, 399–403.

Oberhammer, F., Fritsch, G., Schmied, M., Pavelka, M., Printz, D., Purchio, T., Lassman, H., and Schulte-Hermann, R. (1993). Condensation of the chromatin at the membrane of an apoptotic nucleus is not associated with activation of an endonuclease. *J. Cell Sci.* **104,** 317–326.

Ojeda, J. L., and Hurlé, J. M. (1975). Cell death during the formation of tubular heart of the chick embryo. *J. Embryol. Exp. Morphol.* **33,** 523–534.

O'Leary, D. D., Fawcett, J. W., and Cowan, W. M. (1986). Topographic targeting errors in the retinocollicular projection and their elimination by selective ganglion cell death. *J. Neurosci.* **6,** 3692–3705.

Oppenheim, R. W. (1989). The neurotrophic theory and naturally occurring motoneuron death. *Trends Neurosci.* **12,** 252–255.

Oppenheim, R. W. (1991). Cell death during development of the nervous system. *Annu. Rev. Neurosci.* **14,** 453–501.

Oppenheim, R.W., and Núñez, R. (1982). Electrical stimulation of hindlimb increases neuronal cell death in chick embryo. *Nature (London)* **295,** 57–59.

Oppenheim, R. W., Maderdrut, J. L., and Wells, D. J. (1982). Cell death of motoneurons in the chick embryo spinal cord. VI. Reduction of naturally occurring cell death in the thoracolumbar column of Terni by nerve growth factor. *J. Comp. Neurol.* **210,** 174–189.

Oppenheim, R. W., Haverkamp, L. J., Prevette, D., McManaman, J. L., and Appel, S. H. (1988). Reduction of naturally occurring motoneuron death *in vivo* by a target-derived neurotrophic factor. *Science* **240,** 919–922.

Oppenheim, R. W., Prevette, D., Tytell, M., and Homma, S. (1990). Naturally occurring and induced neuronal death in the chick embryo *in vivo* requires protein and RNA synthesis, evidence for the role of cell death genes. *Dev. Biol.* **138,** 104–113.

Oppenheim, R. W., Qin-Wei, Y., Prevette, D., and Yan, Q. (1992a). Brain-derived neurotrophic factor rescues developing avian motoneurons from cell death. *Nature (London)* **360,** 755–757.

Oppenheim, R.W, Schwartz, L. M., and Shatz, C. J. (1992b). Neuronal death, a tradition of dying. *J. Neurobiol.* **23,** 1111–1115.

Orrenius, S., McConkey, D. J., and Nicotera, P. (1991). Role of calcium in toxic and programmed cell death. *Adv. Exp. Med. Biol.* **283,** 419–425.

Owen-Schaub, L. B., Yonehara, S., Crump, W. L., and Grimm, E. A. (1992). DNA fragmentation. *Cell. Immunol.* **140,** 197–205.

Parr, E. L., Tung, H. N., and Parr, M. B. (1987). Apoptosis as the mode of uterine epithelial cell death during embryo implantation in mice and rats. *Biol. Reprod.* **36,** 211–225.

Peitsch, M. C., Polzar, B., Stephan, H., Crompton, T., MacDonald, H. R., Mannherz, H. G., and Tschopp, J. (1993). Characterization of the endogenous deoxyribonuclease involved in nuclear DNA degradation during apoptosis (programmed cell death). *EMBO J.* **12,** 371–377.

Perez-Polo, J. R., Foreman, P. J., Jackson, G. R., Shan, D., Taglialatela, G., Thorpe, L. W., and Werrbach-Perez, K. (1990). Nerve growth factor and neuronal cell death. *Mol. Neurobiol.* **4,** 57–91.

Pesce, M., Farrace, M. G., Piacentini, M., Dolci, S., and De Felici, M. (1993). Stem cell factor and leukemia inhibitory factor promote primordial germ cell survival by suppressing programmed cell death (apoptosis). *Development (Cambridge, UK)* **118,** 1089–1094.

Pexieder, T. (1975). Cell death in the morphogenesis and teratogenesis of the heart. *Adv. Anat. Embryol. Cell Biol.* **51,** 1–100.

Piatigorsky, J. (1981). Lens differentiation in vertebrates: A review of cellular and molecular features. *Differentiation (Berlin)* **9,** 134–153.

Pierce, G. B., Lewellyn, A. L., and Parchment, R. E. (1989). Mechanism of programmed cell death in the blastocyst. *Proc. Natl. Acad. Sci. U.S.A.* **86,** 3654–3658.

Pittman, R. H., and Oppenheim, R. W. (1978). Neuromuscular blockade increases motoneurone survival during normal cell death in the chick embryo. *Nature (London)* **271,** 364–365.

Pittman, R. H., and Oppenheim, R. W. (1979). Cell death of motoneurons in the chick embryo spinal cord. IV. Evidence that a functional neuromuscular interaction is involved in the regulation of naturally occurring cell death and the stabilization of synapses. *J. Comp. Neurol.* **187**, 425–446.

Poelmann, R. E. (1980). Differential mitosis and degeneration patterns in relation to the alterations in the shape of the embryonic ectoderm of early post-implantation mouse embryos. *J. Embryol. Exp. Morphol.* **55**, 33–51.

Poelmann, R. E., and Vermeij-Keers, C. (1976). Cell degeneration in the mouse embryo: A prerequisite for normal development. In "Progress in Differentiation Research" (N. Müller-Bérat, ed.), pp. 93–102. North-Holland Pub., Amsterdam.

Portera-Cailliau, C., Sung, C. H., Nathans, J., and Adler, R. (1994). Apoptotic photoreceptor cell death in mouse models of retinitis pigmentosa. *Proc. Natl. Acad. Sci. U.S.A.* **91**, 974–978.

Pratt, R. M. (1980). Involvement of hormones and growth factors in the development of the secondary palate. In "Development in Mammals" (M. H. Johnson, ed.), Vol. 4, pp. 203–231. Elsevier North-Holland Biomedical Press, Amsterdam.

Pratt, R. M., and Greene, R. M. (1976). Inhibition of palatal epithelial cell death by altered protein synthesis. *Dev. Biol.* **54**, 135–145.

Pratt, R. M., and Martin, G. R. (1975). Epithelial cell death and cyclic AMP increase during palatal development. *Proc. Natl. Acad. Sci. U.S.A.* **72**, 874–877.

Prestige, M. C. (1970). Differentiation, degeneration, and the role of the periphery, quantitative considerations. In "The Neurosciences" (F. O. Schmitt, ed.), pp. 73–82. Rockefeller Univ. Press, New York.

Prestige, M. C. (1976). Evidence that at least some of the motor nerve cells that die during development have first made peripheral connections. *J. Comp. Neurol.* **170**, 123–134.

Price, J. M., Donahoe, P. K., Ito, Y., and Hendren, W. H. (1977). Programmed cell death in the Müllerian duct induced by Müllerian inhibiting substance. *Am. J. Anat.* **149**, 353–376.

Prigent, P., Blanpied, C., Aten, J., and Hirsch, F. (1993). A safe and rapid method for analyzing apoptosis-induced fragmentation of DNA extracted from tissues or cultured cells. *J. Immunol. Methods* **160**, 139–140.

Rabelo, E. M., Baker, B. S., and Tata, J. R. (1994). Interplay between thyroid hormone and estrogen in modulating expression of their receptor and vitellogenin genes during *Xenopus* metamorphosis. *Mech. Dev.* **45**, 49–57.

Raff, M. C. (1992). Social controls on cell survival and cell death. *Nature (London)* **356**, 397–400.

Raff, M. C. (1994). Cell death genes, *Drosophila*, enters the field. *Science* **264**, 668–669.

Raff, M. C., Barres, B. A., Burne, J. F., Coles, H. S., Ishizaki, Y., and Jacobson, M. D. (1993). Programmed cell death and the control of cell survival, lessons from the nervous system. *Science* **262**, 695–699.

Ramos, R. G. P., Igloi, G. L., Lichte, B., Baumann, U., Maier, D., Schneider, T., Brandstätter, J. H., Fröhlich, A., and Fischbach, K.-F. (1993). The *irregular chiasm C-roughest* of *Drosophila*, which affects axonal projections and programmed cell death, encodes a novel immunoglobulin-like protein. *Genes Dev.* **7**, 2533–2547.

Re, F., Zanetti, A., Sironi, M., Polentarutti, N., Lanfrancone, L., Dejana, E., and Colotta, F. (1994). Inhibition of anchorage-dependent cell spreading triggers apoptosis in cultured human endothelial cells. *J. Cell Biol.* **127**, 537–546.

Reed, J. C. (1994). Bcl-2 and the regulation of programmed cell death. *J. Cell Biol.* **124**, 1–6.

Renaud F., Oliver L., Desset S., Tassin J., Romquin N., Courtois Y., and Laurent M. (1994). Up-regulation of aFGF expression in quiescent cells is related to cell survival. *J. Cell. Physiol.* **158**, 435–443.

Robinow, S., Talbot, W. S., Hogness, D. S., and Truman, J. W. (1993). Programmed cell death in the *Drosophila* CNS is ecdysone-regulated and coupled with a specific ecdysone receptor isoform. *Development (Cambridge, UK)* **119**, 1251–1259.

Rohrer, H., Hofer, M., Hellweg, R., Korschung, S., Stehle, A. D., Saadat, S., and Thoenen, H. (1988). Antibodies against mouse nerve growth factor interfere *in vivo* with the development of avian sensory and sympathetic neurones. *Development (Cambridge, UK)* **103**, 545–552.

Rong, P. M., Teillet, M.-A., Ziller, C., and Le Douarin, N. M. (1992). The neural tube/notochord complex is necessary for vertebral but not limb and body wall striated muscle differentiation. *Development (Cambridge, UK)* **115**, 657–672.

Rotello, R. J., Fernandez, P.-A., and Yuan, J. (1994). Anti-apogens and anti-engulfens, monoclonal antibodies reveal specific antigens on apoptotic and engulfment cells during chicken embryonic development. *Development (Cambridge, UK)* **120**, 1421–1431.

Rowe, D. A., Cairns, J. M., and Fallon, J. F. (1982). Spatial and temporal patterns of cell death in limb bud mesoderm after apical ectodermal ridge removal. *Dev. Biol.* **93**, 83–91.

Rubin, L. L., Philpott, K. L., and Brooks, S. F. (1993). The cell cycle and cell death. *Curr. Biol.* **3**, 391–394.

Rubin, L. L., Gatchalian, C. L., Rimon, G., and Brooks, S. F. (1994). The molecular mechanisms of neuronal apoptosis. *Curr. Opin. Neurobiol.* **4**, 696–702.

Ruoslahti, E., and Reed, J. C. (1994). Anchorage dependence, integrins, and apoptosis. *Cell (Cambridge, Mass.)* **77**, 477–478.

Ryan, J. J., Prochownik, E., Gottlieb, C. A., Apel, I. J., Merino, R., Nuñez, G., and Clarke, M. F. (1994). *c-myc* and *bcl-2* modulate p53 function by altering p53 subcellular trafficking during the cell cycle. *Proc. Natl. Acad. Sci. U.S.A.* **91**, 5878–5882.

Salzgeber, B., and Weber, R. (1966). La régression du mésonéphros chez l'embryon de poulet. *J. Embryol. Exp. Morphol.* **15**, 397–419.

Sanders, E. J., Khare, M. K., Ooi, V. C., and Bellairs, R. (1986). An experimental and morphological analysis of the tail bud mesenchyme of the chick embryo. *Anat. Embryol.* **174**, 179–185.

Saunders, J. W. (1966). Death in embryonic systems. *Science* **154**, 604–612.

Saunders, J. W., Gasseling, M. T., and Saunders, L. C. (1962). Cellular death in morphogenesis of the avian wing. *Dev. Biol.* **5**, 147–178.

Savill, J., Fadok, V., Henson, P., and Haslett, C. (1993). Phagocyte recognition of cells undergoing apoptosis. *Immunol. Today* **14**, 131–136.

Schlüter, G. (1973). Ultrastructural observations on cell necrosis during formation of the neural tube in mouse embryos. *Z. Anat. Entwicklungsgesch.* **141**, 251–264.

Schnellmann, R. G., Swagler, A. R., and Compton, M. M. (1993). Absence of endonuclease activation during acute cell death in renal proximal tubules. *Am. J. Physiol.* **265**, C485–C490.

Schoenwolf, G. C. (1981). Morphogenetic processes involved in the remodelling of the tail region of the chick embryo. *Anat. Embryol.* **162**, 183–197.

Schook, P. (1980). Morphogenetic movements during the early development of the chick eye. An ultrastructural and spatial reconstructive study. B. Invagination of the optic vesicle and fusion of its walls. *Acta Morphol. Neerl.-Scand.* **18**, 159–180.

Schuler, C. F., Guo, Y., Majumder, A., and Luo, R. Y. (1991). Molecular and morphologic changes during the epithelial-mesenchymal transformation of palatal shelf medial edge epithelium *in vitro*. *Int. J. Dev. Biol.* **35**, 463–472.

Schulze-Osthoff, K., Walczak, H., Dröge, W., and Krammer, P. H. (1994). Cell nucleus and DNA fragmentation are not required for apoptosis. *J. Cell Biol.* **127**, 15–20.

Schwartz, L. M. (1991). The role of cell death genes during development. *BioEssays* **13**, 389–395.

Schwartz, L. M. (1992). Insect muscle as a model for programmed cell death. *J. Neurobiol* **23**, 1312–1326.

Schwartz, L. M., and Osborne, B. A. (1993). Programmed cell death, apoptosis and killer genes. *Immunol. Today* **14**, 582–590.

Schwartz, L. M., Kosz, L., and Kay, B. K. (1990a). Gene activation is required for developmentally programmed cell death. *Proc. Natl. Acad. Sci. U.S.A.* **87,** 6594–6598.

Schwartz, L. M., Myer, A., Kosz, L., Engelstein, M., and Maier, C. (1990b). Activation of polyubiquitin gene expression during developmentally programmed cell death. *Neuron* **5,** 411–419.

Schwartz, L. M., Smith, S. W., Jones, M. E. E., and Osborne, B. A. (1993). Do all programmed cell deaths occur via apoptosis? *Proc. Natl. Acad. Sci. U.S.A.* **90,** 980–984.

Schwartzman, R. A., and Cidlowski, J. A. (1993). Apoptosis, the biochemistry and molecular biology of programmed cell death. *Endocr. Rev.* **14,** 133–151.

Selvakumaran, M., Lin, H. K., Sjin, R. T., Reed, J. C., Liebermann, D. A., and Hoffman, B. (1994). The novel primary response gene MyD118 and thew proto-oncogenes myb, myc, and bcl-2 modulate transforming growth factor beta1-induced apoptosis of myeloid leukemia cells. *Mol. Cell. Biol.* **14,** 2352–2360.

Sen, S. (1992). Programmed cell death, concept, mechanism and control. *Biol. Rev. Cambridge Chilos. Soc.* **67,** 287–319.

Sendtner, M., Arakawa, Y., Stockli, K. A., Kreutzberg, G. W., and Thoenen, H. (1991). Effect of ciliary neurotrophic factor (CNTF) on motoneuron survival. *J. Cell Sci., Suppl.* **15,** 103–109.

Sengelaub, D. R., and Finlay, B. L. (1982). Cell death in the mammalian visual system during normal development. I. Retinal ganglion cells. *J. Comp. Neurol.* **204,** 311–317.

Server, A. C., and Mobley, W. C. (1991). Neuronal cell death and the role of apoptosis. *In* "Apoptosis: The Molecular Basis of Cell Death" (L. D. Tomei and F. O. Cope, eds.), pp. 263–278. Cold Spring Harbor Lab. Press, Plainview, NY.

Shah, R. M. (1984). Morphological, cellular, and biochemical aspects of differentiation of normal and teratogen-treated palate in hamster and chick embryos. *Curr. Top. Dev. Biol.* **19,** 103–135.

Shah, R. M., Cheng, K. M., Suen, R., and Wong, A. (1985). An ultrastructural and histochemical study of the development of secondary palate in Japanese quail, *Coturnix coturnix japonica. J. Craniofacial Gen. Dev. Biol.* **5,** 41–57.

Shaw, P., Bovey, R., Tardy, S., Sahli, R., Sordat, B., and Costa, J. (1992). Induction of apoptosis by wild-type p53 in a human colon tumor-derived cell line. *Proc. Natl. Acad. Sci. U.S.A.* **89,** 4495–4499.

Shi, Y., Glynn, J. M., Guilbert, L. J., Cotter, T. G., Bissonnette, R. P., and Green, D. R. (1992). Role for c-myc in activation-induced apoptotic cell death in T cell hybridomas. *Science* **257,** 212–214.

Shiota, K., Fujita, S., Akiyama, T., and Mori, C. (1990). Expression of the epidermal growth factor receptor in developing mouse palates, an immunohistochemical study. *Am. J. Anat.* **188,** 401–408.

Smeyne, R. J., Vendrell, M., Hayward, M., Baker, S. J., Miao, G. G., Schilling, K., Robertson, L. M., Curran, T., and Morgan, J. I. (1993). Continuous c-fos expression precedes programmed cell death *in vivo. Nature (London)* **363,** 166–169.

Snow, M. H. L. (1987). Cell death in embryonic development. *In* "Perspectives on Mammalian Cell Death" (C. S. Potten, ed.), pp. 202–228. Oxford Univ. Press, Oxford.

Sohal, G. S. (1992). The role of target size in neuronal survival. *J. Neurobiol.* **23,** 1124–1130.

Sorenson, C. M., Barry, M. A., and Eastman, A. (1990). Analysis of events associated with cell cycle arrest at G2 phase and cell death induced by cisplatin. *J. Natl. Cancer Inst.* **82,** 749–755.

Steller, H., Abrams, J. M., Grether, M. E., and White, K. (1994). Programmed cell death in *Drosophila. Philos. Trans. R. Soc. London, Ser. B* **345,** 247–250.

Sternberg, P. W. (1991). Control of cell lineage and cell fate during nematode development. *Curr. Top. Dev. Biol.* **25,** 177–225.

Stockdale, F. E. (1992). Myogenic cell lineages. *Dev. Biol.* **154**, 284–298.

Sugimoto, A., Friesen, P. D., and Rothman, J. H. (1994). Baculovirus *p35* prevents developmentally programmed cell death and rescues a *ced-9* mutant in the nematode *Caenorhabditis elegans. EMBO J.* **13**, 2023–2028.

Sulik, K. K., and Dehart, D. B. (1988). Retinoic-acid-induced limb malformations resulting from apical ectodermal ridge cell death. *Teratology* **37**, 527–537.

Sulston, J. E. (1976). Post embryonic development in the ventral cord of *Caenorhabditis elegans. Philos. Trans. R. Soc. London, Ser. B.* **275**, 287–297.

Sulston, J. E., Albertson, D. G., and Thomson, J. N. (1980). The *Caenorhabditis elegans* male, postembryonic development of nongonadal structures. *Dev. Biol.* **78**, 542–576.

Summerbell, D. (1977). Reduction of the rate of outgrowth, cell density, and cell division following removal of the apical ectodermal ridge of the chick limb-bud. *J. Embryol. Exp. Morphol.* **40**, 1–21.

Svendsen, C. N., Kew, J. N., Staley, K., and Sofroniew, M. V. (1994). Death of developing septal cholinergic neurons following NGF withdrawal *in vitro,* protection by protein synthesis inhibition. *J. Neurosci.* **14**, 75–87.

Tang, J., and Landmesser, L. (1993). Reduction of intramuscular nerve branching and synaptogenesis is correlated with decreased motoneuron survival. *J. Neurosci.* **13**, 3095–3103.

Tata, J. R. (1966). Requirement for RNA and protein synthesis for induced regression of the tadpole tail in organ culture. *Dev. Biol.* **12**, 77–94.

Tata, J. R. (1993). Gene expression during metamorphosis, an ideal model for post-embryonic development. *BioEssays* **15**, 239–248.

Tata, J. R., Kawahara, A., and Baker, B. S. (1991). Prolactin inhibits both thyroid hormone-induced morphogenesis and cell death in cultured amphibian larval tissues. *Dev. Biol.* **146**, 72–80.

Tenniswood, M. P., Guenette, R. S., Lakins, J., Mooibroek, M., Wong, P., and Welsh, J. E. (1992). Active cell death in hormone-dependent tissues. *Cancer Metastasis Rev.* **11**, 197–220.

Thaler, C. D., Suhr, L., Ip, N., and Katz, D. M. (1994). Leukemia inhibitory factor and neurotrophins support overlapping populations of rat nodose sensory neurons in culture. *Dev. Biol.* **161**, 338–344.

Toné, S., Tanaka, S., Minatogawa, Y., and Kido, R. (1994). DNA fragmentation during the programmed cell death in the chick limb buds. *Exp. Cell Res.* **215**, 234–236.

Torrey, T. W. (1965). Morphogenesis of the vertebrate kidney. *In* "Organogenesis" (R. L. DeHaan and H. Ursprung, eds.), pp. 559–579. Holt, Rinehart & Winston, Toronto.

Tran, D., Picard, J. Y., Campargue, J., and Josso, N. (1987). Immunocytochemical detection of anti-mullerian hormone in Sertoli cells of various mammalian species including human. *J. Histochem. Cytochem.* **35**, 733–743.

Treier, M., Seufert, W., and Jentsch, S. (1992). *Drosophila UbcD1* encodes a highly conserved ubiquitin-conjugating enzyme involved in selective protein degradation. *EMBO J.* **11**, 367–372.

Trelstad, R. L., Hayashi, A., Hayashi, K., and Donahoe, P. K. (1982). The epithelial-mesenchymal interface of the male rat Mullerian duct, loss of basement membrane integrity and ductal regression. *Dev. Biol.* **92**, 27–40.

Truman, J. W. (1983). Programmed cell death in the nervous system of an adult insect. *J. Comp. Neurol.* **216**, 445–452.

Truman, J. W. (1984). Cell death in invertebrate nervous systems. *Annu. Rev. Neurosci.* **7**, 171–188.

Truman, J. W. (1990). Metamorphosis of the central nervous system of *Drosophila. J. Neurobiol.* **21**, 1072–1084.

Truman, J. W., and Schwartz, L. M. (1984). Steroid regulation of neuronal death in the moth nervous system. *J. Neurosci.* **4**, 274–280.

Truman, J. W., Fahrbach, S. E., and Kimura, K.-I. (1990). Hormones and programmed cell death, insights from invertebrate studies. *Prog. Brain Res.* **86,** 25–35.

Truman, J. W., Thorn, R. S., and Robinow, S. (1992). Programmed neuronal death in insect development. *J. Neurobiol.* **23,** 1295–1311.

Trump, B. F., and Berezesky, I. K. (1992). The role of cytosolic calcium ions in cell injury, necrosis and apoptosis. *Curr. Opin. Cell Biol.* **4,** 227–232.

Turley, E. A., Hollenberg, M. D., and Pratt, R. M. (1985). Effect of epidermal growth factor/urogastrone on glycosaminoglycan synthesis and accumulation *in vitro* in the developing mouse palate. *Differentiation (Berlin)* **28,** 279–285.

Tyler, M. S., and Pratt, R. M. (1980). Effect of epidermal growth factor on secondary palate epithelium *in vitro*, tissue isolation and recombination studies. *J. Embryol. Exp. Morphol.* **58,** 93–106.

Uckun, F. M., Tuel-Ahlgren, L., Song, C. W., Waddick, K., Myers, D. E., Kirihara, J., Ledbetter, J. A., and Schieven, G. L. (1992). Ionizing radiation stimulates unidentified tyrosine-specific protein kinases in humanB-lymphocyte precursors, triggering apoptosis and clonogenic cell death. *Proc. Natl. Acad. Sci. U.S.A.* **89,** 9005–9009.

van der Hoeven, F., Schimmang, T., Volkmann, A., Mattei, M.-G., Kyewski, B., and Rüther, U. (1994). Programmed cell death is affected in the novel mouse mutant Fused toes (Ft). *Development (Cambridge, UK)* **120,** 2601–2607.

Vaux, D. L. (1993). Toward an understanding of the molecular mechanisms of physiological cell death. *Proc. Natl. Acad. Sci. U.S.A.* **90,** 786–789.

Vaux, D. L., Weissman, I. L., and Kim, S. K. (1992). Prevention of programmed cell death in *Caenorhabditis elegans* by human *bcl-2*. *Science* **235,** 1955–1957.

Vaux, D. L., Haecker, G., and Strasser, A. (1994). An evolutionary perspective on apoptosis. *Cell (Cambridge, Mass.)* **76,** 777–779.

Veis, D. J., Sorenson, C. M., Shutter, J. R., and Korsmeyer, S. J. (1993). Bcl-2-deficient mice demonstrate fulminant lymphoid apoptosis, polycystic kidneys, and hypopigmented hair. *Cell (Cambridge, Mass.)* **75,** 229–240.

Vermeij-Keers, C., and Poelmann, R. E. (1980). The neural crest, a study on cell degeneration and the improbability of cell migration in mouse embryos. *Neth. J. Zool.* **30,** 74–81.

Villa, P., Miehe, M., Sensenbrenner, M., and Pettman, B. (1994). Synthesis of specific proteins in trophic factor-deprived neurons undergoing apoptosis. *J. Neurochem.* **62,** 1468–1475.

Voris, B. P., and Young, D. A. (1981). Glucocorticoid-induced proteins in rat thymus cells. *J. Biol. Chem.* **256,** 11319–11329.

Wadewitz, A. G., and Lockshin, R. A. (1988). Programmed cell death, dying cells synthesize a co-ordinated, unique set of proteins in two different episodes of cell death. *FEBS Lett.* **241,** 19–23.

Walker, N. I., Harmon, B. V., Gobe, G. C., and Kerr, J. F. R. (1988). Patterns of cell death. *Methods Achiev. Exp. Pathol.* **13,** 18–54.

Walker, P. R., Kokileva, L., LeBlanc, J., and Sikorska, M. (1993a). Detection of the initial stages of DNA fragmentation in apoptosis. *BioTechniques* **15,** 1032–1036.

Walker, P. R., Kwast-Welfeld, J., and Sikorska, M. (1993b). Relationship between apoptosis and the cell cycle. *In* "Programmed Cell Death: The Cellular and Molecular Biology of Apoptosis" (M. Lavin and D. Watters, eds.), pp. 59–72. Harwood Academic Publishers, Chur, Switzerland.

Wang, L., Miura, M., Bergeron, L., Zhu, H., and Yuan, J. (1994). *Ich-1,* an *Ice/ced-3*-related gene, encodes both positive and negative regulators of programmed cell death. *Cell (Cambridge, Mass.)* **78,** 739–750.

Waring, P. (1990). DNA fragmentation induced in macrophages by gliotoxin does not require protein synthesis and is preceded by raised inositol triphosphate levels. *J. Biol. Chem.* **265,** 14476–14480.

Webb, J. N. (1972). The development of human skeletal muscle with particular reference to muscle cell death. *J. Pathol.* **106,** 221–228

Webb, J. N. (1974). Muscular dystrophy and muscle cell death in normal foetal development. *Nature (London)* **252,** 233–234.

Weber, R. (1964). Ultrastructural changes in regressing tail muscles of *Xenopus* larvae at metamorphosis. *J. Cell Biol.* **22,** 481–487.

Weeks, J. C., Davidson, S. K., and Debu, B.H. G. (1992). Effects of a protein synthesis inhibitor on the hormonally mediated regression and death of motoneurons in the tobacco hornworm, *Manduca sexta. J. Neurobiol.* **24,** 125–140.

Weill, C. L. (1991). Somatostatin (SRIF) prevents natural motoneuron cell death in embryonic chick spinal cord. *Dev. Neurosci.* **13,** 377–381.

Welsh, A. O., and Enders, A. C. (1987). Trophoblast-decidual cell interactions and the establishment of maternal blood circulation in the parietal yolk sac placenta of the rat. *Anat. Rec.* **217,** 203–219.

White, K., Grether, M. E., Abrams, J. M., Young, L., Farrell, K., and Steller, H. (1994). Genetic control of programmed cell death in *Drosophila. Science* **264,** 677–683.

Wijsman, J. H., Jonker, R. R., Keizer, R., van de Velde, C.J. H., Cornelisse, C. J., and van Dierendonck, J. H. (1993). A new method to detect apoptosis in paraffin sections, *in situ* end-labeling of fragmented DNA. *J. Histochem. Cytochem.* **41,** 7–12.

Williams, R. W., and Rakic, P. (1988). Elimination of neurons from the Rhesus monkey's lateral geniculate nucleus during development. *J. Comp. Neurol.* **272,** 424–436.

Williams, G. T., Smith, C. A., Spooncer, E., Dexter, T. M., and Taylor, D. R. (1990). Haemopoietic colony stimulating factors promote cell survival by suppressing apoptosis. *Nature (London)* **343,** 76–79.

Wilson, C. A., di Clemente, N., Ehrenfels, C., Pepinsky, R. B., Josso, N., Vigier, B., and Cate, R. L. (1993). Mullerian inhibiting substance requires its N-terminal domain for maintenance of biological activity, a novel finding within the transforming growth factor-beta superfamily. *Mol. Endocrinol.* **7,** 247–257.

Wolff, T., and Ready, D. F. (1991). Cell death in normal and rough eye mutants of *Drosophila. Development (Cambridge, UK)* **113,** 825–839.

Wolszon, L. R., Rehder, V., Kater, S. B., and Macagno, E. R. (1994). Calcium wave fronts that cross gap junctions may signal neuronal death during development. *J. Neurosci.* **14,** 3437–3448.

Wong, R. O., and Hughes, A. (1987). Role of cell death in the topogenesis of neuronal distributions in the developing cat retinal ganglion cell layer. *J. Comp. Neurol.* **262,** 496–511.

Wride, M. A., and Sanders, E. J. (1993). Expression of tumor necrosis factor-α (TNFα)-cross-reactive proteins during early chick embryo development. *Dev. Dyn.* **198,** 225–239.

Wride, M. A., and Sanders, E. J. (1995). Potential roles for tumour necrosis factor-α during embryonic development. *Anat. Embryol.* **191,** 1–10.

Wride, M. A., Lapchak, P. H., and Sanders, E. J. (1994). Distribution of TNFα-like proteins correlates with some regions of programmed cell death in the chick embryo. *Int. J. Dev. Biol.* **38,** 673–682.

Wu, X., and Levine, A. J. (1994). p53 and E2F-1 cooperate to mediate apoptosis. *Proc. Natl. Acad. Sci. U.S.A.* **91,** 3602–3606.

Wyllie, A. H. (1980). Glucocorticoid-induced thymocyte apoptosis is associated with endogenous endonuclease activation. *Nature (London)* **284,** 555–556.

Wyllie, A. H. (1981). Cell death, a new classification separating apoptosis from necrosis. *In* "Cell Death in Biology and Pathology" (I. D. Bowen and R. A. Lockshin, eds.), pp. 9–34. Chapman & Hall, London.

Wyllie, A. H. (1992). Apoptosis and the regulation of cell numbers in normal and neoplastic tissues: An overview. *Cancer Metastasis Rev.* **11,** 95–103.

Wyllie, A. H., and Morris, R. G. (1982). Hormone-induced cell death. *Am. J. Pathol.* **109,** 78–87.

Wyllie, A. H., Kerr, J.F. R., and Currie, A. R. (1980). Cell death, the significance of apoptosis. *Int. Rev. Cytol.* **68,** 251–306.

Wyllie, A. H., Morris, R. G., Smith, A. L., and Dunlop, D. (1984). Chromatin cleavage in apoptosis, association with condensed chromatin morphology and dependence on macromolecular synthesis. *J. Pathol.* **142,** 67–77.

Wyllie, A. H., Rose, K. A., Morris, R. G., Steel, C. M., Foster, F., and Spandidos, D. A. (1987). Rodent fibroblast tumours expressing human myc and ras genes, growth, metastasis and endogenous oncogene expression. *Br. J. Cancer* **56,** 251–259.

Yonish-Rouach, E., Resnitzky, D., Lotem, J., Sachs, L., Kimchi, A., and Oren, M. (1991). Wild-type p53 induces apoptosis of myeloid leukaemic cells that is inhibited by interleukin-6. *Nature (London)* **352,** 345–347.

Young, R. W. (1984). Cell death during differentiation of the retina in the mouse. *J. Comp. Neurol.* **229,** 362–373.

Yuan, J., and Horvitz, H. R. (1990). The *Caenorhabditis elegans* genes ced-3 and ced-4 act cell autonomously to cause programmed cell death. *Dev. Biol.* **138,** 33–41.

Yuan, J., and Horvitz, H. R. (1992). The *Caenorhabditis elegans* cell death gene ced-4 encodes a novel protein and is expressed during the period of extensive programmed cell death. *Development (Cambridge, UK)* **116,** 309–320.

Yuan, J., Shaham, S., Ledoux, S., Ellis, H. M., and Horvitz, H. R. (1993). The *C. elegans* cell death gene ced-3 encodes a protein similar to mammalian interleukin-1β-converting enzyme. *Cell (Cambridge, Mass.)* **75,** 641–652.

Zakeri, Z. F., Quaglino, D., Latham, T., and Lockshin, R. A. (1993). Delayed internucleosomal DNA fragmentation in programmed cell death. *FASEB J.* **7,** 470–478.

Zakeri, Z. F., Quaglino, D., and Ahuja, H. S. (1994). Apoptotic cell death in the mouse limb and its suppression in the hammertoe mutant. *Dev. Biol.* **165,** 294–297.

Zheng, L. M., Zychlinsky, A., Liu, C. C. L., Ojcius, D. M., and Young, J. D.-E. (1991). Extracellular ATP as a trigger for apoptosis or programmed cell death. *J. Cell Biol.* **112,** 279–288.

Biochemistry and Molecular Biology of Chromoplast Development

Bilal Camara,* Philippe Hugueney,* Florence Bouvier,* Marcel Kuntz,*
and René Monégert
*Institut de Biologie Moléculaire des Plantes du Centre National de la
Recherche Scientifique and Université Louis Pasteur, 67087
Strasbourg Cedex, France; and †Laboratoire de Pathologie et Biochimie
Végétales, Université Pierre et Marie Curie, 75025 Paris Cedex, France

Plant cells contain a unique class of organelles, designated the plastids, which distinguish them from animal cells. According to the largely accepted endosymbiotic theory of evolution, plastids are descendants of prokaryotes. This process requires several adaptative changes which involve the maintenance and the expression of part of the plastid genome, as well as the integration of the plastid activity to the cellular metabolism. This is illustrated by the diversity of plastids encountered in plant cells. For instance, in tissues undergoing color changes, i.e., flowers and fruits, the chromoplasts produce and accumulate excess carotenoids. In this paper we attempt to review the basic aspects of chromoplast development.

KEY WORDS: Chromoplast differentiation, Biogenesis, Molecular biology.

I. Introduction

Plants, in contrast to animals, contain a unique plastid organelle which, according to the endosymbiotic theory, derives from Cyanobacteria. Due to the totipotency of plant cells, the plastid is subject to considerable structural and biochemical adaptative changes, which reflect the prevailing physiological state of the cell. Consequently, several interconvertible types of plastids with specialized functions have been described (Whatley, 1978; Thomson and Whatley, 1980). Among them, chromoplasts observed in flowers, fruits, roots, and stressed leaves are characterized by an increased

175

accumulation of carotenoids (Schimper, 1885; Whatley and Whatley, 1987). Partial reviews covering special facets of chromoplast development are available (Sitte *et al.*, 1980; Camara *et al.*, 1989; Marano *et al.*, 1993; Price *et al.*, 1993). The aim of this review is to offer an integrated view of chromoplast biology. A detailed account is given of previous observations as a background for understanding the information derived from recent biochemical and molecular approaches and for raising areas that deserve further studies.

II. Consistent Organization
   of Chromoplast Morphotypes

According to the concept of plastid continuity (Schimper, 1885), plastids arise from preexisting plastids that are transmitted from parent cells to daughter cells during sexual or vegetative reproduction. The number of plastids for a given plant cell is generally constant, and plastid division usually is correlated to the growing phase of the cell (Possingham *et al.*, 1988). Chromoplasts generally occur in mature tissue and derive from preexisting mature plastids (Zurzycki, 1962), as one could note during the chloroplast to chromoplast differentiation in pepper fruit (*Capsicum annuum*) (Fig. 1). Consequently, it has usually been inferred that chromoplasts do not divide. The constriction of the central region of *C. annuum* chromoplast, shown in Fig. 2, is characteristic of dividing plastids (Leech and Pyke, 1988). This observation, in connection with similar data described for *Forsythia suspensa* petals (Sitte, 1987), demonstrates that chromoplasts also divide and prefigures their dynamic roles.

Data concerning chromoplast differentiation can be inferred from light microscopy observations (Kraus, 1872) indicating that chlorophyll bodies in green tomato turn yellow during the ripening period. Subsequent studies (Schimper, 1885; Courchet, 1888) allowed the classification of chromoplasts into four main types: (1) chromoplasts composed of proteic stroma with granules, (2) chromoplasts composed of protein crystals and amorphous pigment granules, (3) chromoplasts composed of protein and pigment crystals, (4) chromoplasts containing only crystals. Later, with the advent of electron microscopy and improvements in methods, the configuration of chromoplast substructures (globules, crystals, membranes, fibrils, tubules) was better resolved; it was observed that not a single chromoplast substructure exists in the same plastid, a phenomenon illustrated by *Aglaonema commutatum* chromoplasts which contain membranous and crystalline substructures (Knoth, 1981). Subsequently, the data compiled from several studies (Steffen, 1955; Steffen and Walter, 1955; Simpson, 1974; Sitte, 1974;

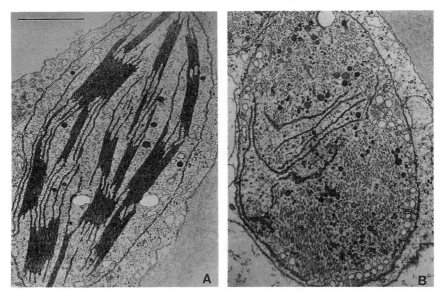

FIG. 1   Electron micrograph of chloroplast (A) and chromoplast (B) from normal *Capsicum* fruits. The chloroplast thylakoid is totally disorganized in the chromoplast. Bar = 1 μm.

Sitte *et al.*, 1980) allowed the distinction of five main categories based on the frequency of different chromoplast substructures encountered within the same plastid.

A. Globular Chromoplasts

Globular chromoplasts developing in several plants (Table I) are mainly characterized by the accumulation of plastoglobules which appear in the plastid stroma as spheroidal osmiophile bodies (Lichtenthaler, 1968). Their chemical composition, which is considered later, allows their distinction from plastoglobules usually observed in normal and senescing chloroplasts. The simplicity of the organization of plastoglobules can be taken as an argument to classify the globular chromoplasts as the most primitive morphotype. This view is reinforced by the fact that, in some groups of plants such as alga, bryophytes, pteridophytes, and gymnosperms, primitive chromoplast-like plastids are found in which excess carotenoids are dissolved in lipid globules (Whatley, 1983). In addition, similar substructures develop in *Flavobacterium* mutants producing excess carotenoids (Bauer *et al.*, 1974).

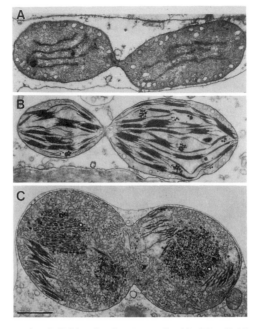

FIG. 2 Aspects of constricted division in *Capsicum* plastids. The divisions are observed at the proplastid (A), chloroplast (B), and chromoplast (C) stages. Bar = 1 μm.

B. Crystalline Chromoplasts

Crystalline chromoplasts which are widely represented in plants (Table II) can be subdivided into three categories based on their aspect after electron microscopy.

1. Crystalline Chromoplasts Containing Large Crystals of β-Carotene

In these chromoplasts the crystalline substructures are generally characterized by the fact that after fixation using glutaraldehyde and osmium tetroxide, they appear as empty spaces surrounded by a membrane structure. These structures have been observed in the root of *Daucus carota* (Ben-Shaul *et al.,* 1968; Israel and Steward, 1967). A very similar structure has been described for the fungus *Clathrus cancellatus,* which sequesters excess carotene in the cytoplasm (Eymé and Parriaud, 1970). On the other hand, when the fixation is carried out using potassium permanganate or glutaraldehyde/permanganate, the β-carotene crystals appear as elongated and frequently electron-dense spindles (Trabucci, 1964; Grilli, 1965a,b), probably due to the deposition of insoluble manganese dioxide (Sitte, 1977).

TABLE I

Occurrence of Globular Chromoplasts in Plants

Species	Organ	References[a]
Allamanda neriifolia	Aril	1
Aloë plicatis	Petal	2
Alyssum saxatile	Petal	1
Berberis vulgaris	Petal	1
Brassica napus	Petal	1
Caltha palustris	Petal	1
Capsicum annuum, yellow and orange cultivar	Fruit	3
Cheiranthus allionii	Petal	4,5
Chimonanthus praecox	Petal	1
Chrysanthemum segetum	Petal	6
Chrysosplenium alternifolium	Sepal	1
Chrysosplenium oppositifolium	Sepal	1
Citrus aurantium	Fruit	7
Citrus sinensis	Fruit	8–14
Convallaria majalis	Fruit	15
Cotoneaster microphylla	Fruit	16
Cucurbita pepo	Fruit	17–21
Daphne alpina	Anther	22
Daphne altaica	Anther	22
Daphne arbuscula	Anther	22
Daphne laureola	Anther	22
Daucus carota, yellow and white mutants	Root	23
Doronicum latifolium	Petal	24
Dorinuccum orientale	Petal	1
Eranthis hiemalis	Petal	1
Euphorbia cyparissias	Leaves	1
Ficaria ranunculoides	Nectary	25
Forsythia ovata	Petal	26
Gerbera jamesonii	Bract	27
Hamamelis mollis	Petal	1
Heliopsis helianthoides	Petal	28,29
Hemerocallis flava	Tepal	1
Iris germancia	Tepal	30
Iris pseudacorus	Tepal	1
Jasminum nudiflorum	Petal	1,31
Lamium galeobdolon	Petal	1

(*continued*)

TABLE I (*Continued*)

Species	Organ	References[a]
Lilium longiflorum	Anther	32
Lilium martagon	Tepal	1
Lycopersicon esculentum	Petal	33
Lycopersicon esculentum, yellow cultivar	Fruit	34
Lycopersicon esculentum, ghost mutant	Fruit	35
Mahonia aquifolium	Petal	1
Mahonia ilicifolia	Petal	31
Malpighia glabra	Fruit	36
Melampyrum nemorosum	Petal	1
Mimulus luteus	Petal	1
Musa sp	Fruit	37
Nuphar lutea	Sepal	27
Oenothera suaveolens	Petal	38
Orobanche fuliginosa	Stem	39
Orobanche hederae	Stem	40
Orobanche ramosa	Stem	41
Oxalis stricta	Petal	42
Oxalis succulenta	Petal	1
Physalis alkekengi	Fruit, sepal	28
Physalis franchetii	Fruit	14
Pyrus malus	Fruit	43
Ranunculus acer	Petal	44
Ranunculus repens	Petal	45
Rhinanthus alectorolophus	Petal	1
Sarothamus scoparius	Petal	28,29
Solanum luteum	Fruit	46
Solanum tuberosum	Fruit	7
Spartium junceum	Petal	47
Tagetes erecta	Petal	48
Taraxacum officinale	Petal	1
Taxus baccata	Aril	49
Tulipa gesneriana	Tepal	50
Tulipa sylvestris	Tepal	1
Tussilago farfara	Petal	1
Vinca rosea	Petal	4
Viola tricolor	Petal	28
Zantedeschia elliottiana	Spathe	27

(*continued*)

TABLE I (Continued)

[a] 1. Sitte et al. (1980); 2. Steffen and Walter (1958); 3. Kirk and Juniper (1967); 4. Gourret (1971); 5. Gourret and Le Normand (1971); 6. Lance-Nougarède (1960); 7. Sitte (1974); 8. Salema (1968); 9. Thomson (1965); 10. Thomson (1966b); 11. Thomson (1966a); 12. Thomson (1969); 13. Thomson et al. (1967); 14. Osumi (1961); 15. Steffen (1964a); 16. Simpson (1974); 17. Grilli (1965a); 18. Devidé (1970); 19. Devidé and Ljubesic (1972); 20. Ljubesic (1970); 21. Matienco (1967); 22. Garagaty-Feissly (1970); 23. Klein and Ben Shaul (1967); 24. Schimper (1885); 25. Eymé (1967); 26. Ledbetter and Porter (1970); 27. Grönegress (1974); 28. Lichtenthaler (1969); 29. Lichtenthaler (1971); 30. Buvat (1969); 31. Courchet (1888); 32. Dickinson (1973); 33. Laborde and Spurr (1973); 34. Harris and Spurr (1969a); 35. Scolnik et al. (1987); 36. Arnott and Smith (1967); 37. Mesquita et al. (1980); 38. Schötz (1961); 39. Kollmann et al. (1969); 40. Dodge and Lawes (1974); 41. Laudi and Albertini (1967); 42. Guilliermond (1919); 43. Catesson (1970); 44. Frey-Wyssling and Kreutzer (1958a); 45. Lichtenthaler (1966); 46. Simpson et al. (1975b); 47. Nougarède (1964); 48. Steffen (1964b); 49. Camefort (1964); 50. Lichtenthaler (1970b).

2. Crystalline Chromoplasts Containing Small Crystals of β-Carotene

In some cases, the chromoplast substructures consist of small crystals of β-carotene which are apparently better preserved after osmium tetroxide fixation. These substructures can be observed in the chromoplasts of the tomato mutant *high-beta* (Harris and Spurr, 1969a), in carrot root (Wrischer, 1972), and in *Cucurbita pepo* fruits (Devidé, 1970).

3. Crystalline Chromoplasts Containing Lycopene Crystals

Lycopenic crystals are not well preserved during osmium tetroxide fixation, and appear in the chromoplast stroma as an undulating structure enclosed in an electron-transparent space surrounded by a membrane structure (Ben-Shaul and Naftali, 1969; Harris and Spurr, 1969a). Potassium permanganate fixation reveals the lycopenic structures as electron-dense sheets or as laminated structures (Rosso, 1968; Walles, 1971).

C. Fibrillar and Tubular Chromoplasts

The ultrastructures of fibrillar and tubular chromoplast substructures observed in several plants (Table III) were first described by Steffen (1955) and Steffen and Walter (1955) from the fruits of *Rosa canina* and *Solanum capsicastrum*. Data compiled from several systems allow subdivision of this group into three categories.

TABLE II

Occurrence of Crystalline Chromoplasts in Plants

Species	Organ	References[a]
Citrullis vulgaris	Fruit	1
Clivia nobilis	Tepal	2
Cucurbita maxima	Fruit	3,4
Cucurbita pepo cv. ovifera	Fruit	5
Cucurbita pepo cv. small sugar	Fruit	3,4,6
Daucus carota	Root	7–20
Hemerocallis fulva	Tepal	2
Hordeum vulgare (*tig-o*[34] mutant)	Leaf	21
Lycopersicon esculentum	Fruit	3, 22–34
Lycopersicon esculentum (*high-beta* mutant)	Fruit	22,23,35
Lycopersicon esculentum (*high-delta* mutant)	Fruit	22,23
Lycopersicon esculentum (*tangerine* mutant)	Fruit	24,25
Narcissus poeticus	Petal	36–39
Physalis pubescens	Fruit	40
Prunus persica	Receptacle	41
Solanum aviculare	Fruit	16
Solanum pseudocapsicum	Fruit	40
Zea mays (lycopenic mutant)	Leaves	42,43

[a] 1. Knoth (1981); 2. Matienco *et al.* (1968); 3. Sitte *et al.* 1980); 4. Grilli (1964); 5. Grilli (1965a); 6. Devidé (1970); 7. Grilli (1965b); 8. Roberts (1946); 9. Roberts and Southwick (1948); 10. Ben-Shaul (1962); 11. Trabucci (1964); 12. Frey-Wyssling and Schwegler (1965); 13. Tôyama (1967); 14. Israel *et al.* (1969); 15. Nedukha (1969); 16. Wu and Salunkhe (1971); 17. Grönegress (1971); 18. Mühlethaler (1971); 19. Wrischer (1972); 20. Jordan and Chapman (1973); 21. von Wettstein *et al.* (1971); 22. Grilli (1965c); 23. Harris (1967); 24. Harris (1968); 25. Rosso (1967a); 26. Rosso (1967b); 28. Rosso (1968); 28. Ben-Shaul and Shmueli (1968); 29. Ben-Shaul and Naftali (1969); 30. Harris and Spurr (1969a); 31. Spurr (1969); 32. Khudairi (1972); 33. Laval-Martin (1969); 34. Laval-Martin (1974); 35. Harris (1970); 36. Frey-Wyssling (1967); 37. Kuhn (1970); 38. Frey-Wyssling and Kuhn (1969); 39. Kuhn *et al.* (1969); 40. Salema (1968); 41. Eymé (1971); 42. Walles (1971); 43. Walles (1972).

1. Fibrillar Chromoplast Substructures Organized into Bundles

Representative types are illustrated by chromoplasts encountered in *R. canina* fruits, in *S. capsicastrum* fruits (Steffen, 1955; Steffen and Walter, 1955) in *Strelitzia reginae* sepals (Bornman, 1968), or in *Nuphar luteum* sepals (Grönegress, 1974). In the latter case, the fibrils appear as spindle-shaped bundles which are aligned in parallel, and sometimes show a swell-

TABLE III

Occurrence of Fibrillar and Tubular Chromoplasts in Plants

Species	Organ	References[a]
Agrimonia eupatoria	Petal	1
Asparagus officinalis	Fruit	2,3
Capsicum annuum	Fruit	4–15
Celastrus scandens	Aril	16
Chelidonium majus	Petal	1
Citrus unshiu	Fruit	17,18
Crataegus ellwangeriana	Fruit	19
Cucumis sativus[b]	Petal	20
Cucurbita moschata	Fruit	21
Cucurbita pepo[b]	Petal	22–24
Cucurbita pepo[b]	Fruit	21
Euonymus europaeus	Aril	1
Hemerocallis liliousphodelus	Petal	19
Oxalis europaea	Petal	1
Palisoata barteri[b]	Petal	25
Physalis peruviana	Fruit	3
Physalis pubescens	Fruit	8
Pittosporum rhombifolium	Fruit	19
Rhododendron flavum	Petal	11
Rosa canina	Fruit	26
Rosa rugosa	Fruit	27
Solanum capsicastrum	Sepal	28
Solanum capsicastrum	Fruit	29
Solanum pseudocapsicum	Fruit	8
Sorbus aucuparia	Fruit	19
Sorbus intermedia	Fruit	1
Strelitzia reginae	Sepal	30
Tropaeolum majus[b]	Petal	31

[a] 1. Sitte *et al.* (1980); 2. Rezende-Pinto and Salema (1959); 3. Simpson *et al.* (1977c); 4. Frey-Wyssling and Kreutzer (1958b); 5. Grob (1963); 6. Kirk and Juniper (1967); 7. Clowes and Juniper (1968); 8. Salema (1968); 9. Spurr and Harris (1968); 10. Laborde (1969); 11. Lichtenthaler (1969); 12. Spurr (1970); 13. Laborde and Spurr (1973); 14. Suzuki (1974); 15. Camara and Brangeon (1981); 16. Bornman (1968); 17. Yuasa (1960); 18. Osumi (1961); 19. Simpson (1974); 20. Smith and Butler (1971); 21. Ljubesic (1977); 22. Matienco (1965); 23. Grilli (1964); 24. Grilli, (1965c); 25. Knoth *et al.* (1986); 26. Steffen and Walter (1955); 27. Wuttke (1976); 28. Steffen and Walter (1958); 29. Rezende-Pinto and Salema (1959); 30. Simpson *et al.* (1975a); 31. Falk (1976).

[b] Tubular chromoplast.

ing protuberance due to elongated plastoglobules which probably initiate their formation. At higher magnification, one can observe that the fibrillar substructures are composed of microfibrils (Simpson and Lee, 1976a).

2. Fibrillar Chromoplasts Organized into Dispersed Substructures

This second type, observed in *Celastrus scandens* aril (Bornman, 1968), in *Physalis pubescens* fruits (Salema, 1968), and in *Asparagus officinalis* fruits (Simpson *et al.,* 1977c), is characterized by the persistence of large plastoglobules connected to fibrils in the mature chromoplast. In addition, the fibrils are not organized in bundles.

3. Tubular Chromoplasts

The tubular chromoplast morphotype, at lower magnification, resembles the fibrillar morphotype. However, at higher magnification no microfibrillar structure is observed in tubular chromoplasts. The tubules are generally shorter than fibrils. Representative types have been described in *Cucumis sativus* petals (Smith and Butler, 1971), in *Cucurbita pepo* fruits and petals (Grilli, 1965c), in *Cucurbita maxima* fruits (Ljubesic, 1977), in *Tropaeolum majus* petals (Falk, 1976), and in *Palisota barteri* fruits (Knoth *et al.,* 1986) (Table III).

D. Membranous Chromoplasts

These chromoplast morphotypes are characterized by an extended development of concentric membranes and a low plastoglobule content. Typical examples are observed in *Narcissus pseudonarcissus* corona (Mollenhauer and Kogut, 1968; Kowallik and Herrmann, 1972; Mesquita, 1976; Liedvogel *et al.,* 1976) and in *Calceolaria rugosa* petals (Wrischer and Ljubesic, 1984). In both cases, the concentric membrane substructures apparently derive from the inner envelope membrane of rudimentary chloroplasts (Wrischer and Ljubesic, 1984), as shown by their positive reaction to caffeic acid (Vaughn and Wilson, 1981) and their negative reaction after diaminobenzidine treatment (Wrischer, 1978). Surprisingly, the internal chromoplast substructures can develop in the absence of carotenoid (Stead and Duckett, 1980; Liedvogel and Falk, 1980).

E. Reticulo-Tubular Chromoplasts

These chromoplasts display a network of twisted fibrils which fill the stroma, in addition to few plastoglobules. Two cases have been described which include the spadice-appendix chromoplasts of *Typhonium divaricatum* (Schnepf and Czygan, 1966) and the petal chromoplasts of *Liriondendron tulipifera* L. (Ljubesic, 1979).

III. Organization of Chromoplast-like Plastids

A. Senescing Plastids or Gerontoplasts

The plastids of senescing leaves (Ikeda and Ueda, 1964; Thomson *et al.*, 1964; Barton, 1966; Butler and Simon, 1971) or "autumn" leaves (Grob and Eichenberger, 1962; Tôyama and Ueda, 1965) have been considered as chromoplasts. In fact, the observed phenotypes are due to the degradation of the chlorophylls and the subsequent unmasking of preexisting carotenoids. They no longer represent true chromoplasts and have been designated gerontoplasts (Sitte *et al.*, 1980).

B. Chlorochromoplasts

In contrast to senescing plastids, the term chlorochromoplast is used for plastids which, in the absence of any mutation, synthesize chromoplast carotenoids such as rhodoxanthin, while preserving part of their chlorophylls (Buscalioni and Bruno, 1928; Savelli, 1938a,b, 1960). Their distinction is facilitated by the fact that they are largely confined to gymnosperm families (Ida, 1981). Some representative species are listed in Table IV. The plastid stroma of chlorochromoplasts contain numerous plastoglobules, like the globular chromoplast morphotypes. These plastoglobules could represent the site of rhodoxanthin deposit.

C. Chromoplast-like Plastids in Pathogenic or Symbiotic Tissues

The normal pattern of chloroplast differentiation is altered during pathogenic infection. Under these conditions the presence of chromoplast-like plastids has been described after viral infection (Arnott and Smith, 1967; Arnott *et al.*, 1969; Kitajima and Costa, 1973), fungal infection (Orcival, 1968; Heath, 1974), or *Mycoplasma* infection (Gourret, 1971; Gourret and Le Normand, 1971). The plastids of these infected tissues, like normal chromoplasts, are characterized by disorganization of the thylakoid membranes, grana loss, accumulation of plastoglobules, and degradation of chlo-

TABLE IV

Plant Leaves Containing Chlorochromoplasts

Species	References[a]
Adoxa sp	1
Aloë sp	2,3
Apica sp	1
Biota orientalis	4
Bulbine sp	5
Buxus sp	3
Chamaecyparis obtusa	4
Cryptomeria sp	4
Equisetum sp	4
Gasteria sp	6
Hawortia sp	3
Juniperus sp	5
Kniphofia sp	3
Marsilia sp	3,4
Metasequoia glyptostroboides	4
Potamogeton sp	6,7
Reseda sp	6
Sciadopitys verticillata	4
Scirpus sp	5
Selaginella sp	4
Sequoiadendron giganteum	4
Shishindenia oricoides	4
Taxus sp	4,5
Thuja sp	7

[a] 1. Geitler (1937); 2. Buscalioni and Bruno (1928); 3. Savelli (1938b); 4. Ueda and Momose (1968); 5. Lippmaa (1926b); 6. Lippmaa (1926a); 7. Prat (1924).

rophyll. However, increased carotenoid synthesis, which is a characteristic of true chromoplasts, is not usually observed. Indeed, Arnott *et al.* (1969) suggested that the disappearance of chlorophyll is due to photooxidation which follows the inhibiton of the carotenoid biosynthetic pathway by tobacco mosaic viruses. Very similar observations have been reported in the case of *Tussilago farfara* infected by the fungus *Puccinia poarum* (Orcival, 1968). During the interaction between the leaves of *Vigna sinensis* and the fungus *Uromyces phaseoli,* Heath (1974) observed, in addition to the accumulation of plastoglobules, the formation of new membrane structures resembling the lycopene crystals observed in normal chromoplasts. This phenomenon has been correlated with the emission of ethylene, which is a potential inducer of chromoplast formation. Also during endomycorrhizal root formation in *Ornithogalum umbellatum,* one of the plant responses observed involves the induction of chromoplast differentiation. This phenomenon is paralleled by the reduction of starch accumulation and by the induction of carotenoid synthesis concomitant to the accumulation of plastoglobules (Scannerini and Bonfante-Fasolo, 1977). The signaling mechanism involved in these changes is not known, but in the case of fungi, this could involve the endogenous liberation of polysaccharides (Dexheimer *et al.,* 1990). Overall these phenomena clearly indicate that pathogens can interfere with the normal signaling pathway involved in chromoplast differentiation.

D. Chromoplast-like Plastids in Lycopenic or ζ-Carotenic Plant Mutants

Several lycopenic maize (Faludi-Daniel *et al.,* 1968; Walles, 1971, 1972) and barley mutants (von Wettstein *et al.,* 1971; Nielsen, 1974a,b) or ζ-carotenic maize mutants (Faludi-Daniel, 1975) grow on the autotrophic mode under low light intensity (below 50 lux) and accumulate less than one-third of their total chlorophyll content (Faludi-Daniel *et al.,* 1966). When these mutants are subjected to higher intensities, one can note the bleaching of the chlorophylls and the subsequent differentiation of the chloroplasts, either into crystalline chromoplasts accumulating lycopene or into rudimentary membranous chromoplasts containing ζ-carotene.

IV. General Characteristics of Isolated Chromoplasts

A. Isolation of Chromoplasts and Chromoplast Substructures

Our present understanding of chromoplast biochemistry owes much to the development of techniques for the isolation of intact chromoplasts. The *Narcissus* (Liedvogel *et al.,* 1976) and the pepper (Camara, 1985a; 1993) systems have been widely used for this purpose. In both cases, it has been realized that intact chromoplasts band at low buoyant density, compared to chloroplasts, during centrifugation in sucrose gradients. In contrast, chromoplasts do not enter or poorly enter Percoll gradient, unless extreme dilutions of the Percoll solution are achieved (Hansmann and Sitte, 1984). For detailed analysis, disrupted chromoplasts subfractionated by ultracentrifugation allow the isolation of different chromoplast substructures, which include globules (Hansmann and Sitte, 1982), tubules (Winkenbach *et al.,* 1976; Knoth *et al.,* 1986), fibrils (Deruère *et al.,* 1994), and internal membranes (Liedvogel *et al.,* 1976; Bouvier *et al.,* 1994).

B. Biochemical Composition

1. Carotenoids and Prenyllipids

During chloroplast to chromoplast transformation, there is an enhanced synthesis of carotenoids, which usually involves the apparition of new carotenoid structures typical of chromoplasts (Table V). this phenomenon is usually paralleled by an increased esterification of xanthophylls by fatty acids (Eilati *et al.,* 1975; Camara and Monéger, 1978), thus enhancing their lipophilicity and their subsequent dissolution or sequestration into various plastid substructures. These later modifications of the carotenoid molecules are largely restricted to chromoplasts, but have been described in senescing chloroplasts of leaves of deciduous trees (Goodwin, 1958) and in drought-stressed barley seedlings (Barry *et al.,* 1992). The accumulation of massive amounts of carotenoids in chromoplasts is usually paralleled by an increased synthesis of tocopherol (Lichtenthaler, 1977; Camara *et al.,* 1982b), which probably exerts an antioxidant role.

2. Acyl Lipids

As for chloroplasts, the monogalactosyldiglyceride and the digalactosyldiglyceride species represent the major polar lipid (Table VI). The galactolipid composition of chromoplasts is reminiscent of plastid envelope lipids (Douce and Joyard, 1990), and reflects the degradation of the thylakoid lipids, in the case where the differentiation process involves chloroplast to chromoplast transition. This trend is further supported by the absence of the fatty acid Δ3-trans-hexadecenoate in the phosphatidylglycerol (Camara and Monéger, 1977; Whitaker, 1986). The phosphatidylethanolamine content is very low, arguing against the presence of significant cytosolic contaminations in purified chromoplasts so far analyzed.

3. Biochemical Composition of Chromoplast Substructures

The lipophilic nature of plastoglobules was noted during the first studies which demonstrated that they could be isolated as a cream suspension in

TABLE V

Percentage Carotenoid Composition of Different Chromoplast Morphotypes

Carotenoids	Chloroplast (1)	Globular chromoplast (2)	Fibrillar chromoplast (3)	Crystalline chromoplast (4)	Membranous chromoplast (5)	Chlorochromoplast (6)
Phytoene			3.5			
Phytofluene			1.5	7.2		
ζ-Carotene			2.0			
Lycopene				82.9		
α-Carotene			tr			4.7
β-Carotene	24.0	29.0	7.0	6.3	tr	15.3
β-Carotene-5,6-epoxide		23.6				
β-Crytoxanthin	9.1		6.2			
Cryptocapsin			3.2			
Rhodoxanthin						32.0
Lutein	34.9	31.1	2.6	1.1	>56.6	33.1
Lutein epoxide		7.5	3.4			
Zeaxanthin			5.2			
Antheraxanthin			3.8			
Violaxanthin	14.8		8.5	0.9	24.4	14.8
Capsanthin			35.3			
Capsanthin epoxide			1.8			
Capsorubin			9.3			
Neoxanthin	17.1	8.6	5.3		9	
Mutatoxanthin			1.4			
Chlorophylls	+[b]					+

[a] Data from: (1) green pepper fruits, Camara and Monéger (1978); (2) Solanum luteum, Simpson et al. (1975b); (3) pepper chromoplasts, Camara et al. (1983); (4) tomato fruit, Laval-Martin et al. (1975); (5) Narcissus pseudonarcissus, Liedvogel et al. (1976); and (6) Taxus cuspida, Ida (1981).

[b] The presence of chlorophylls is indicated by the plus sign.

TABLE VI

Acyl Lipid Composition (Percentage of Total Lipid) of Isolated *Capsicum* and *Narcissus* Chromoplasts

Lipid molecule	*Capsicum* (1)[a]	*Narcissus* (2)
Monogalactosyldiglyceride	39.1	63.1
Digalactosyldiglyceride	38.7	18.3
Sulfoquinovosyldiglyceride	6.1	5.4
Phosphatidylcholine	9.8	1.6
Phosphatidylethanolamine	2.0	0.6
Phosphatidylinositol	2.0	1.1
Phosphatidylglycerol	2.3	9.6
Phosphatidic acid	—	0.5

[a] 1. Camara *et al.* (1983); 2. Kleinig and Liedvogel (1978).

the upperlayer of osmotically shocked chloroplast subjected to ultracentrifugation. Biochemical analysis revealed that they consist of 95% lipid and 0.88% nitrogen on a dry weight basis (Bailey and Whyborn, 1963). Although the molecular species of the different lipids are not clearly established, due to the problem of cross-contamination with other subplastidial structures, photosynthetic carotenoid are not normal constituents of chloroplast plastoglobules (Bailey and Whyborn, 1963). An apparent exception to this trend has been observed in chloroplast plastoglobules of *Dunaliella* which accumulate neutral lipid and β-carotene when cultivated under high salt concentration (Ben-Amotz *et al.,* 1982). On the other hand, chromoplast plastoglobules contain carotenoids (Table VII), and become electron-translucent (Lichtenthaler, 1970a). This latter property has been confirmed using phytoene-enriched plastoglobule fractions, which derive from wheat seedlings treated with the herbicide Norflurazon (Dahlin *et al.,* 1983; Dahlin and Ryberg, 1986), and also in the case of cotton seedlings (Stegink and Vaughn, 1988) treated with Norflurazon. In general, there is a positive correlation between the increase of carotenoids and that of tocopherols in chromoplast plastoglobules (Lichtenthaler, 1971). The fact that plastoglobules contain mainly neutral lipids (Steinmüller and Tevini, 1985) argues for the absence of a bordering membrane in plastoglobules, in contrast to a previously held view (Grönegress, 1974; Laval-Martin, 1974). The absence of coalescence and the presence of nitrogen in plastoglobules, in connection with the aggregation of plastoglobules induced by pH below 5, indicates that protein may participate in their organization (Bailey and Whyborn, 1963). The involvement of protein is much more prominent in the organiza-

TABLE VII

Biochemical Composition of Chromoplast Substructures

Components	Globules (1)[a]	Tubules (2)	Fibrils (3)	Membranes (4)
Acyl lipids				
Triglyceride	53.8–69.2			
Galactolipids	1.15	0.8	0.5	2.1
Sulfolipid	—	—	—	0.1
Phospholipids	0.4	0.5	0.03	0.7
Prenyl lipids	0.8	tr	0.03	0.1
Carotenoids	5.4–9.2	0.6	0.4	0.2

[a] Data from *Caltha, Viola* and *Tulipa* (1), *Tropaeolum* and *Rosa* (2), and *Narcissus* (4) chromoplasts are expressed in milligrams dry weight per milligram of protein (Sitte, 1977), while data from pepper chromoplasts (3) are expressed in nanomoles per milligram protein (Doruòre *et al.*, 1994).

tion of fibrils, tubules and internal chromoplast membranes, since they have lower lipid to protein ratios (Table VII).

V. Control of Chromoplast Differentiation

A. Genetic Factors Affecting Chromoplast Formation

1. Tomato Chromoplasts

The presence of regulatory genes controlling chromoplast differentiation in plants is suggested by several lines of evidence. For instance, genes which control the formation of acyclic carotenoids accumulating in ripe tomato do not affect the carotenoid composition of the green fruits and leaves, except for the high-pigment (hp^+/hp) and the ghost (gh^+/gh) mutant genes (Mackinney et al., 1956; Baker and Tomes, 1964). The hp allele increases the chlorophyll and carotenoid content in both leaf and fruit plastids (Baker and Tomes, 1964). In the case of *ghost* mutant, phytoene accumulates in both leaf and fruit plastids (Mackinney *et al.*, 1956). Homozygous plants (*ghgh*) for the recessive *ghost* allele (*gh*) yield white fruits, in which the structure of the typical crystalline chromoplast of tomato fruit is changed into a globular type (Scolnik *et al.*, 1987). The fact that *ghost* mutation maps to chromosome 11 of tomato and not to chromosome 3 as phytoene desaturase (Giuliano *et al.*, 1993), and the failure of CPTA to induce lycopene synthesis in *ghost* tomato fruits (Raymundo, 1971), may suggest that *ghost* mutation regulates the biosynthetic steps downstream phytoene, according to an unknown mechanism. The grafting of normal scion onto *ghost* plants induces *ghost* phenotype in the emerging shoot (Mackinney *et al.*, 1956), suggesting that a diffusible substance could be transported from the *ghost* tissue to the graft.

In addition, genetic analyses have demonstrated that in tomato three nonallelic genes r^+/r, at^+/at, and *hp/hp*, regulate the level of the total carotenoid in the fruit chromoplasts (Table VIII). For example, tomato fruits with the genotype *rr* or *at*, i.e., homozygous for the recessive allele *r* or *at*, have very low levels of total carotenoids (5 to 10% compared to the normal red fruit). Recent data indicate that the *r* locus is equivalent to phytoene synthase (Fray and Grierson, 1993). In contrast, tomato fruits with the genotype *hphp*, i.e., homozygous for the recessive high-pigment allele *hp*, have an increased total carotenoid content (Table VIII). Further analysis of other tomato fruit mutants has revealed the existence of several other regulatory genes whose mechanisms of action are presently unknown (Table VIII). Furthermore, several other allele mutants have pleiotropic effects on chromoplast development in tomato fruits (Robinson and Tomes, 1968; Darby, 1978; Grierson *et al.*, 1987). Among these, the *never ripe* mutant induces, in homozygous lines, the formation of a dirty orange color

TABLE VIII

Genes Affecting Carotenoid Accumulation and Chromoplast Development in Tomato

Phenotype	Genotype (locus)	Carotenes[a]	References[b]
Red	r^+	a, b, c, d, e	1
Yellow	*r*	a, b, d, e	2
High-pigment	*hp*	a, b, c, d, e	3
Crimson	og^c	a, b, c, d, e	4
Tangerine	*t*	a, b, d, e, f, g, j	1
Apricot	*at*	a, b, d, e	2
Intermediate-beta	*B*	a, b, c, d, e	1
High-beta	Bmo_B	a, b, c, d, e	1
Delta	*Del*	a, b, c, d, e, g, h, i, j	1
High-delta	*Delhp*	a, b, c, d, h, i, j	1
Ghost	*gh*	a, b, d, e	2
Verkerk 377-2 $\alpha\alpha$	*vo*	a, b, c, d, e, g	5
Sherry	*sh*	a, b, e, j	6

[a] a, lycopene; b, β-carotene; c, γ-carotene; d, phytoene; e, phytofluene; f, prolycopene; g, ζ-carotene; h, α-carotene; i, γ-carotene; j, neurosporene.

[b] 1. Tomes (1963); 2. Tomes (1969); 3. Baker and Tomes (1964); 4. Thompson *et al.* (1967); 5. Tomes and Verkerk (1965); 6. Zscheile and Lesley (1967).

instead of the normal red color. Recent physiological and molecular analyses based on the exploitation of the triple response (Kieber and Ecker, 1993) and the epinastic response (Ursin and Bradford, 1989) induced by ethylene have revealed that this mutant is probably impaired in the transduction of ethylene signal (Lanahan *et al.*, 1994) which bears strong resemblance to the two-component signal-transducing mechanism of prokaryotes (Chang *et al.*, 1993). In the case of the *green flesh* mutant, part of the chlorophyll content is preserved after the final stage of chromoplast differentiation (Ramirez and Tomes, 1964). This phenotype is due to a single recessive mutation located on chromosome 8 (Kinzer *et al.*, 1990). The cause of this mutation is not due to an impairment of chlorophyllase activity in the *green flesh* fruit since the activity detected in the *green flesh* mutant is higher than that in the normal fruit (Ramirez and Tomes, 1964).

2. Pepper Chromoplasts

The pigments accumulating in pepper chromoplasts are under the control of three genes, y, c_1, and c_2, which regulate the formation of ketocarotenoids, and an additional gene, *cl*, which determines the retention of chlorophylls (Hurtado-Hernandez and Smith, 1985) according to the combinations shown in Table IX.

3. Genes Inducing Chromoplast-like Structures in Maize

Several maize mutants in which the chloroplasts develop into chromoplasts have been described (Walles, 1971; Robertson, 1975; Robertson *et al.*, 1978). The putative *loci* implicated in these mutants are indicated (Table X).

TABLE IX

Genes Affecting Carotenoid Accumulation and Chromoplast Development in Pepper Fruits

Phenotype[a]	Genotype (*locus*)[a]	Carotenoid composition[b]
Red	$y^+ c1^+ c2^+ cl^+$	1–8,10,11,14–19
Orange	$y^+ c1^+ c2\ cl^+$	1–11,14–19
Yellow	$y\ c1\ c2\ ^+cl^+$	1–14
Brown-chocolate	$y^+ c1^+ c2^+ cl$	1–8,10,11,14–19
Cream	$y\ c1\ c2\ cl^+$	5,6,9,11,14
Green	$y\ c1\ c2\ cl$	5,6,9,11,14

[a] Smith (1950); Hurtado-Hernandez and Smith (1985).

[b] 1. phytoene; 2. phytofluene; 3. ζ-carotene; 4. neurosporene; 5. β-carotene; 6. α-carotene; 7. cryptoxanthin; 8. zeaxanthin; 9. lutein; 10. antheraxanthin; 11. violaxanthin; 12. auroxanthin; 13. luteoxanthin; 14. neoxanthin; 15. cryptocapsin; 16. capsanthin; 17. capsanthin-5,6-epoxide; 18. capsanthin isomer; 19. capsorubin isomer (Simpson *et al.*, 1977b), B. Camara (unpublished results).

TABLE X

Genes Inducing Chromoplast-like Structures in Maize

Endosperm phenotype	Leaf phenotype	Genotype (*locus*)	Site of inhibition
White	Albino	*vp2*	Phytoene desaturation
White	Albino	*vp5*	Phytoene desaturation
White	Albino	*w3*	Phytofluene desaturation
Pale yellow	Albino	*vp9*	ζ-Carotene desaturation
Pink	Albino to green	*vp7*	Lycopene cyclization
Pale yellow	Green tips	*y9*	Steps after ζ-carotene

4. Genes Inducing Chromoplast Formation in *Cucurbita pepo* L. (Squash, Pumpkin, Ornamental Gourds)

In several cultivars belonging to the edible group of *C. pepo*, chloroplasts present in the ovary differentiate into chromoplasts just before or in most case after anthesis (Shifriss, 1981). On the other hand, in the case of ornamental gourds, the first signs of chromoplast differentiation are visible long before the anthesis period. The development of chromoplast is under the control of a dominant gene *B* and a nonallelic gene *Y* which control the retention of chlorophylls (Boyer, 1989). The expression of gene *B* is extremely variable. In the presence of modifier genes *Ep-1* and *Ep-2* (Shifriss and Paris, 1981) the effect of gene *B* can extend to leaves which subsequently differentiate chromoplasts (Shifriss, 1981). These data clearly indicate that leaf tissues have the potential to develop a typical chromoplast distinct from gerontoplasts. In this context one could observe that *in vitro* culture of green tomato calyx induces the differentiation of lycopenic chromoplasts (Ishida, 1991).

B. Regulation of Photosynthetic Genes during Chromoplast Differentiation

1. General Organization and Topological Change of Chromoplast DNA

Since the initial report on the presence of nucleic acids in carrot chromoplasts (Straus, 1954), detailed analysis has been carried out using *N. pseudonarcissus* chromoplasts (Kowallik and Herrmann, 1972; Hermann, 1972). Analysis of the DNA profile obtained after restriction endonuclease digestion reveals that chloroplast and chromoplast DNA are identical and plastid genes have the same organization in different plants, including *N. pseudo-*

narcissus (Thompson, 1980; Hansmann, 1987), pepper (Gounaris *et al.,* 1986), tomato (Iwatsuki *et al.,* 1985; Hunt *et al.,* 1986), and pumpkin (Lim *et al.,* 1989).

Owing to the endosymbiotic origin of plastids, the genetic information of chromoplasts is located in a nucleoid, i.e., a DNA–protein complex which may contain RNA (Kowallik and Herrmann, 1972; Hansmann *et al.,* 1985), a feature characteristic of procaryotes which show coupling of transcription and translation (Miller *et al.,* 1970). Using the DNA-specific fluorochrome DAPI (4′,6-diamidino-2-phenylindole), it has been shown that the number of nucleoids decreases during chloroplast to chromoplast differentiation in *Narcissus* petals (Hansmann *et al.,* 1985). During the disorganization of the photosynthetic membranes in differentiating *Capsicum* chromoplasts, the nucleoids previously connected to thylakoid, as in mature chloroplasts (Kuroiwa *et al.,* 1981), are located mainly at the periphery of the stroma and remain associated to specific proteins of the chromoplast envelope membrane (Hansmann *et al.,* 1985; Carde *et al.,* 1988). In the classification of nucleoids (Kuroiwa *et al.,* 1981), this distribution of chromoplast nucleoid belongs to the PS-type (i.e., peripheral scattered). In addition, the protein composition of plastid nucleoid is subject to change during chromoplast differentiation (Hansmann *et al.,* 1985), as shown previously during chloroplast development (Nemo *et al.,* 1990; Reiss and Link, 1985). Obviously, topological changes and modifications of the protein content of the nucleoid could play a role in the expression of plastid genes during chromoplast differentiation. This assumption is supported by the fact that in pea, the plastid nucleoids bind to the plastid envelope of young leaves via a 130-kDa protein, while no binding could be observed in mature leaves (Sato *et al.,* 1993).

2. Transcriptional Activity during Chromoplast Development

Available data from tomato fruit (Piechulla *et al.,* 1985, 1986) and pepper fruit (Kuntz *et al.,* 1989) indicate a general reduction in the accumulation of the transcripts of plastid and nuclear-encoded proteins involved in photosynthesis. The loss of the light-harvesting chlorophyll a/b-binding protein (LHCP) mRNA represents an early marker of this phenomenon (Piechulla *et al.,* 1985, 1986; Kuntz *et al.,* 1989). In tomato, the parallelism between the down-regulation of the expression of LHCP gene and the increased activity of superoxide dismutase (Livne and Gepstein, 1988) suggests that oxidative stress prevailing in the ripening fruit could trigger this phenomenon. On the other hand, the gene *psbA* encoding the 32-kDa "herbicide binding" protein of the photosystem II is expressed until the last period of chromoplast development in tomato (Piechulla *et al.,* 1986) and in pepper (Kuntz *et al.,* 1989).

The down-regulation of the expression of photosynthetic plastid genes during chromoplast differentiation has been analyzed directly using run-on experiments, which involve the incubation of isolated plastid with nucleoside triphosphates so as to elongate initiated transcripts. The final products are then hybridized with the DNA probe to be analyzed (Mullet and Klein, 1987; Deng *et al.*, 1987). Data reported indicate low transcriptional activity in chromoplasts compared to leaves (Deng *et al.*, 1987; Kuntz *et al.*, 1989; Kobayashi *et al.*, 1990; Marano and Carrillo, 1992). Furthermore, the transcriptional rates vary according to the different genes; for example, in pepper chromoplasts the decreasing order $psbA > rbcL$ (large subunit of ribulose diphosphate carboxylase oxygenase) $> psaA$ (apoprotein A_1 of P700) was noted (Kuntz *et al.*, 1989). Therefore, the down-regulation of photosynthetic plastid genes during chloroplast to chromoplast transformation is not due to the dismantling of the plastid transcription apparatus. The regulation of plastid gene expression during chromoplast differentiation in tomato has been explained on the basis of plastid DNA methylation (Kobayashi *et al.*, 1990; Ngernprasirtsiri *et al.*, 1988). However, no DNA methylation could be detected during further analysis using chromoplast DNA from tomato (Marano and Carrillo, 1991). DNase I footprinting assays to characterize the different sites of protein interactions with the different *rbcS* (small subunit of ribulose diphosphate carboxylase oxygenase) promoters revealed that in developing tomato fruits most of the DNA–protein interactions disappear, in good agreement with the low expression of these genes (Manzara *et al.*, 1991).

Indeed, detailed analysis of plastid gene expression is hampered by our limited knowledge on plastid promoters. Plastid promoters generally display features of procaryote promoters, i.e., conserved sequence in the nucleotide region -35 and -10 relative to the transcription start (Helmann and Chamberlin, 1988; Gruissem and Tonkyn, 1993). In addition, the promoters of some genes like *psbA* have additional elements similar to the eucaryotic TATA box of nuclear genes transcribed by RNA polymerase II (Link, 1984; Klein *et al.*, 1992). Evidence based on the study of the *pbsA* promoter indicates that in etioplasts and chloroplasts of mustard seedlings, the -35 and the TATA-like elements function differently (Eisermann *et al.*, 1990). The picture is further complicated by the fact that the catalytic binding requirement of the sigma factors to the RNA polymerase are different (Tiller and Link, 1993) for both types of plastids. In addition, a second RNA polymerase, nuclear encoded, is involved in the control of the transcription of plastid genes. Its existence is further reinforced by the fact that barley plastid mutants deficient in plastid ribosomes have a competent transcription system (Hess *et al.*, 1993), and by the fact that nongreen plastids of *Epifagus* accumulate rRNA in the absence of *rpo*B, *rpo*C1, and *rpo*C2, which encode the subunits of the plastid-encoded RNA polymerase

(Morden et al., 1991). Further resolution of these problems is essential to understanding the transcriptional regulation of gene expression during chromoplast development.

3. In Organello Translation and Import of Proteins during Chromoplast Differentiation

From the limited number of published results, one could note that the *in vitro* translation capacity of chromoplast is generally low or negligible (Bathgate et al., 1986; Marano et al., 1993; Price et al., 1989). This phenomenon can be explained by the fact that electron microscope observations indicate that plastoribosomes are visible during the initial period of development and disappear in fully differentiated chromoplasts (Suzuki, 1974; Carde et al., 1988). In Cucurbita pepo this change is reversible since plastid ribosomes reappear during the regreening process (Devidé and T jubesic, 1974) In addition, data based on ultracentrifugation analysis reveal the absence of polysomes in Narcissus chromoplasts (Hansmann et al., 1987) and in pepper chromoplasts (Carde et al., 1988). Finally, the inefficiency of the translation of plastid RNAs could not be ascribed to the absence of editing which operates in chromoplasts (Kuntz et al., 1992).

Consequently, during chromoplast differentiation massive import of nuclear-encoded protein must occur, as shown by the accumulation of specific proteins in tomato fruit chromoplasts (Bathgate et al., 1985), Cucumis sativus petal chromoplasts (Smirra et al., 1993), and pepper fruit chromoplasts where two proteins of 58 and 35 kDa molecular mass, designated ChrA and ChrB (Hadjeb et al., 1988), have been identified. Further analysis revealed that ChrA is a carotenoid binding protein (Cervantes-Cervantes et al., 1990).

The detailed study of plastid protein import is hampered by the fragility of isolated chromoplasts. However, available results from non-green plastids (Boyle et al., 1986, 1990; Strzalka et al., 1987; Klösgen et al., 1989; Halpin et al., 1989) suggest that a similar mechanism operates in chromoplasts, i.e., the import process requires the recognition of receptor sites on the plastid envelope and the involvement of exogenous ATP. Until now, only the correct import and processing of the precursors of stromal proteins such as the small subunit of rubisco have been observed *in vitro* in nongreen plastids, while the precursor of the membrane protein LHCP was apparently not imported in chlorophyll free amyloplasts isolated from sycamore cells (Strzalka et al., 1987). The intraplastidial routing of imported proteins probably must follow different pathways than chloroplasts. The energy requirement for the integration of thylakoid destined has allowed the distinction of three routes characterized by their energy requirements: ATP alone, proton gradient alone, or ATP plus proton gradient (Cline *et al.,* 1993). Obviously, in chromoplasts, additional mechanisms must operate due to the impossibility of generating a proton gradient in the absence of photosynthetic electron transfer. This point is indirectly supported by the observation that nonphotosynthetic leucoplasts from castor bean endosperm can import the precursor of plastocyanin, a thylakoid lumen protein, but are unable to carry out the final maturation steps (Halpin *et al.,* 1989). In contrast, using transgenic tomato plant expressing a plastocyanin, De Boer *et al.* (1988) were able to demonstrate that import and complete maturation occurred in the chromoplasts of the petals. Overall, the situation of chromoplasts is reminiscent of that of the albino rice plant, which has a large deletion in plastid DNA and still develops without translation in the plastidial compartment (Harada *et al.,* 1992).

In addition to the above situation, two nonexclusive mechanisms can be postulated to explain the disparition of photosynthetic-related proteins during normal chromoplast development. The first mechanism could be due to a developmental regulation of the import capacity during chromoplast differentiation. The observation that in wheat seedlings the import capacity of the plastids is age-dependent represents a key precedence to this end (Dahlin and Cline, 1991). Alternatively, one may postulate that selective proteases triggered during chromoplast differentiation dismantle preexisting photosynthetic proteins. Very few data have been reported concerning proteolytic activity during plastid differentiation (Debesh and Apel, 1983; Hampp and De Filipps, 1980; Hauser *et al.,* 1984). Based on available data one can hypothesize that normal proteases (Feller and Fischer, 1994) and ATP-dependent proteolytic systems (Liu and Jagendorf, 1984; Malek *et al.,* 1984) operate during chromoplast differentiation. ATP-dependent proteolysis has been reported from pea chloroplasts (Liu and Jagendorf, 1984; Malek *et al.,* 1984). In *Escherichia coli,* of which detailed studies have been conducted, the activity of the ATP-dependent proteolytic system requires the cooperation of two subunits which include the ATP-dependent regulator named Clp A (Seol *et al.,* 1994) and the proteolytic subunit designated Clp P, which is a serine protease. Analysis of the plastid genome of tobacco revealed the presence of an open reading frame homologous to the proteolytic subunit of *E. coli* ClpP (Shinozaki *et al.,* 1986). Finally, the chloroplast *ClpP* gene has been cloned recently from *Pinus contorta* (Clarke *et al.,* 1994). The situation is much less clear for the plant regulatory subunit, since several Clp-like proteins exist (Moore and Keegstra, 1993; Kiyosue *et al.,* 1993). We have characterized recently from pepper chromoplast stroma three types of Clp proteins whose sequences are much more similar to *Mycobacterium leprae* ClpC (Nath and Laal, 1990) than to ClpA, thus demonstrating that they probably act as molecular chaperones or possess a function not yet identified. In preliminary tests we have observed negligible ATP-dependent proteolysis during chromoplast differentiation in pepper fruits. On the other hand, the aminopeptidase activity tested with different artificial substrates is particularly up-regulated (B. Camara, unpublished results). These observations corroborate the N-end rules, which assume that the half-life of a protein is determined mainly by its N-terminal residues (Bachmair *et al.*, 1986; Varshavsky, 1992).

C. Environmental and Nutritional Factors Affecting Chromoplast Differentiation

1. Light Regulation of Chromoplast Development

Light is an important environmental factor for chloroplast differentiation. When grown in the dark, most plants become etiolated and the characteristic chloroplast thylakoid membranes are not formed but are replaced by prola mellar bodies and prothylakoids. Although light is not strictly necessary for chromoplast development (Smith and Smith, 1931, Denison, 1948), there is evidence that light speeds and amplifies this process in tomato fruit (Duggar, 1913; Smith, 1936; Denison, 1951; Shewfelt and Halpin, 1967; Raymundo *et al.*, 1976) and in pepper fruit (Simpson *et al.*, 1977a). Further studies using tomato fruits revealed that red and blue lights were the most potent inducers of chromoplast development (Jen, 1974a,b). Indeed, the induction of chromoplast formation in tomato fruit, as judged by the accumulation of lycopene, is proportional to the logarithm of red light intensity up to the saturation point (Thomas and Jen, 1975a). The red light effect is inhibited by far red light illumination (Thomas and Jen, 1975b), thus demonstrating a typical phytochrome response as suggested previously (Khudairi and Arboleda, 1971). The stimulating effect of blue light is also worth considering since this phenomenon has been documented previously for nonphotosynthetic bacteria (Nelson *et al.*, 1989), fungi, and plants, in relation to carotenoid formation (Rau, 1985; Kaldenhoff and Richter, 1990).

2. Temperature Regulation of Chromoplast Development

Duggar, (1913) first observed that temperatures above 30°C inhibit chloroplast to chromoplast transformation in tomato fruits, as shown by the inhibition of lycopene accumulation, whereas β-carotene is unaffected. This phenomenon was subsequently confirmed by detailed biochemical analysis (Goodwin and Jamikorn, 1952; Tomes, 1963; Baqar and Lee, 1978), and by electron microscope studies which demonstrated the disappearance of lycopene crystals and the persistence of β-carotene sequestered in lipid globules (Ben-Shaul and Naftali, 1969). This phenomenon is reversible (Went *et al.*, 1942) and can be overcome by CPTA treatment (Rabinowitch and Rudich, 1972; Baqar and Lee, 1978). This has been taken as evidence that two spatially distinct pathways of carotenoid biosynthesis operate in developing tomato chromoplasts. The first pathway corresponds to the preexisting chloroplast pathway which affords β-carotene, lutein, violaxanthin, and neoxanthin, while the second pathway is triggered at the onset of chromoplast development and yields lycopene as the main final product. The validity of this assumption is further supported by the fact that in the tomato mutant *high-beta,* in which the chromoplasts preferentially accumulate β-carotene, ripening at 30°C drastically inhibits the formation of β-carotene (Tomes, 1963; Baqar and Lee, 1978). In addition, it has been shown in normal tomato that dimethylsulfoxide inhibits the accumulation of lycopene and not β-carotene (Raymundo *et al.,* 1967). It has been shown in tomato fruits that the high-temperature effect can be partially reversed by CPTA (Rabinowitch and Rudich, 1972; Baqar and Lee, 1978), which also acts as a derepressor of the gene regulating the biosynthesis of carotenogenic enzymes (Coggins *et al.,* 1970; Seyama and Splittstoesser, 1975; Kirk, 1978). One could therefore hypothesize that the action of high temperature proceeds through the inhibition of the synthesis or the inactivation of an endogenous derepressor of carotenogenic genes.

3. Nutritional Regulation of Chromoplast Development

Experiments on *Citrus* pericarp discs (Huff, 1983) address the question of whether nutritional factors control chromoplast development. It can be inferred from data gained from this model that the carbon to nitrogen ratio (C/N) plays a key role during chromoplast formation. According to this postulate, a high ratio induces chromoplast differentiation, with the typical loss of chlorophylls and chloroplast structures, while a low ratio favors the reversion process (Huff, 1983, 1984; Mayfield and Huff, 1986). Similarly, in *Euglena* a high C/N ratio represses chlorophyll synthesis (Schwelitz *et al.,* 1978). Consistent with the mechanism of nutritional control was the observation that excess carbohydrates (Dalton and Street, 1977) inhibit chlorophyll synthesis and plastid development (Pasqua *et al.,* 1989), while nitrogen availability or starvation regulates plastid biogenesis (Horum and Schwartzbach, 1980) and the expression of photosynthetic genes (Plumpley and Schmidt, 1989). Though this view is attractive and supported by indirect evidence, we are unaware of any study at the enzymatic level. Extrapolation of these results would predict that plastid enzymes of nitrogen assimilation are partially inactive or subject to strict control during chromoplast differentiation. One may expect that glutamine synthetase, which is a key enzyme for the synthesis of nitrogen, could be a main target. Several features make this assumption reasonable. First, chloroplast glutamine synthetase activity, like many aspects of plastid differentiation, is activated or repressed by several elicitors of plastid differentiation. Second, through the action of glutamate synthase, glutamine affords glutamic acid, which is a specific precursor of the pyrrole ring of chlorophylls in higher plants (Gough and Kannangara, 1976; Beale, 1984). Finally, it has been shown that the inhibition of glutamine synthetase by tabtoxinine-β-lactam led to a chlorophyll decrease of 35% in 48 h (Sinden and Durbin, 1988). Based on these observations, a partial answer to this question can be deduced from the data described for tomato fruits, demonstrating that during chloroplast to chromoplast differentiation, the plastidial glutamine synthetase is progressively degraded (Gallardo et al., 1988).

The mechanisms inducing these changes are unknown. However, in many respects, the phenotypic events that occur during chromoplast differentiation, i.e., the chlorophyll loss and the decreased activity of the Calvin cycle enzymes, are identical to those reported to occur when excess glucose is supplied to mature leaves (Krapp et al., 1991) or in transgenic plants expressing yeast invertase (von Schaewen et al., 1990; Stitt et al., 1990). Detailed studies using maize protoplasts and a transient expression assay reveal that carbohydrates repress the transcriptional activity of several photosynthetic gene promoters (Sheen, 1990; Jang and Sheen, 1994), according to a signaling pathway in which hexokinase exerts a key role (Jang and Sheen, 1994). This conclusion reinforces the previous observation showing that hexose kinase activity increases within 1 min following artificial wounding of leaf tissue (MacNicol, 1976). Obviously, further studies are required to delineate the role of hexose kinase in the regulation of the cellular homeostasis occurring during chromoplast development.

4. Hormonal Regulation of Chromoplast Development

Attempts to understand the cellular mechanism underlying the differentiation or the reversion process between chloroplasts and chromoplasts have led to postulate hormonal determinants (Coggins et al., 1962, 1980; Lewis and Coggins, 1964; Thomson et al., 1967; Gemmrich and Kayser, 1984; Eilati et al., 1969; Goldschmidt, 1988). Analysis of available data provides an accurate picture of the overall effects of different plant hormones on chromoplast development, but little direct information on how they operate. Ethylene has been used widely to induce chromoplast differentiation in tomato fruits (Khudairi, 1972). The specificity of ethylene is supported by the fact that silver nitrate, an inhibitor of ethylene action, inhibits the degradation of chlorophylls (Purvis, 1980). The involvement of ethylene is also demonstrated by the analysis of transgenic plants in which ethylene biosynthesis is inhibited by expressing antisense aminocyclopropane carboxylic acid synthase or oxidase (Theologis et al., 1993). In general, the first symptom following ethylene application is the loss of chlorophylls and the disintegration of thylakoid membranes (Denny, 1924; Eaks, 1977; Purvis, 1980; Purvis and Barmore, 1981). The degradation of chlorophylls is catalyzed by chlorophyllase, which cleaves the phytyl side chain and yields a water soluble chlorophyllide which is subject to oxidative attack according to unknown mechanisms (Amir-Shapira et al., 1987). Chlorophyllase activity of Citrus fruits is enhanced after ethylene treatment (Shimokawa et al., 1978; Purvis, 1980; Barmore, 1975), according to a mechanism which presumably occurs through de novo synthesis of chlorophyllase (Trebitsh et al., 1993). In general these changes induced by ethylene are concomitant to increased synthesis (Stewart and Wheaton, 1971; Young and Jahn, 1972) and esterification (Eilati et al., 1972) of carotenoids. Very similar effects have also been observed with abscisic acid (Khudairi, 1972). The ethylene control of chromoplast differentiation in fruit is reversed by cytokinins (Khudairi and Arboleda, 1971; Abdel-Gawad and Romani, 1974; Goldschmidt et al., 1977) and gibberellins (Lewis and Coggins, 1964; Dostal and Leopold, 1967; Eilati et al., 1969; Goldschmidt et al., 1977). Thus gibberellins have been used frequently to elicit the reversion of chromoplasts into chloroplasts (Coggins et al., 1962).

Though these data are based on exogenous application of plant hormones, they are generally corroborated by quantitative determination of their endogenous level during the different stages of chromoplast differentiation under normal conditions (Goldschmidt et al., 1973; Adato et al., 1976; Davey and Van Staden, 1978). However, caution must be exercised since it has been observed that in transgenic tomato fruits expressing Agrobacterium cytokinine biosynthetic gene ipt, the pericarp tissue containing fully differentiated chromoplasts had at least six times more cytokinin than the green portion of the pericarp (Martineau et al., 1994). In addition, callus cultures derived from the endosperm of Ricinus communis grown in the absence of auxin but in the presence of kinetin as the sole hormone differentiates chromoplast containing the red carotenoid rhodoxanthin (Kayser and Gemmrich, 1983; Gemmrich and Kayser, 1984). This apparent chromoplast-inducing effect of cytokinins deviates from the classical view that cytokinins activate chloroplast differentiation (Parthier, 1979) and delay the induction of ripening and chromoplast differentiation. Parallel to these findings one could note that in Picea abies, cytokinins repress the expression of the photosynthetic genes encoding LHCP and the small subunit of ribulose diphosphate carboxylase oxygenase (Stabel et al., 1991). These apparent contradictions can be reconcilied when considering the fact that in Arabidopsis, the same endogenous concentrations of cytokinins can lead to plastids with different states of differentiation, thus indicating the prevalence of the mechanism of the transduction of cytokinin signal (Stabel et al., 1991) over the effect of steady concentration of the endogenous hormone.

This view is reinforced by the fact that in *Funaria hygrometrica,* the biological response elicited by cytokinins is mediated by an increase of the intracellular calcium concentration (Saunders and Helper, 1982, 1983).

5. Stress Regulation of Chromoplast Development

The signaling mechanisms during chromoplast differentiation are unknown. One can observe that any perturbation of the homeostasis of chlorophyllous cells, rapidly induces the down-regulation of chloroplast metabolism or activity. This can be observed during pathogenic interactions which selectively alter the activity of the photosystem II of tobacco chloroplasts after viral infection (Hodgson *et al.,* 1989), according to a mechanism which could be due to interference with the formation of the nuclear-encoded plastid protein synthesized in the cytosol and transported into the plastids (Lindbeck *et al.,* 1991). Similar mechanisms operate during the down-regulation of the expression of ribulose diphosphate carboxylase oxygenase genes *rbcL* and *rbcS* in infected leaves (Walters and Ayres, 1984; Roby *et al.,* 1988), and during the initial steps of tobacco protoplast culture (Fleck *et al.,* 1979). Further analysis using leaf strips and isolated protoplasts incubated in the presence of sucrose or mannitol clearly demonstrated that the mechanical stress during these treatments severely inhibited the transcription of the genes *rbcS* and the light-harvesting chlorophyll a/b binding protein of photosystem II (LHCII), while the expression of the gene *rbcL* was less affected (Criqui *et al.,* 1992). All these changes are diagnostic features of chromoplast development, and allow the hypothesise that stress-derived signals play a key role in the regulation of the expression genes during chromoplast differentiation. For the moment, it is important to note that the "stress hormone" abscisic acid activates the formation of the chromoplast specific protein fibrillin (Deruère *et al.,* 1994) and that mechanical stress induces the expression of the capsanthin-capsorubin synthase gene (Hugueney *et al.,* 1995a), which encodes the last enzyme involved in the biogenesis of chromoplast carotenoids in pepper fruits (Bouvier *et al.,* 1994).

6. Chemical Regulation of Chromoplast Development

Chromoplast development in plant tissues can be induced by several chemicals which alter the normal plastid carotenogenic pathway or induce the formation of a new carotenogenic pathway, which both lead to the accumulation of chromoplast carotenoids. Most of these compounds are tertiary amines having the general formula RCH_2NEt_2 (Coggins *et al.,* 1970).

One of the most potent compounds is 2-(chlorophenylthio) triethyl-ammonium chloride (CPTA) which induces the formation of lycopene-accumulating chromoplasts in several plants (Coggins *et al.,* 1970; Simpson *et al.,* 1974a,b, 1977a). Several related analogs induce the accumulation of poly-*cis* carotenoids (Poling *et al.,* 1980). Two mechanisms of action of CPTA have been proposed. The first mechanism involves lycopene cyclase inhibition (Beyer *et al.,* 1991). The second mechanism involves the induction of a new chromoplast carotenogenic pathway, which results in the accumulation of lycopene and its precursors phytoene, phytofluene, ξ-carotene, and γ-carotene. This assumption is supported by the fact that CPTA can increase the accumulation of lycopene up to 100-fold in tissue where the original β-carotene content is extremely low (Radin, 1986). The inhibition of lycopene cyclase is difficult to understand under these conditions, unless β-carotene exerts a feedback regulatory role, since the stoichiometric amount of lycopene formed *de novo* and the preexisting amount of β-carotene preclude a precursor–product relationship. Furthermore, the induction of lycopene synthesis by CPTA is blocked by protein translation and transcription inhibitors (Seyama and Splittstoesser, 1975; Fosket and Radin, 1983; Benedict *et al.,* 1985). These data suggest that CPTA may act via gene expression. This trend is further supported by the fact that CPTA did not inhibit lycopene cyclase *in vitro* (Bucholtz *et al.,* 1977). The genetic mediation of CPTA activity is also strengthened by its ability to enhance lycopene accumulation in tomato fruit mutants (Raymundo, 1971; Sink *et al.,* 1974) and in normal tomato fruits (Salunkhe *et al.,* 1974; Wiley *et al.,* 1972), and also by its ability to partially overcome the high-temperature inhibition of lycopene formation in normal tomato chromoplasts (Rabinowitch and Rudich, 1972; Baqar and Lee, 1978). These data could be explained by the existence of genes involved in the negative regulation of carotenoid biosynthesis and whose expression is induced by CPTA and related derivatives.

Similar but less pronounced phenomena have been observed with different compounds. For instance, nicotine induces the accumulation of lycopene in greening *Cucurbita ficifolia* cotyledons in a pH-dependent manner (Howes, 1974a). This allows the conclusion that the nonionic form of nicotine is the active species (Howes, 1974b). The growth retardant Chlormequat (2-chloroethyltrimethylammonium chloride) (Tolbert, 1960) also induces the formation of lycopenic chromoplasts in *Cucurbita pepo* cotyledons (Knypl, 1969), in *Cucumis sativus* cotyledons (Mikulska *et al.,* 1973), and in tobacco suspension cultures (Gamburg, 1978). The same trend has been observed in maize leaves treated with the herbicide amitrole (3-amino-1,2,4-triazole) (Guillot-Salomon *et al.,* 1967).

7. Regulation of Chromoplast Reversion

The pathway of chromoplast differentiation is not always monotropic since, in several plant tissues, a reversion has been observed (Table XI). The reversion is favored by gibberellin treatment in *Citrus* fruits (Coggins *et al.*, 1962; Ismail *et al.*, 1967; Eilati *et al.*, 1969), while in carrot roots cultivated *in vitro*, the reversion strictly requires the addition of auxin and sucrose (Kumar *et al.*,1984). The reversion is also stimulated by nutritional factors as discussed previously and by several environmental factors, including high light intensity (Grönegress, 1971; Wrischer, 1972) and temperature changes (Caprio, 1956). It is worthwhile to note that the chromoplast plastoglobules play a key role during the reversion process. Indeed, in *Citrus* fruit, one could note the remobilization of the lipids initially sequestered in the plastoglobules (Miller *et al.*, 1940; Lee *et al.*, 1971), which progressively become less dense to electron and smaller (Thomson *et al.*, 1967), while connected to the nascent thylakoid system.

TABLE XI

Occurrence of Chromoplast Reversion in Plants

Species	References[a]
Aphelandra squarrosa	1
Chrysosplenium alternifolium	2
Chrysosplenium oppositifolium	2
Citrus natsundaidai	3
Citrus paradisi	4
Citrus sinensis	5,6
Cucurbita pepo cv. ovifera	7,8
Daucus carota	9,10
Narcissus poeticus	11
Nuphar luteum	1
Oenothera suaveolens (mutant)	12
Zantedeschia elliottiana	1

[a] 1. Grönegress (1974); 2. Sitte (1974); 3. Daito and Hirose (1970); 4. Coggins *et al.* (1962); 5. Thomson *et al.* (1967); 6. Huff (1983); 7. Devidé and Ljubesic (1972); 8. Devidé and Ljubesic (1974); 9. Israel *et al.* (1969); 10. Grönegress (1971); 11. Kuhn (1970); 12. Schötz and Senser (1961).

VI. Chromoplast Biogenesis

A. Plastid Envelope Membranes Budding in Relation to the Acquisition of New Biosynthetic Capacities

Results obtained from several mutants including tomato (Ramirez and Tomes, 1964; Cheung *et al.,* 1993) and pepper fruits (Laborde and Spurr, 1973; Deruère *et al.,* 1994), where the chloroplast thylakoids are preserved during chromoplast differentiation, indicate that the new set of chromoplast substructures is formed *de novo* and does not derive from preexisting materials released during the degradation of chloroplast thylakoids. However, the sequence of events is not understood. Based on ultrastructural data, it has been proposed that the plastid envelope plays a key role during chloroplast thylakoid ontogeny in higher plants (Mühlethaler and Frey-Wyssling, 1959; Wettstein, 1958; Menke, 1962; Berger and Feierabend, 1967; Whatley, 1974; Wozny and Szweykowska, 1975; Brangeon and Nato, 1981; Oliveira, 1982), in *Chlamydomonas reinhardtii* (Hoober *et al.,* 1992), or in *Euglena* (Ben-Shaul *et al.,* 1964; Stern *et al.,* 1964a,b). By analogy with chloroplasts, similar observations can be made during chromoplast differentiation since the inner chromoplast envelope membrane is often subject to vesiculation fission and fusion during chloroplast to chromoplast differentiation (Rosso, 1968; Harris, 1970; Cran and Possingham, 1973; Shimokawa and Horiba, 1975; Ljubesic, 1977; Shimokawa *et al.,* 1978; Sitte *et al.,* 1980; Camara and Brangeon, 1981; Knoth, 1981) or during the reversion of chromoplasts to chloroplasts (Thomson *et al.,* 1967; Pryer *et al.,* 1992; Devidé and Ljubesic, 1974; Whatley and Whatley, 1987). These sequential events suggest that vesicles derived from the inner plastid envelope membrane represent assembly platforms for the new biochemical potentialities and allow the formation of the different chromoplast substructures. This phenomenon is reminiscent of vesicle fission and fusion (Pryer *et al.,* 1992), which have been demonstrated in animals (Wilson *et al.,* 1989; Rothman and Orci, 1992) and yeasts (Eakle *et al.,* 1988) and involve a cytosolic protein designated N-ethylmaleimide sensitive factor (NSF), which participates in the intracellular transport of proteins from membrane donors to membrane acceptors. It can be inferred from these studies that the fusion process must be assisted by "fusogen proteins" since it is an energetically unfavorable process due to the need for counteracting repulsive hydratation forces that inhibit fusion, in addition to the need of promoting hydrophobic forces that favor fusion (Helm *et al.,* 1992; Rand *et al.,* 1988). We devised a procedure for isolating putative components of the vesicle transport system (Hugueney *et al.,* 1995b). Data gained from these studies allowed the resolution of several proteins based on their ATP-binding capacity. One of these proteins has been purified from *Capsicum* fruit and *Crocus* stigma chromoplasts, and designated "plastid fusion and/or translocation factor" (Pftf) (Hugueney *et al.,* 1995b). The purified protein has a molecular mass of 72 kDa (Hugueney *et al.,* 1995c). The sequence of the mature part of this protein deduced from the cDNA displays significant homology to the NSF of Chinese hamster (Wilson *et al.,* 1989; 51% similarity and 28% identity), to the yeast Sec 18p (Eakle *et al.,* 1988; 53% similarity and 27% identity), and to FtsH (Ogura *et al.,* 1991; 73% similarity and 56% identity), which represents a protein involved in bacterial protein translocation (Akiyama *et al.,* 1994a,b). In addition Pftf has significant homology to Bs, a bacterial division protein (Accession No. D 26185, from EMBL nucleotide database; 75% similarity and 60% identity). Analysis of the deduced peptide sequence of Pftf reveals only one typical ATP-binding domain (Erdmann *et al.,* 1991; Tagaya *et al.,* 1993; Whiteheart *et al.,* 1994). Northern blot analysis further revealed that the corresponding gene is constitutively expressed during plastid development in pepper. These data reinforce the long outstanding hypothesis of vesicle fusion in plastids.

In fact, this dynamic role of plastid envelope membrane vesiculation can be followed from photosynthetic bacteria, where the periplasmic membrane budding gives photosynthetic substructures (De Greef, 1992). This trend is further strengthened by the fact that the periplasmic membrane of *Cyanobacteria* and its higher plant homolog, i.e., the plastid envelope membrane, have very similar galactolipid composition (Murata *et al.,* 1981). In addition, nonbilayer lipids (Gruner, 1985) such as monogalactosyldiacylglycerol (Gounaris and Barber, 1983) have greater propensity to form membrane vesicles. In this context, it is worth noting that Morré *et al.* (1991) have demonstrated that envelope galactolipids are transferred to thylakoid membranes. Finally, the role of such proteins in plastid membrane biogenesis is supported by the fact that the plastid genomes of several red Alga (Valentin, 1993; Scaramuzzi *et al.,* 1992; Reith and Munholland, 1993) display open reading frames encoding NSF homologs.

B. General Metabolism of Chromoplasts

1. ATP/ADP and NADPH Sources

Though an adenylate (ATP/ADP) translocator has been characterized from *Narcissus* chromoplasts (Liedvogel and Kleinig, 1980), pea root plastids (Schünemann *et al.,* 1993), and tobacco proplastids (Ardila *et al.,* 1993), its exact physiological role is not clear. In mitochondria the adenylate translocator allows the import of ADP involved in oxidative phosphorylation and the export of ATP formed in the mitochondria into the cytosol.

It has been shown that during either light or dark cycle treatment the ATP/ADP ratio is higher in the cytosol compared to the mitochondrial matrix (Heineke et al., 1991). Consequently, one can assume that in the absence of photosynthesis, the major function of the chromoplast translocator is to import the ATP generated in the mitochondria. This hypothesis is supported by available results from nongreen plastids, which indicate that externally supplied ATP sustains starch biosynthesis in plastids of pea embryos (Hill and Smith, 1991) and in cauliflower bud plastids (Neuhaus et al., 1993b). Very similar conclusions can be drawn from the stimulating role of external ATP supply during fatty acid biosynthesis in Narcissus chromoplasts (Kleinig and Liedvogel, 1980), in castor bean endosperm leucoplasts (Smith et al., 1992), and in pea root plastids (Stahl and Sparace, 1991). Further studies using plastids isolated from cauliflower flower buds demonstrated the presence of a high-affinity ATP/ADP translocator with a K_m of 12 μM for ATP (Neuhaus et al., 1993b). Alternatively, ATP could be generated endogenously in the plastids through the plastid glycolytic sequence at the step of pyruvate kinase and phosphoglycerate kinase. For instance, in castor bean leucoplasts the rate of fatty acid synthesis in the presence of phospho-enolpyruvate is 3.5 times higher than the rate obtained in the presence of an external ATP supply (Boyle et al., 1990), thus suggesting that ATP formed through the plastidial pyruvate kinase is a better source. In tobacco proplastids, phosphoglycerate kinase is regulated by the energy charge of adenylate and preferentially functions in the ATP-forming direction as soon as the adenylate charge is below 0.85 (Washitani and Sato, 1981). It is clear, therefore, that further studies addressing the degree of overlap between the different potential sources of ATP and the molecular characterization of the adenylate translocator in heterotrophic plastids are required.

The source of NADPH in nongreen plastids has been assessed indirectly. Data obtained based on the reductive assimilation of nitrite by tobacco proplastids (Washitani and Sato, 1977) and glutamate synthesis by pea root plastids (Bowsher et al., 1992) indicate that the oxidative pentosephosphate pathway generates the required reductants in nongreen plastids. This trend is reinforced by the fact that in developing pepper chromoplasts, the activity of glucose-6-phosphate dehydrogenase, a key enzyme of the pathway, is increased and remains insensitive to light (Camara, 1981; Ziegler et al., 1983), which down-regulates the chloroplast enzyme (Ziegler et al., 1983). However, due to redox potential requirements, the true electron donors in nonphotosynthetic plastid have not been assessed in detail. The characterization of several forms of ferredoxin and ferredoxin-NADP$^+$ reductase proteins from nonphotosynthetic tissues (Morigasaki et al., 1990; Green et al., 1991; Bowsher et al., 1993) and their in vitro reconstitution with target enzymes could increase our understanding of the basic organization of the electron transfer system.

2. Carbohydrate Metabolism

During chromoplast differentiation the starch content of the plastid is generally degraded. However, in a limited number of cases starch retention has been observed in chromoplasts of *Solanum capsicastrum* (Steffen and Walter, 1958), *Citrus* (Thomson, 1966b), *Zantedeschia elliottiana* (Gröne-gress, 1974), and *Liriodendron tulipefera* (Ljubesic, 1979). Evidence that amylogenic enzymes are still active in chromoplasts is provided by the fact that excised discs from ripening *Solanum pseudocapsicum* fruits (Salema, 1968) and red pepper fruits (B. Camara, unpublished), when incubated with exogenous sucrose, synthesize and accumulate detectable amounts of starch inside the chromoplasts. Similar findings have been observed using *Forsythia suspensa, Laburnum anagyroides,* and *Ranunculus acer* petals, *Lilium croceum* tepals (Keresztes and Schroth, 1979), *Citrus limon* fruits (Kordan, 1971), and *Croton* sp leaves which develop chromoplast-like struc tures (Mesquita and Dias, 1990).

During chloroplast to chromoplast transformation in pepper fruits the activity of ribulose diphosphate caboxylase oxygenase and phosphoribulo-kinase, two key enzymes of the Calvin cycle, decrease up to 80% (Ziegler *et al.,* 1983). In addition, the characteristic post-translational inhibition of the plastidial glucose-6-phosphate dehydrogenase by light (Lendzian and Ziegler, 1970) is lost, while its activity is increased by 140% (Camara, 1981; Ziegler *et al.,* 1983). This phenomenon is a diagnostic feature of the prevailing function of the oxydative pentose phosphate cycle over the reduc-tive cycle. An interesting question therefore regards the origin of carbohy-drates metabolized, and the nature of the precursors imported during chlo-roplast to chromoplast differentiation, in the absence of photosynthetic activity. Our studies (B. Camara, unpublished) reveal that during this period the activity of invertase is increased at least fivefold, and that the resulting glucose and fructose residues represent the main carbohydrates metabo-lized in the sink fruit, illuminating the nature of hexose entering the chromo-plast compartment. We have observed in pepper chromoplasts very high hexose kinase activities (B. Camara, unpublished), suggesting that free hexoses might be transported. The possibility of hexose import into the chloroplast compartment has been raised previously in studies using spinach (Schäfer *et al.,* 1977), and *Sedum* (Piazza *et al.,* 1982), and potato (Kobmann *et al.,* 1992); more recently, functional demonstration of a glucose trans-porter in *Arabidopsis thaliana* chloroplasts has been reported (Trethewey and ap Rees, 1994). Data from developing wheat grain indicate that glucose-1-phosphate is the main source of carbon imported in the endosperm amy-loplasts for starch synthesis (Keeling *et al.,* 1988). This is supported by the fact that endosperm amyloplasts lack fructose-1,6-bisphosphatase (Entwis-tle and ap Rees, 1988), which is a key enzyme for the conversion of triose phosphates into starch in nongreen plastids. Further analysis confirmed that import of hexose phosphate into the plastid is a general characteristic of nonphotosynthetic plastids (Table XII).

Using purified pepper fruit chloroplasts and chromoplasts, we have made three key observations (B. Camara, unpublished). First, glucose-6-phosphate is a better source for starch synthesis in pepper chloroplasts and chromoplasts. Second, fructose-1,6-bisphosphatase is immunologically detectable in pepper chloroplast and chromoplast stroma. Third, the specific activity of fructose-1,6-bisphosphatase is twofold higher in the chromoplasts compared to the chloroplasts isolated from pepper fruits. Collectively these data indicate that in pepper chromoplasts triose phosphates as well as glucose-6-phosphate are imported into the plastid compartment. Preliminary data gained from tomato fruit chloroplasts and chromoplasts reveal that the phosphate translocator catalyzes in addition to inorganic phosphate and 3-phosphoglycerate, the transport of glucose-1-phosphate, and phosphoenolpyruvate (Schünemann and Borchert, 1994). These data suggest that the chromoplast translocator is probably different from that of mesophyll leaf chloroplasts having a C_3 type of photosynthesis, since in that case, phosphoenolpyruvate is poorly transported, while glucose-1-phosphate is

TABLE XII

Substrate Specificity for Starch Synthesis in Nonphotosynthetic Plastids

Plastid	Species	Tissue	Substrate specificity[a]	References[b]
Amyloplast	*Triticum aestivum*	Endosperm	Glc 1 P	1
Amyloplast	*Pisum sativum*	Root	Glc 6 P	2,3
Amyloplast	*Pisum sativum*	Cotyledons	Glc 6 P	4
Amyloplast	*Brassica oleracea*	Floral bud	Glc 6 P	5
Amyloplast	*Glycine max*	Cell culture	Glc 6 P	6
Amyloplast	*Zea mays*	Endosperm	Glc 1 P and Glc 6 P	7
Amyloplast	*Brassica napus*	Embryo	Glc 6 P	8
Amyloplast	*Solanum tubersosum*	Tuber cell culture	Glc 1 P	9
Etioplast	*Hordeum vulgare*	Leaf	DHAP	10
Chromoplast	*Capsicum annuum*	Fruit	Glc 6 P > Glc 1 P	11

[a] Glucose 1 or 6 phosphate (Glc 1P or 6 P), as well as dihydroxy acetone phosphate (DHAP) were used as substrates.

[b] 1. Tyson and ap Rees (1988); 2. Borchert *et al.* (1989); 3. Borchert *et al.* (1993); 4. Hill and Smith (1991); 5. Neuhaus *et al.* (1993b); 6. Coates and ap Rees (1993); 7. Neuhaus *et al.* (1993a); 8. Kang and Rawsthorne (1994); 9. Kosegarten and Mengel (1994); 10. Batz *et al.* (1992); 11. B. Camara, unpublished.

not transported (Fliege *et al.*, 1978). Indeed, the diversity of the phosphate translocator (Heldt *et al.*, 1991) illustrates the specific integration of the plastid metabolism to the rest of the cell, this phenomenon can be inferred from the fact that the phosphate translocator of mesophyll chloroplasts of C_4 plants transports phosphoenolpyruvate and dihydroxyacetone phosphate, in addition to inorganic phosphate, 3-phosphoglycerate, and phosphoenolpyruvate (Gross *et al.*, 1990), while the amyloplast translocator from pea roots transports glucose-6-phosphate in addition to inorganic phosphate, 3-phosphoglycerate, and dihydroxyacetone phosphate (Borchert *et al.*, 1989, 1993). Interestingly, a putative hexose phosphate translocator has been identified from cauliflower bud amyloplasts (Batz *et al.*, 1993).

Further analysis of the fate of imported triose or hexose phosphates into developing pepper chromoplasts has been evaluated by determining the activity of the different enzymes involved in starch metabolism, the glycolytic and the oxidative pentose pathways (B. Camara, unpublished). These data are summarized in Fig. 3, which indicates that pepper chromoplasts have an extended set of enzymes involved in carbohydrate metabolism, in comparison to previous data described for nongreen plastids (Simcox *et al.*, 1977; Journet and Douce, 1985; Frehner *et al.*, 1990; Trimming and Emes, 1993; Entwistle and ap Rees, 1988). One could conclude that these activities could sustain the basic requirements for the different biosynthetic activities operating in the developing chromoplasts. In the future, it will be essential to evaluate the flux of precursors engaged in each plastid pathway and how this process is regulated.

3. Isoprenoid Pathways

a. Carotenoids   One of the crucial issues is whether or not the distal precursor isopentenyl pyrophosphate (IPP) is synthesized in the plastid. Available data indicate that isolated chloroplast plastids have low or no hydroxymethylglutaryl CoA reductase, which, affords the formation of mevalonic acid, the precursor of IPP (Bach, 1987). Until now, the different plant HMG CoA reductase genes cloned encode enzymes localized in the endoplasmic reticulum (Weissenborn *et al.*, 1995). These results reinforce the hypothetic mechanism in which IPP derived from the cytosol is transported through the plastid envelope and used for plastid isoprenoid synthesis (Kreuz and Kleinig, 1984). One major unresolved element of this concept is the need for characterizing the plastid envelope carrier involved in the transport of polar IPP, which has been tentatively adressed (Soler *et al.*, 1993). Furthermore, overexpression of a hamster HMG-CoA reductase cDNA under the control of the constitutive promoter (CaMV 35S) resulted in ten-fold higher HMG-CoA reductase, and a 3- to 5-fold increase in sterol content while the carotenoid content remained unchanged (Chappell *et al.*,

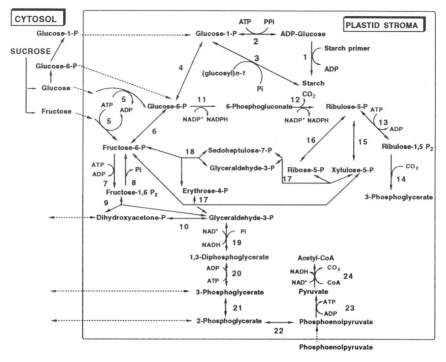

FIG. 3 Pathway of the utilization of sucrose-derived metabolites in sink chromoplasts. Numbers refer to the following enzymes: (1) starch synthase, (2) ADP-glucose pyrophosphorylase, (3) α-glucan phosphorylase, (4) phosphoglucomutase, (5) hexokinase, (6) hexose phosphate isomerase, (7) ATP-phosphofructokinase, (8) fructose,1,6-bisP phosphatase, (9) aldolase, (10) triose-P isomerase, (11) glucose 6-P dehydrogenase, (12) gluconate 6-P dehydrogenase, (13) ribulose 5-P kinase, (14) ribulose 1,5-bisP carboxylase, (15) ribulose 5-P 3-epimerase, (16) ribose 5-P isomerase, (17) transcetolase, (18) transaldolase, (19) glyceraldehyde 3-P dehydrogenase, (20) phosphoglycerate kinase, (21) phosphoglycerate mutase, (22) enolase, (23) pyruvate kinase, (24) pyruvate dehydrogenase.

1991). In contrast with the hypothesis of the cytosolic origin of IPP, it has been demonstrated that the autonomy of plastids to synthesize IPP is subject to developmental regulation, i.e., only immature chloroplasts are capable of synthesizing isoprenoids from CO_2 (Heintze et al., 1990; Hoppe et al., 1993; Gölz and Feierabend, 1993). As the chloroplasts reach their mature stage, the autonomy to synthesize IPP is progressively lost, and IPP imported from the cytosol becomes the prevailing source for plastid isoprenoid synthesis. Based on the procaryotic origin of plant plastids, one may consider an alternate pathway of IPP synthesis from pyruvate which occurs in bacteria (Rohmer et al., 1993). However, recent data (Heintze et al., 1994) indicate a limited contribution of the pyruvate pathway to the plastid pathway of isoprenoid synthesis. It is worth pointing out that a better understanding of plastid autonomy should represent a major topic for future studies.

In contrast, IPP is readily metabolized by isolated chromoplasts to afford carotenoids (Fig. 4) and several geranylgeranyl derivatives (Camara and Monéger, 1980; Beyer et al., 1980). Subsequent studies using pepper (Camara et al., 1982a) and Narcissus chromoplasts (Kreuz et al., 1982) demonstrated that the pathway of carotenoid biosynthesis in chromoplasts is compartmentalized (Camara et al., 1982a; Camara and Monéger, 1982; Kreuz, et al., 1982), i.e., the stroma is the site of phytoene synthesis, as shown in other type of plastids (Dogbo and Camara, 1987a; Dogbo et al., 1987),

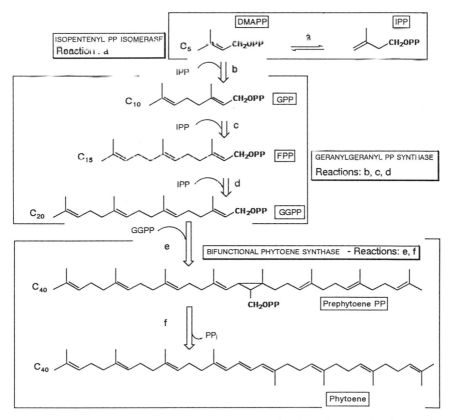

FIG. 4 Pathway of isopentenyl pyrophosphate conversion into phytoene by chromoplast stroma. The three individual enzymes (IPP isomerase, GGPP synthase, and phytoene synthase) and the reactions catalyzed are boxed. The intermediates include DMAPP (dimethylallyl pyrophosphate), GPP (geranyl pyrophosphate), FPP (farnesyl pyrophosphate), GGPP (geranylgeranyl pyrophosphate), and prephytoene pyrophosphate.

while the later steps which involve desaturations of phytoene (Hugueney *et al.*, 1992), cylization of lycopene (Camara and d'Harlingue, 1985; Camara *et al.*, 1985; Camara and Dogbo, 1986), and xanthophyll formation (Camara, 1980; Camara and Monéger, 1981; Bouvier *et al.*, 1994) are catalyzed by the chromoplast membranes.

Earlier studies focused on tomato (Porter and Spurgeon, 1979) and pepper (Camara, 1985b) chromoplasts demonstrated that the different enzymes involved in phytoene synthesis behave as an operationally soluble complex of 160 to 200 kDa. Subsequently, methods were developed to purify and characterize the individual enzymes. A monomeric isopentenyl pyrophosphate isomerase affords dimethylallyl pyrophosphate (Dogbo and Camara, 1987b), which allows subsquent polymerization, catalyzed by a dimeric prenyltransferase designated geranylgeranyl pyrophosphate synthase, to yield geranylgeranyl pyrophosphate (GGPP) (Dogbo and Camara, 1987b). Ultimately a bifunctional phytoene synthase synthase affords the first C_{40} carotenoid phytoene, by dimerization of GGPP (Dogbo *et al.*, 1988) (Fig. 4). The enzymology of the later steps is less known, except for phytoene-phytofluene desaturase (Hugueney *et al.*, 1992) and capsanthin-capsorubin synthase (Bouvier *et al.*, 1994) which have been characterized recently (Fig. 5). Data gained from these studies paved the way for cloning the cDNAs encoding several carotenogenic enzymes which are listed in Table XIII. Steady state analysis of expression of carotenogenic genes by Northern blot reveals a marked difference in the regulation of carotenogenic genes during chromoplast development. The pattern observed is summarized in Table XIII. We are unaware of the signals inducing these changes. One could note that in tomato fruit expressing antisense RNA to ethylene biosynthetic enzymes, the expression of phytoene-phytofluene desaturase is ethylene-independent, and is probably developmentally regulated (Theologis *et al.*, 1993). The preponderance of the developmental regulation over a putative direct ethylene action is further strengthened by the fact that ethylene has an inhibitory role during the greening of etiolated tissues, which occurs concomitant to the activation of carotenoid synthesis (Kang and Stanley, 1972). An unusual effect of methyl jasmonate is the induction of the synthesis of β-carotene in Golden Delicious apple fruits (Pérez *et al.*, 1993) and in tomato fruits (Saniewski and Czapski, 1983; Czapski and Saniewski, 1985). Using a more direct approach, in which the expression of lycopene cyclase gene was analyzed by Northern blot, we could not detect any stimulatory effect of methyl jasmonate (B. Camara, unpublished). Alternatively, it has been shown that photooxidative stress stimulates the expression of phytoene synthase and phytoene-phytofluene desaturase in tomato (Giuliano *et al.*, 1993; Bartley *et al.*, 1994). Obviously, questions about the control of carotenoid accumulation in chromoplasts have been approached in a piecemeal way. Given the current state of our knowledge, it would now seem possible to approach directly the analysis

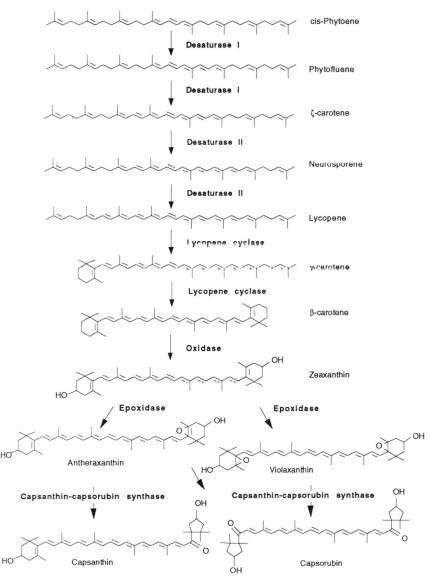

FIG. 5 Pathway of capsanthin and capsorubin synthesis in *C. annuum* chromoplasts.

about the signals and the molecular regulation of carotenoid biosynthesis during chromoplast development.

One of the prominent features of developing chromoplasts is the enhanced synthesis and accumulation of carotenoids. This raises the question of the mechanism of sequestration of excess carotenoids in chromoplasts.

TABLE XIII

Expression of Carotenogenic Genes during Chromoplast Development in Pepper and Tomato

Enzyme	Plant gene	Stage I[a]	Stage II	Stage III
Geranylgeranyl pyrophosphate synthase	GGPS			
	Pepper (1)[b]	−[c]	++	
	Tomato (2)	−	−	
Phytoene synthase	PSY			
	Pepper (3)	−	+	++
	Tomato (4)	−	++	+++
Phytoene-phytofluene desaturase	PDS			
	Pepper (5)	−	+	+
	Tomato (6 or 7)	−	+	+ or ++
Lycopene cyclase	LCY (8)			
	Pepper	−	−	+
Capsanthin-capsorubin synthase	CCS			
	Pepper (9)	−	++	+++

[a] The different stages correspond to green fruits (I), turning fruits (II), and red fruits (III).

[b] 1. Kuntz et al. (1992); 2. M. Kuntz, unpublished results; 3. Gray et al. (1992); 4. Römer et al. (1993); 5. Hugueney et al. (1992); 6. Giuliano et al. (1993); 7. Pecker et al. (1992); 8. Hugueney et al., 1995b; 9. Bouvier et al. (1994).

[c] The plus sign (+) denotes an enhanced expression of the gene, in contrast to the barely detectable level of expression (−).

A suggestion comes from the consideration of the morphological and biochemical consequences observed in several procaryotes and eucaryotes whose lipid metabolism is deregulated. In animals, when the level of cholesterol is too high for its incorporation into membranes or further metabolism, the macrophage differentiates into cells characterized by intense accumulation of lipid bodies containing esterified cholesterol (Fruchart, 1992). A very similar phenomenon occurs in yeast overexpressing hydroxymethylglutaryl, coenzyme A reductase, which develops an intense accumulation of lipid droplets in the cytosol (Saunders, 1991). In plants, a very similar situation has been described, for the first time, in tobacco mutants overproducing sterols (Maillot-Vernier et al., 1991). An extreme situation is more frequently observed in seeds in which excess triacylglycerol is sequestered within oleosomes, which consist of a triacylglycerol core surrounded by a specific protein designated oleosin (Tzen and Huang, 1990). In fact, a very similar phenomenon occurs in carotenogenic alga (Sprey, 1970; Ben-Amotz et al., 1982; Whithers and Haxo, 1978; Lang, 1968; Santos and Mesquita, 1984) and in carotenogenic fungi (Heim, 1946; Eymé and Parriaud, 1970), as well as in developing chromoplasts (Steffen, 1955; Steffen and Walter, 1955; Frey-Wyssling and Kreutzer, 1958b; Ben-Shaul and Klein, 1965; Kuhn,

1970; Ben-Shaul and Naftali, 1969; Harris and Spurr, 1969a,b); in all these cases the overaccumulation of carotenoids at a level far above that which could be incorporated into the membranes may be correlated with specific deposit structures. In developing chromoplasts the most simple structure is the globule, which has been described for the chromoplasts of *Viola tricolor* (Hansmann and Sitte, 1982) and *Caltha palustris* petals and for *Tulipa sylvestris* tepals (Sitte, 1977). In this case the carotenoid to protein ratio is very low (Sitte, 1977). In crystalline chromoplasts, for which the biochemical composition has not been described, one can easily observe that the lycopene crystal (Rosso, 1968; Knoth, 1981) or the β-carotene crystal is sequestered within a membrane (Ben-Shaul *et al.*, 1968). In fibrillar chromoplasts, the presence of a specific protein has been described (Wuttke, 1976; Winkenbach, *et al.*, 1976; Emter *et al.*, 1990; Knoth *et al.*, 1986). Comparative analysis of these different situations provided the framework for a detailed investigation of fibril formation at the biochemical and molecular level in pepper chromoplasts (Deruère *et al.*, 1994). In pepper chromoplasts, it has been shown that the overaccumulated carotenoids are sequestered within the chromoplast fibrils. This process involves the recruitment of a specific 35-kDa protein, designated fibrillin, which induces the formation of a subplastidial fibrillar architecture. Topological and biochemical analysis revealed that the core of fibrils is composed of carotenoids (up to 95% of the chromoplast carotenoids) surrounded by a layer of acyl lipids and finally by fibrillin which is located at the periphery (Deruère *et al.*, 1994). cDNA cloning of fibrillin revealed the presence of the adhesion motif RGD (D'Souza *et al.*, 1991) which could be involved in the alignment of fibrils observed *in vivo*. Analysis of the expression fibrillin gene indicates that its pattern of expression is paralleled by the accumulation of the carotenoids (Deruère *et al.*, 1994). A very similar type of sequestration has been described in the case of *Tropaeolum majus* chromoplasts (Emter *et al.*, 1990).

The overaccumulation of carotenoid in chromoplast can lead to dramatic cellular modification. This trend is observed during chromoplast development in the stigma of saffron (*Crocus sativus*), where increased synthesis of crocetin occurs (B. Camara, unpublished). During this process, the proplastids or amyloplasts develop into chromoplasts. At the very beginning of the development of the stigma, histochemical analysis reveals the presence of plastids staining positively for polysaccharides, as shown by the dark deposit due to starch. Subsequently, the degradation of starch is followed by chromoplast differentiation and the induction of crocetin accumulation. During this period the plastids exhibit a marked provacuolation, similar to that described during the senescing of *Sinapis alba* cotyledons (Hudak, 1981), and which resembles subplastidial organelles designated plastosomes (Arnott and Harris, 1973; Harris and Arnott, 1973) (Fig. 6). Following this step, the chromoplast content is digested, and the degraded chromoplast fuse with the large vacuole. Though correlation of dynamic events and static electron micrograph must be tentative, the membrane continuities observed in *Crocus* stigma point to the fact that crocetin is sequestered into vacuole. This is reinforced by the fact that the polarity of crocetin glycosides found in *Crocus* style follows the essential criterion of water solubility reported for vacuolar sequestered solutes in plant cells (Frey-Wyssling, 1942; Matile, 1990).

Although association between plastids and endoplasmic reticulum is implicated in the transfer of secretory products (Wooding and Northcote, 1965; Whatley *et al.,* 1991), convincing associations between plastids and vacuoles are very rare. So far, this type of association has been observed in the gametophyte of the fern *Pteris vittata* (Crotty and Ledbetter, 1973) and occasionally in a limited group of higher plants, including leaves of *Lycopersicon esculentum* (Crotty and Ledbetter, 1973) and *Hypoestes sanguinolenta* (Vaughn and Duke, 1981) and pepper leaves (Simpson and Lee,

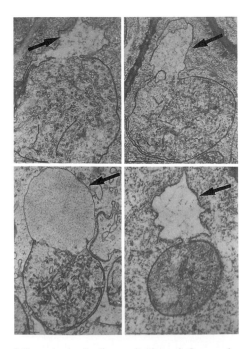

FIG. 6   Initial steps of the autophagic disorganization of *Crocus* chromoplasts during the formation of crocetin in the stigma. The plastid envelope shows an extensive budding indicated by the arrows, before digestion of the plastid content. Bar = 1 μm.

1976b), or more frequently in reproductive tissues (Dickinson, 1973; Pacini *et al.*, 1992; Yu and Russell, 1994) where the lipid material released from the degenerating chromoplasts of the tapetum participates in the synthesis of pollen sporopollenin (Dickinson and Bell, 1972). The mechanism involved in the autophagy of the chromoplasts is reminiscent of the lytic disorganization observed in senescing organs (Matile and Winkenbach, 1971; Hudak, 1981; Wittenbach *et al.*, 1982), in the course of the formation of root sieve elements (Buvat, 1990) and laticifers (Marty, 1978), and during the differentiation of protein bodies (Van der *et al.*, 1980). The mechanism inducing these changes is unknown. Since all these tissues are heterotrophic one could postulate that carbohydrate starvation could induce the autophagic process (James, 1994; El Amrani, 1994). Recent demonstration of an extensive degradation of amyloplasts in rice suspension cells starved of sucrose (Chen *et al.*, 1994) suggests that the availability of sucrose or sucrose derived metabolites could act as an inducer of plastid autophagy.

b. Other Geranylgeranyl Derivatives   Once synthesized in the chromoplast stroma (Kreuz *et al.*, 1982; Camara *et al.*, 1982a), geranylgeranyl pyrophosphate is also engaged in the formation of the side chains of different prenyl lipids, i.e., tocopherols and plastoquinones (Camara *et al.*, 1982b; Gaudillère *et al.*, 1984; Camara and d'Harlingue, 1985; d'Harlingue and Camara, 1985) or chlorophylls (Kreuz and Kleinig, 1981; Dogbo *et al.*, 1984), according a pathway which is practically similar to that occurring in chloroplasts (Soll *et al.*, 1985). Due to the central role of plastids in the synthesis of geranylgeranyl pyrophosphate in plant cells, one may suggest that part of the products is exported in the cytosol for the synthesis of diterpene derivatives which could include geranylgeranylated proteins. Presently, the available data do not give insight into the regulatory mechanisms. Characterizing the individual proteins involved in this process and cloning their cDNAs will provide the basis from which to address the important question of how geranylgeranyl pyrophosphate is channeled into these alternative pathways.

4. Acyl Lipid Metabolism

Previous studies demonstrated that isolated chromoplasts are competent for the activation of acetate into acetyl CoA, which is subsequently engaged in the synthesis of fatty acids (Kleinig and Lievogel, 1978; Camara and Brangeon, 1981). The origin of acetate in the plant cell is still unresolved. The plastidial pyruvate dehydrogenase complex which catalyzes the oxidative decarboxylation of pyruvate with the formation of acetyl CoA is a likely route for acetyl CoA supply (Denyer and Smith, 1988). However, the mitochondrial pyruvate dehydrogenase complex is much more active than its plastid counterpart (Williams and Randall, 1979). Under these conditions, it has been proposed that acetyl CoA formed in the mitochondria is hydrolyzed by an acetyl thioesterase to free acetate which diffuses out of the mitochondria and enters the plastid compartment for further activation into acetyl CoA (Murphy and Stumpf, 1981). Alternatively, the transfer of the acetate anion could be mediated by an acyl carnitine transferase (Masterson *et al.,* 1990). Whatever the origin of acetyl CoA, two enzyme complexes designated acetyl CoA carboxylase and fatty acid synthetase afford fatty acids of different chain lengths (Slabas and Fawcett, 1992). With respect to the formation of complex lipids, one could note that *Narcissus* chromoplast membranes efficiently catalyze the last step of galactolipid synthesis, i.e., the incorporation of galactose residues into monogalactosyldiglyceride and digalactosyldiglyceride (Liedvogel and Kleinig, 1976). Molecular analysis of fatty acid synthesized in *Narcissus* chromoplasts reveals prominent synthesis of saturated fatty acids (Kleinig and Lievogel, 1978). It is worth pointing out that deficiencies in the plastidial (prokaryotic) desaturase involved in the formation of Δ3-*trans*-hexadecenoic acid and Δ7,10,13-hexadecatrienoic acid at the *sn*-2 of glycerol in phosphatidylglycerol and monogalactosyldiglyceride can be inferred from previous analysis (Camara and Monéger, 1977; Whitaker, 1992).

5. Nitrogen Metabolism

The key involvement of nongreen plastids in the reduction of nitrite to ammonia is well known (Washitani and Sato, 1977; Emes and Fowler, 1983). On the other hand, data obtained from pepper chromoplasts indicate that nitrite reductase activity is barely detectable (B. Camara, unpublished results). The additional step in which the first carbon to nitrogen bond (C–N) is formed involves the assimilation of ammonia. This step is catalyzed by glutamine synthetase which is apparently absent from tomato chromoplasts (Gallardo *et al.,* 1988). In addition, the activity of glutamate synthase is 80% decreased during tomato fruit ripening (Gallardo *et al.,* 1993). One may conclude that, under these conditions, the plastid glutamine synthetase–glutamate synthase cycle which plays a major role in the synthesis of amino acids is no longer operating in tomato chromoplasts, and that most of the amino acids required for chromoplast metabolism probably are imported from the cytosol and ultimately from the photosynthetic leaves. Interestingly, these transports could be facilitated by plasmalemma amino acid carriers and the plastid glutamine–glutamate transporters (Yu and Woo, 1988; Flügge *et al.,* 1988). In marked contrast, cysteine synthase, which catalyzes the formation of the first carbon to sulfur bond (C–S), operates actively in pepper chromoplasts (Römer *et al.,* 1992). Further analysis revealed that cysteine synthase gene is up-regulated during chromoplast development in pepper fruits (Römer *et al.*, 1992). This phenomenon occurs in unison with increased glutathione accumulation (Römer *et al.*, 1992), suggesting that the increased cysteine synthase activity might participate in the biosynthesis of glutathione.

VII. Conclusion

In contrast to chloroplasts, which have been the subject of considerable studies devoted to photosynthetic activity, the physiology of chromoplasts has been explored only recently. The accumulated data display the multiple facets and the amazing flexibility of plastid development. Despite the major advances that have been made, no clear picture has yet emerged in understanding the signaling mechanism involved during chromoplast differentiation. The discovery that some critical factors and genes for regulation of cell development have been conserved in the eukaryotic kingdom, with a certain amount of redundancy, raises the possibility that the basic features of chromoplast development will operate in a very similar way. Given the number of signaling pathways described so far, it is necessary to dissect those specifically triggering the development of chromoplasts.

Acknowledgments

We thank past and present colleagues who contributed to our study on chromoplast development over the past 20 years.

References

Abdel-Gawad, H., and Romani, R. J. (1974). Hormone-induced reversal of colour change and related respiratory effects in ripening apricot fruits. *Physiol. Plant.* **32**, 161–165.

Adato, I., Gazit, S., and Blumenfeld, A. (1976). Relationship between changes in abscisic acid and ethylene production during ripening of avocado fruits. *Aust. J. Plant Physiol.* **3**, 553–559.

Akiyama, Y., Ogura, T., and Ito, K. (1994a). Involvement of FtsH in protein assembly into and through the membrane. I. Mutations that reduce retention efficiency of cytoplasmic reporter. *J. Biol. Chem.* **269**, 5218–5224.

Akiyama, Y., Shirai, Y., and Ito, K. (1994b). Involvement of FtsH in protein assembly into and through the membrane. II. Dominant mutations affecting FtsH functions. *J. Biol. Chem.* **269**, 5225–5229.

Amir-Shapira, D., Goldschmidt, E. E., and Altman, A. (1987). Chlorophyll catabolism in senescing plant tissues: *In vivo* breakdown intermediates suggest different pathways for *Citrus* fruit and parsley leaves. *Proc. Natl. Acad. Sci. U.S.A.* **84**, 1901–1905.

Ardila, F., Pozueta-Romero, J., and Akazawa, T. (1993). Adenylate uptake by proplastids from cultured cells of tobacco (*Nicotiana tabacum* L. cv. BY2) indicates that an adenylate translocator is present in all types of plastid. *Plant Cell Physiol.* **34,** 237–242.

Arnott, H. J., and Harris, J. B. (1973). Development of chloroplast substructures with apparent secretory roles in the young tobacco leaf. *Tissue Cell* **5,** 337–347.

Arnott, H. J., and Smith, K. M. (1967). Electron microscopy of virus-infected sunflower leaves. *J. Ultrastruct. Res.* **19,** 173–195.

Arnott, H. J., Rosso, S. W., and Smith, K. M. (1969). Modification of plastid ultrastructure in tomato leaf cells infected with tobacco mosaic virus. *J. Ultrastruct. Res.* **27,** 149–167.

Bach, T. J. (1987). Synthesis and metabolism of mevalonic acid in plants. *Plant Physiol. Biochem.* **25,** 163–178.

Bachmair, A., Finley, D., and Varshavsky, A. (1986). *In vivo* half-life of a protein is a function of its amino-terminal residue. *Science* **234,** 179–186.

Bailey, J. L., and Whyborn, A. G. (1963). The osmiophilic globules of chloroplasts: II. Globules of the spinach beet, chloroplast. *Biochim. Biophys. Acta* **78,** 163–174.

Baker, L. R., and Tomes, M. L. (1964). Carotenoids and chlorophylls in two tomato mutants and their hybrid. *Proc. Am. Soc. Hortic. Sci.* **85,** 507–513.

Baqar, M. R., and Lee, T. H. (1978). Interaction of CPTA and high temperature on carotenoid synthesis in tomato fruit. *Z. Pflanzenphysiol.* **88,** 431–435.

Barmore, C. R. (1975). Effect of ethylene on chlorophyllase activity and chlorophyll content in calamondin rind tissue. *HortScience* **106,** 595–596.

Barry, P., Evershed, R. P., Young, A., Prescott, M. C., and Britton, G. (1992). Characterization of carotenoid acyl esters produced in drought-stressed barley seedlings. *Phytochemistry* **31,** 3163–3168.

Bartley, G. E., Scolnik, P. A., and Giuliano, G. (1994). Molecular biology of carotenoid biosynthesis in plants. *Annu. Rev. Plant Physiol. Plant Mol. Biol.* **45,** 287–301.

Barton, R. (1966). Fine structure of mesophyll cells in senescing leaves of *Phaseolus*. *Planta* **71,** 314–325.

Bathgate, B., Purton, M. E., Grierson, D., and Goodenough, P. W. (1985). Plastid changes during the conversion of chloroplasts to chromoplasts in ripening tomatoes. *Planta* **165,** 197–204.

Bathgate, B., Goodenough, P. W., and Grierson, D. (1986). Regulation of the expression of the *psb A* gene in tomato fruit chloroplasts and chromoplasts. *J. Plant Physiol.* **124,** 223–233.

Batz, O., Scheibe, R., and Neuhaus, H. E. (1992). Transport processes and corresponding changes in metabolites levels in relation to starch synthesis in barley (*Hordeum vulgare* L.). *Plant Physiol.* **100,** 184–190.

Batz, O., Scheibe, R., and Neuhaus, H. E. (1993). Identification of the putative hexose-phosphate translocator of amyloplasts from cauliflower buds. *Biochem. J.* **294,** 15–17.

Bauer, H., Shepherd, D., and Sigarlakie, E. (1974). Osmiophilic structure in pigment-producing *Flavobacterium* mutants. *J. Bacteriol.* **118,** 747–755.

Beale, S. I. (1984). Biosynthesis of photosynthetic pigments. *In* "Chloroplast Biogenesis" (N. R. Baker and J. Barber, eds.), pp. 133–205. Elsevier North-Holland, Amsterdam.

Ben-Amotz, A., Katz, A., and Avron, M. (1982). Carotenes in halotolerant algae: Purification and characterization of β-carotene-rich globules from *Dunaliella bardawill* (chlorophyceae). *J. Phycol.* **18,** 529–537.

Benedict, C. R., Rosenfield, C. L., Mahan, J. R., Madhavan, S., and Yokoyama, H. (1985). The chemical regulation of carotenoid biosynthesis in *Citrus*. *Plant Sci.* **41,** 169–173.

Ben-Shaul, Y. (1962). Studies on the carotene bodies in the carrot root. Ph.D. Thesis, Hebrew University, Jerusalem.

Ben-Shaul, Y., and Klein, S. (1965). Development and structure of carotene bodies in carrot roots. *Bot. Gaz.* (*Chicago*) **126,** 79–85.

Ben-Shaul, Y., and Naftali, Y. (1969). The development and ultrastructure of lycopene bodies in chromoplast of *Lycopersicon esculentum*. *Protoplasma* **67**, 333–344.

Ben-Shaul, Y., Shmueli, U. (1968). Electron diffraction of carotenoid pigments in chromoplasts. *Proc. Eur. Reg. Conf. Electron. Microsc., 4th,* Rome, 1968, pp. 411–412.

Ben-Shaul, Y., Schiff, J. A., and Epstein, H. T. (1964). Studies of chloroplast development in *Euglena*. VI. Fine structure of the developing plastid. *Plant Physiol.* **39**, 213–240.

Ben-Shaul, Y., Treffry, T., and Klein, S. (1968). Fine structure of carotene body development. *J. Microsc (Paris)* **7**, 265–274.

Berger, C., and Feierabend, J. (1967). Plastidenentwicklung und Bildung von Photosynthese-Enzymen in etiolierten Roggenkeimlingen. *Physiol. Vég.* **5**, 109–122.

Beyer, P., Kreuz, K., and Kleinig, H. (1980). β-carotene synthesis in isolated chromoplasts from *Narcissus pseudonarcissus*. *Planta* **150**, 435–438.

Beyer, P., Kröncke, U., and Nievelstein, V. (1991). On the lycopene isomerase/cyclase reaction in *Narcissus pseudonarcissus* L. chromoplasts. *J. Biol. Chem.* **266**, 17072–17078.

Borchert, S., Grobe, H., and Heldt, H. W. (1989). Specific transport of inorganic phosphate and 3-phosphoglycerate into amyloplasts from pea roots. *FEBS Lett.* **253**, 183–186.

Borchert, S., Harborth, J., Schünemann, D., Hoferichter, P., and Heldt, H. W. (1993). Studies of the enzymic capacities and transport properties of pea root plastids. *Plant Physiol.* **101**, 303–312.

Bornman, C. R. (1968). Observations on the ultrastructure of chromoplasts in *Celastrus scandens*. *J. S. Afr. Bot.* **34**, 295–301.

Bouvier, F., Hugueney, P., d'Harlingue, A., Kuntz, A., and Camara, B. (1994). Xanthophyll biosynthesis in chromoplast: Isolation and molecular cloning of an enzyme catalyzing the conversion of 5,6-epoxycarotenoid into ketocarotenoid. *Plant J.* **6**, 45–54.

Bowsher, C. G., Boulton, E. L., Rose, J., Nayagam, S., and Emes, M. J. (1992). Reductant for glutamate synthase is generated by the oxidative pentose phosphate pathway in non-photosynthetic root plastids. *Plant J.* **2**, 893–898.

Bowsher, C. G., Hucklesby, D. P., and Emes, M. J. (1993). Induction of ferredoxin-NADP$^+$ oxidoreductase and ferredoxin synthesis in pea root plastids during nitrate assimilation. *Plant J.* **3**, 463–467.

Boyer, C. D. (1989). Genetic control of chromoplast formation during fruit development of *Cucurbita pepo* L. *In* "Physiology, Biochemistry, and Genetics of Nongreen Plastids" (C. D. Boyer, J. Shannon, and R. Hardison, eds.), pp. 241–252. Am. Soc. Plant Physiol., Rockville, MD.

Boyle, S. A., Hemmingsen, S. M., and Dennis, D. T. (1986). Uptake and processing of the precursor to the small subunit of ribulose 1,5-bisphosphate carboxylase by leucoplasts from the endosperm of developing castor oil seeds. *Plant Physiol.* **81**, 817–822.

Boyle, S. A., Hemmingsen, S. M., and Dennis, D. T. (1990). Energy requirement for the import of protein into plastids from developing endosperm of *Ricinus communis* L. *Plant Physiol.* **92**, 151–154.

Brangeon, J., and Nato, A. (1981). Heterotrophic tobacco cell cultures during greening. 1. Chloroplast and cell development. *Physiol. Plant.* **53**, 327–334.

Bucholtz, M. L., Maudinas, B., and Porter, J. W. (1977). Effects of *in vivo* inhibitors of carotene biosynthesis on the synthesis of carotenes by a soluble tomato plastid enzyme system. *Chem.-Biol. Interact.* **17**, 359–362.

Buscalioni, L., and Bruno, F. (1928). Sui cloroplasti cromici delle *Aloinee*. *Malpighia* **31**, 50–67.

Butler, R. D., and Simon, E. W. (1971). Ultrastructural aspects of senescence in plants. *Adv. Gerontol. Res.* **3**, 73–129.

Buvat, R. (1969). Plastids. *In* "Plant Cells" (R. Buvat, ed.), pp. 150–198. McGraw-Hill, New York.

Buvat, R. (1990). Histological differentiation of vascular plants. *In* "Ontogeny, Cell Differentiation and Structure of Vascular Plants" (R. Buvat, ed.), pp. 227–557. Springer-Verlag, Berlin and New York.

Camara, B. (1980). Carotenoid biosynthesis. *In vitro* conversion of violaxanthin to capsorubin by a chromoplast enriched fraction of *Capsicum* fruits. *Biochem. Biophys. Res. Commun.* **93,** 113–117.

Camara, B. (1981). Mécanismes et facteurs de la biogenèse des caroténoïdes dans les chromoplastes du fruit de Poivron (*Capsicum annuum* L.). Ph.D. Thesis, Université Pierre et Marie Curie (Paris VI), Paris.

Camara, B. (1985a). Carotene synthesis in *Capsicum* chromoplasts. *In* "Methods in Enzymology" (J. H. Law and H. C. Rilling, eds.), Vol. 110, pp. 244–253. Academic Press, New York.

Camara, B. (1985b). Carotenogenic enzymes from *Capsicum* chromoplasts. *Pure Appl. Chem.* **57,** 675–677.

Camara, B. (1993). Plant phytoene synthase complex: Component enzymes, immunology, and biogenesis. *In* "Methods in Enzymology" (L. Packer, ed.), Vol. 214, pp. 352–365. Academic Press, San Diego, CA.

Camara, B., Bardat, F., and Monéger, R. (1982a). Sites of biosynthesis of carotenoids in *Capsicum* chromoplasts. *Eur. J. Biochem.* **127,** 255–258.

Camara, B., Bardat, F., Sèye, A., d'Harlingue, A., and Monéger, R. (1982b). Terpenoid metabolism in plastids. Localization of α-tocopherol synthesis in *Capsicum* chromoplasts. *Plant Physiol.* **70,** 1562–1563.

Camara, B., Bardat, F., Dogbo, O., Brangeon, J., and Monéger, R. (1983). Terpenoid metabolism in plastids. Isolation and biochemical characteristics of *Capsicum annuum* chromoplasts. *Plant Physiol.* **3,** 94–99.

Camara, B., Bousquet, J., Cheniclet, C., Carde, J. P., Kuntz, M., Evrard, J. L., and Weil, J. H. (1989). Enzymology of isoprenoid biosynthesis and expression of plastid and nuclear genes during chromoplast differentiation in pepper fruits (*Capsicum annuum*). *In* "Physiology, Biochemistry, and Genetics of Nongreen Plastids" (C. D. Boyer, J. C. Shannon, and R. C. Hardison, eds.), pp. 141–156. Am. Soc. Plant Physiol., Rockville, MD.

Camara, B., and Brangeon, J. (1981). Carotenoid metabolism during chloroplast to chromoplast transformation in *Capsicum* annuum. *Planta* **151,** 359–364.

Camara, B., and d'Harlingue, A. (1985). Demonstration and solubilization of S-adenosylmethionine: γ-tocopherol methyltransferase from *Capsicum* chromoplasts. *Plant Cell Rep.* **4,** 31–32.

Camara, B., and Dogbo, O. (1986). Demonstration and solubilization of lycopene cyclase from *Capsicum* chromoplast membranes. *Plant Physiol.* **80,** 172–174.

Camara, B., Dogbo, O., d'Harlingue, A., and Bardat, F. (1985). Inhibition of lycopene cyclization by *Capsicum* chromoplast membranes by 2-aza-2,3-dihydrosqualene. *Phytochemistry* **24,** 2751–2752.

Camara, B., and Monéger, R. (1977). Les lipides du fruit vert et du fruit mûr de Poivron (*Capsicum annuum* L.). *Physiol. Vég.* **15,** 711–722.

Camara, B., and Monéger, R. (1978). Free and esterified carotenoids in green and red fruits of *Capsicum* annuum. *Phytochemistry* **17,** 91–93.

Camara, B., and Monéger, R. (1980). Carotenoid biosynthesis: Biogenesis of capsanthin and capsorubin in pepper fruits (*Capsicum annuum*). *In* "Biogenesis and Function of Plant Lipids" (P. Mazliak, P. Benveniste, C. Costes, and R. Douce, eds.), pp. 363–367. Elsevier/North-Holland Biomed. Press, Amsterdam.

Camara, B., and Monéger, R. (1981). Carotenoid biosynthesis. *In vitro* conversion of antheraxanthin to capsanthin by a chromoplast enriched fraction of *Capsicum* fruits. *Biochem. Biophys. Res. Commun.* **99,** 1117–1122.

Camara, B., and Monéger, R. (1982). Biosynthetic capabilities and localization of enzymatic activities in carotenoid metabolism of *Capsicum annuum* isolated chromoplasts. *Physiol. Vég.* **20,** 757–773.

Camefort, H. (1964). Evolution de la structure des plastes pendant la maturation de l'arille de l'if (*Taxus baccata* L.). *C. R. Hebd. Seances Acad. Sci.* **258,** 1017–1020.

Caprio, J. M. (1956). An analysis of the relation between regreening of Valencia oranges and mean monthly temperatures in southern California. *Proc. Am. Soc. Hortic. Sci.* **67,** 222–235.

Carde, J. P., Camara, B., and Cheniclet, C. (1988). Absence of ribosome in *Capsicum* chromoplasts. *Planta* **173,** 1–11.

Catesson, A. M. (1970). Evolution des plastes de pomme au cours de la maturation du fruit. Modifications ultrastructurales et accumulation de ferritine. *J. Microsc. (Paris)* **9,** 949–974.

Cervantes-Cervantes, M., Hadjeb, N., Newman, L. A., and Price, C. A. (1990). ChrA is a carotenoid-binding protein in chromoplasts of *Capsicum annuum. Plant Physiol.* **93,** 1241–1243.

Chang, C., Kwok, S. F., Bleecker, A. B., and Meyerowitz, E. M. (1993). *Arabidopsis* ethylene-response gene *ETR1*: Similarity of product to two-component regulators. *Science* **262,** 539–544.

Chappell, J., Proulx, J., Wolf, F., Cuellar, R. E., and Saunders, C. (1991). Is HMG-CoA reductase a rate limiting step for isoprenoid metabolism? *Plant Physiol.* **96S,** 127.

Chen, M. H., Liu, L. F., Chen, Y. R., Wu, H. K., and Yu, S. M. (1994). Expression of α-amylases, carbohydrate metabolism, autophagy in cultured rice cells is coordinately regulated by sugar nutrient. *Plant J.* **6,** 625–636.

Cheung, A. Y., McNellis, T., and Piekos, B. (1993). Maintenance of chloroplast components during chromoplast differentiation in the tomato mutant *green flesh. Plant Physiol.* **101,** 1223–1229.

Clarke, A. K., Gustafsson, P., and Lidholm, J. A. (1994). Identification and expression of the chloroplast *Clp P* gene in the conifer *Pinus contorta. Plant Mol. Biol.* **26,** 851–862.

Cline, K., Henry, R., Li, C., and Yuan, J. (1993). Multiple pathways for protein transport into or across the thylakoid membrane. *EMBO J.* **12,** 4105–4114.

Clowes, F. A. L., and Juniper, B. E. (1968). Cytoplasmic organelles. *In* "Cytoplasmic Organelles" (F. A. L. Clowes and B. E. Juniper, eds.), pp. 122–202. Blackwell, Oxford.

Coates, S. A., and ap Rees, T. (1993). Carbohydrate oxidation by leucoplasts from suspension cultures of soybean (*Glycine max* L.). *Planta* **189,** 516–521.

Coggins, C. W., Hield, H. Z., and Burns, R. M. (1962). The influence of potassium gibberellate on grapefruit trees and fruit. *Proc. Am. Soc. Hortic. Sci.* **81,** 223–226.

Coggins, C. W., Henning, G. G., and Yokoyama, H. (1970). Lycopene accumulation induced by 2-(4-chlorophenylthio) triethylamine hydrochloride. *Science* **168,** 1589–1590.

Coggins, C. W., Hield, H. Z., and Garber, M. J. (1980). The influence of potassium gibberellate on valencia orange trees and fruits. *Proc. Am. Soc. Hortic. Sci.* **76,** 193–198.

Courchet, M. (1888). Recherches sur les chromoleucites. *Ann. Sci. Nat. Bot. Biol. Veg.* **1,** 263–374.

Cran, D. G., and Possingham, J. V. (1973). The fine structure of avocado plastids. *Ann. Bot. (London)* [N.S.] **37,** 993–997.

Criqui, M. C., Durr, A., Parmentier, Y., Marbach, J., Fleck, J., and Jamet, E. (1992). How are photosynthetic genes repressed in freshly-isolated mesophyll protoplasts of *Nicotiana sylvestris? Plant Physiol. Biochem.* **30,** 597–601.

Crotty, W. J., and Ledbetter, M. C. (1973). Membrane continuities involving chloroplast and other organelles in plant cells. *Science* **182,** 839–841.

Czapski, J., and Saniewski, M. (1985). Effect of methyljasmonate on carotenoids in tomato fruits. *Gartenbauwissenschaft.* **50,** 35–37.

Dahlin, C., and Cline, K. (1991). Developmental regulation of the plastid protein import apparatus. *Plant Cell* **3,** 1131–1140.

Dahlin, C., and Ryberg, H. (1986). Accumulation of phytoene in plastoglobuli of SAN-9789 (Norflurazon)-treated dark-grown wheat. *Physiol. Plant.* **68,** 39–45.

Dahlin, C., Ryberg, H., and Axelsson, L. (1983). A possible structural role for carotenoids and carotenoid precursors in etioplasts. *Physiol. Plant.* **59,** 562–566.

Daito, H., and Hirose, K. (1970). Studies on acceleration of coloring or degreening of citrus fruit. II. Effects of ethrel (ethylene releasing compound) on the acceleration of coloring and carotenoid pattern of the Natsudaidai (*Citrus natudaidai* Hayata). *Bull. Hortic. Res. Stn., Ser. B* **10**, 35–50.

Dalton, C. C., and Street, H. E. (1977). The influence of applied carbohydrates on the growth and greening of cultured spinach *Spinacia oleracea* L. cells. *Plant Cell Physiol.* **10**, 157–164.

Darby, L. A. (1978). Isogenic lines of tomato fruit colour mutants. *HortScience* **18**, 73–84.

Davey, J. E., and Van Staden, J. (1978). Endogenous cytokinins in the fruits of ripening and non-ripening tomatoes. *Plant Sci. Lett.* **11**, 359–364.

Debesh, K., and Apel, K. (1983). The function of proteases during the light-dependent transformation of etioplasts to chloroplasts in barley (*Hordeum vulgare* L.). *Planta* **157**, 381–383.

De Boer, D., Cremers, F., Teerstra, R., Smits, L., Hille, J., Smeekens, S., and Weisbeek, P. (1988). *In vivo* import of plastocyanin and a fusion protein into developmentally different plastids of transgenic plants. *EMBO J.* **7**, 2631–2635.

De Greef, J. A. (1992). What do we know about the biosynthesis and the development of secondary thylakoids (grana formation) in higher plant systems? *In* "Regulation of Chloroplast Biogenesis" (J. H. Argyroudi-Akoyunoglou, ed.), pp. 551–559. Plenum, New York.

Deng, X. W., Stern, D. B., Tonkyn, J. C., and Gruissem, W. (1987). Plastid-run on transcription: Application to determine the transcriptional regulation of spinach genes. *J. Biol. Chem.* **262**, 9641–9648.

Denison, E. L. (1948). Tomato color as influenced by variety and environment. *Proc. Am. Soc. Hortic. Sci.* **51**, 349–356.

Denison, E. L. (1951). Carotenoid content of tomato fruits as influenced by environment and variety. 1. Effect of temperature and light. *Iowa State Coll. J. Sci.* **25**, 549–564.

Denny, F. E. (1924). Hastening the colorations of lemons. *J. Agric. Res.* **27**, 757–769.

Denyer, K., and Smith, A. M. (1988). The capacity of plastids from developing pea cotyledons to synthesize acetyl CoA. *Planta* **173**, 172–782.

Deruère, J., Römer, S., d'Harlingue, A., Backhaus, R. A., Kuntz, M., and Camara, B. (1994). Fibril assembly and carotenoid over accumulation: A model for supramolecular lipoprotein structures. *Plant Cell* **6**, 119–133.

Devidé, Z. (1970). Ultrastructural changes of plastids in ripe fruit of *Cucurbita pepo* var. ovifera. *Acta Bot. Croat.* **29**, 57–62.

Devidé, Z., and Ljubesic, N. (1972). Plastid tranformations in pumpkin fruits. *Naturwissenschaften* **59**, 39–40.

Devidé, Z., and Ljubesic, N. (1974). The reversion of chromoplasts to chloroplasts in pumpkin fruits. *Z. Pflanzenphysiol.* **73**, 296–306.

Dexheimer, J., Gérard, J., Boudarga, K., and Christine, J. (1990). Les plastes des cellules-hôtes des mycorhizes à vésicules et arbuscules. *Can. J. Bot.* **68**, 50–55.

d'Harlingue, A., and Camara, B. (1985). Plastid enzymes of terpenoid biosynthesis. Purification and characterization of γ-tocopherol methyltransferase from *Capsicum* chromoplasts. *J. Biol. Chem.* **260**, 15200–15203.

Dickinson, H. G. (1973). The role of plastids in the formation of pollen grain coating. *Cytobios* **8**, 25–40.

Dickinson, H. G., and Bell, R. R. (1972). The role of tapetum in the formation of sporopollenin-containing structures during microsporogenesis in *Pinus banksiana*. *Planta* **107**, 205–212.

Dodge, J. D., and Lawes, G. B. (1974). Plastid ultrastructure in some parasitic and semi-parasitic plants. *Cytobiologie* **9**, 1–9.

Dogbo, O., and Camara, B. (1987a). Metabolism of plastid terpenoids. II. Regulation of phytoene synthesis in plastid stroma isolated from higher plants. *Plant Sci.* **49**, 103–109.

Dogbo, O., and Camara, B. (1987b). Purification of isopentenyl pyrophosphate isomerase and geranylgeranyl pyrophosphate synthase from *Capsicum annuum* by affinity chromatography. *Biochim. Biophys. Acta* **920**, 140–148.

Dogbo, O., Bardat, F., and Camara, B. (1984). Terpenoid metabolism in plastids: Activity, localization and substrate specificity of chlorophyll synthetase in *Capsicum annnum* plastids. *Physiol. Vég.* **22**, 75–82.

Dogbo, O., Bardat, F., Laferrière, A., Quennemet, J., Brangeon, J., and Camara, B. (1987). Metabolism of plastid terpenoids. I. Biosynthesis of phytoene in plastid stroma isolated from higher plants. *Plant Sci.* **49**, 89–101.

Dogbo, O., Laferrière, A., d'Harlingue, A., and Camara, B. (1988). Isolation and characterization of a bifunctional enzyme catalyzing the synthesis of phytoene. *Proc. Natl. Acad. Sci. U.S.A.* **85**, 7054–7058.

Dostal, H. C., and Leopold, A. C. (1967). Gibberellin delays ripening of tomatoes. *Science* **158**, 1579–1580.

Douce, R., and Joyard, J. (1990). Biochemistry and function of the plastid envelope. *Annu. Rev. Cell Biol.* **6**, 173–216.

D'Souza, S. E., Ginsberg, M. H., and Plow, E. F. (1991). Arginyl-glycyl-aspartic acid (RGD): A cell adhesion motif. *Trends Biol. Sci.* **16**, 246–250.

Duggar, B. M. (1913). Lycopersicin, the red pigment of the tomato, and the effects of conditions upon its development. *Wash. Univ. Stud., Sci. Ser.* **1**, 22–45.

Eakle, K. A., Bernstein, M., and Emr, S. D. (1988). Characterization of a component of the yeast secretion machinery: Identification of the *SEC*18 gene product. *Mol. Cell. Biol.* **8**, 4098–4109.

Eaks, I. L. (1977). Physiology and degreening—Summary and discussion of related topics. *Proc. Int. Soc. Citric.* **1**, 223–226.

Eilati, S. K., Goldschmidt, E. E., and Monselise, S. P. (1969). Hormonal control of colour changes in orange peel. *Experientia* **25**, 209–210.

Eilati, S. K., Budowski, P., and Monselise, S. P. (1972). Xanthophyll esterification in the flavedo of *Citrus* fruit. *Plant Cell Physiol.* **13**, 741–746.

Eilati, S. K., Budowski, P., and Monselise, S. P. (1975). Carotenoid changes in the "Shamouti" orange peel during chloroplast-chromoplast transformation on and off the tree. *J. Exp. Bot.* **26**, 624–632.

Eisermann, A., Tiller, K., and Link, G. (1990). *In vitro* transcription and DNA binding characteristics of chloroplast and etioplast extracts from mustard (*Sinapis alba*) indicate differential usage of the *psbA* promoter. *EMBO J.* **9**, 3981–3987.

El Amrani, A. (1994). Retard à l'émergence des cotylédons de betterave sucrière (*Beta vulgaris* L.): Mise en évidence d'une carence carbonée et étude de ses effets sur les plastes et la transition hétérotrophie-autotrophie. Ph.D. Thesis, Université Bordeaux I, Bordeaux.

Emes, M. J., and Fowler, M. W. (1983). The supply of reducing power for nitrite reduction in plastids of seedlings pea roots (*Pisum sativum* L.). *Planta* **158**, 97–102.

Emter, O., Falk, H., and Sitte, P. (1990). Specific carotenoids and proteins as prerequisites for chromoplast tubule formation. *Protoplasma* **157**, 128–135.

Entwistle, G., and ap Rees, T. (1988). Enzymatic capacities of amyloplasts from wheat (*Triticum aestivum*) endosperm. *Biochem. J.* **255**, 391–396.

Erdmann, R., Wiebel, F. F., Flessau, A., Rytka, J., Beyer, A., Fröhlich, K. U., and Kunau, W. H. (1991). *PAS1*, a yeast gene required for peroxisome biogenesis, encodes a member of a novel family of putative ATPases. *Cell* (*Cambridge, Mass.*) **64**, 499–510.

Eymé, J. (1967). Nouvelles observations sur l'infrastructure de tissus nectarigènes floraux. *Botaniste* **50**, 169–183.

Eymé, J. (1971). La structure des plastes dans les parenchymes des coupes réceptaculaires florales du Pêcher *Prunus persica* chez les variétés à fruits à chair blanche et à chair jaune. *C. R. Hebd. Seances Acad. Sci., Ser. D* **272**, 1232–1235.

Eymé, J., and Parriaud, H. (1970). Au sujet de l'infrastructure des hyphes de *Clathrus cancellatus* Tournefort, Champignon gastéromycète. *C. R. Seances Acad. Sci., Ser. D* **270**, 1890–1892.

Falk, H. (1976). Chromoplasts of *Tropaeolum majus:* Structure and development. *Planta* **128,** 15–22.

Faludi-Daniel, A. (1975). Pigment synthesis and photosynthetic activity in carotenoid deficient mutants of maize. *In* "Genetic Aspects of Photosynthesis" (Y. E. Nasyrov and Z. Sestak, eds.), pp. 239–245. Junk Publishers, The Hague.

Faludi-Daniel, A., Lang, A., and Fradkin, L. I. (1966). The state of chlorophyll a in leaves of some carotenoid mutant maize. *In* "Biochemistry of Chloroplasts" (T. Goodwin, ed.), pp. 269–274. Academic Press, London.

Faludi-Daniel, A., Fridvalszky, L., and Gyurjan, I. (1968). Pigment composition and plastid structure in leaves of carotenoid mutant maize. *Planta* **78,** 184–195.

Feller, U., and Fischer, A. (1994). Nitrogen metabolism in senescing leaves. *Crit. Rev. Plant Sci.* **13,** 241–273.

Fleck, J., Durr, A., Lett, M. C., and Hirth, L. (1979). Changes in proteins synthesis during the initial stage of life of tobacco protoplasts. *Planta* **145,** 279–285.

Fliege, R., Flügge, N. I., Werdan, K., and Heldt, H. W. (1978). Specific transport of inorganic phosphate, 3-phosphoglycerate, and triose phosphates across the inner membrane of the envelope in spinach chloroplasts. *Biochim. Biophys. Acta* **502,** 232–247.

Flügge, I. U., Woo, K. C., and Heldt, H. W. (1988). Characteristics of 2-oxoglutarate and glutamate transport in spinach chloroplasts. Studies with a double-silicone-layer centrifugation technique and in liposomes. *Planta* **174,** 534–541.

Fosket, D. E., and Radin, D. N. (1983). Induction of carotenogenesis in cultured cells of *Lycopersicon esculentum. Plant Sci.* **30,** 165–175.

Fray, R. G., and Grierson, D. (1993). Identification and genetic analysis of normal and mutant phytoene synthase genes of tomato by sequencing, complementation and co-suppression. *Plant Mol. Biol.* **22,** 589–602.

Frehner, M., Pozueta-Romero, J., and Akazawa, T. (1990). Enzyme sets of glycolysis, gluconeogenesis, and oxidative pentose phosphate pathway are not complete in nongreen highly purified amyloplasts of sycamore (*Acer pseudoplatanus* L.) cell suspension culture. *Plant Physiol.* **94,** 538–544.

Frey-Wyssling, A. (1942). Zur Physiologie der pflanzlichen Glukoside. *Naturwissenchaften* **30,** 500–503.

Frey-Wyssling, A. (1967). Über die Carotinkristalline in der Nebenkrone der Narzissenblute. *An. Edafol. Agrobiol.* **26,** 25–32.

Frey-Wyssling, A., and Kreutzer, E. (1958a). Die submikroskopische Entwicklung der Chromoplasten in den Blüten von *Ranunculus repens* L. *Planta* **51,** 104–114.

Frey-Wyssling, A., and Kreutzer, E. (1958b). The submicroscopic development of chromoplasts in the fruits of *Capsicum annuum. J Ultrastruct. Res.* **1,** 397–411.

Frey-Wyssling, A., and Kuhn, H. (1969). Zur Kristallographie der pflanzlichen Carotinkriställchen. *Mikroskopie* **25,** 78–86.

Frey-Wyssling, A., and Schwegler, J. (1965). Ultrastructure of the chromoplasts in the carrot root. *J. Ultrastruct. Res.* **13,** 543–559.

Fruchart, J. C. (1992). Le transport du cholestérol et sa fixation dans les artères. *Pour Science* **175,** 40–47.

Gallardo, A. R., Galvez, S., Quesda, M. A., Canovas, F. M., and Nunez de Castro, I. (1988). Glutamine synthetase activity during the ripening of tomato. *Plant Physiol. Biochem.* **26,** 747–752.

Gallardo, F., Canton, F. R., Garcia-Gutiérrez, A., and Canovas, F. M. (1993). Changes in photorespiratory enzymes and glutamate synthases in ripening tomatoes. *Plant Physiol. Biochem.* **31,** 189–196.

Gamburg, K. Z. (1978). The influence of 1-naphthaleneacetic acid and (2-chloroethyl)-tri-methylammonium chloride on the carotenoid content of tobacco tissue in suspension culture. *Biol. Plant.* **20,** 93–97.

Garagaty-Feissly, C. (1970). Sur les modifications de l'ultrastructure des cellules tapétales du genre *Daphne* au cours du développement des grains de pollen. *Ber. Schweiz Bot. Ges.* **79**, 221–228.

Gaudillère, J. P., d'Harlingue, A., Camara, B., and Monéger, R. (1984). Prenylation and methylation reactions in phylloquinone (vitamin K1) synthesis in *Capsicum annuum* plastids. *Plant Cell Rep.* **3**, 240–242.

Geitler, L. (1937). Über die Karotinfärbung der Laubbätter von *Adoxa* und Über andere "Karotinpflanzen." *Oesterr. Bot. Ztg.* **86**, 297–301.

Gemmrich, A. R., and Kayser, H. (1984). Hormone induced changes in carotenoid composition in *Ricinus* cell cultures. II. Accumulation of rhodoxanthin during auxin-controlled chromoplast differentiation. *Z. Naturforsch., C: Biosci.* **39C**, 753–757.

Giuliano, G., Bartley, G. E., and Scolnik, P. A. (1993). Regulation of carotenoid biosynthesis during tomato development. *Plant Cell* **5**, 379–387.

Goldschmidt, E. E. (1988). Regulatory aspects of chloro-chromoplast interconversions in senescing *Citrus* fruit peel. *Isr. J. Bot.* **37**, 123–130.

Goldschmidt, E. E., Goren, R., Even-Chen, Z., and Bittner, S. (1973). Increase in free and bound abscisic acid during natural and ethylene-induced senescence in *Citrus* fruit peel. *Plant Physiol.* **51**, 879–882.

Goldschmidt, E. E., Aharoni, Y., Eilati, S. K., Riov, J. W., and Monselise, S. P. (1977). Differential counteraction of ethylene effects by gibberellin A₃ and N₆-benzyladenine in senescing *Citrus* peel. *Plant Physiol.* **59**, 193–195.

Golz, P., and Feierabend, J. (1993). Isoprenoid biosynthesis and stability in developing green and achlorophyllous leaves of Rye (*Secale cereale* L.). *Z. Naturforsch. C: Biosci.* **48C**, 886–895.

Goodwin, T. W. (1958). Studies in carotenogenesis. 24. The changes in carotenoid and chlorophyll pigments in the leaves of deciduous trees during autumn necrosis. *Biochem. J.* **68**, 503–511.

Goodwin, T. W., and Jamikorn, M. (1952). Biosynthesis of carotenes in ripening tomatoes. *Nature (London)* **170**, 104–105.

Gough, S. P., and Kannangara, C. G. (1976). Synthesis of δ-aminolevulinate by isolated plastids. *Carlsberg Res. Commun.* **41**, 184–190.

Gounaris, I., Michalowski, C. B., Bohnert, H. J., and Price, C. A. (1986). Restriction and gene maps of plastid DNA from *Capsicum annuum:* Comparison of chloroplast and chromoplast DNA. *Curr. Genet.* **11**, 7–16.

Gounaris, K., and Barber, J. (1983). Monogalactosyldiacylglycerol: The most abundant polar lipid in nature. *Trends Biochem. Sci.* **8**, 378–381.

Gourret, J. P. (1971). Evolution des plastes dans les végétaux infectés par des mycoplasmes: Problèmes de la jaunesse et des pétales verts. *Physiol. Vég.* **9**, 583–594.

Gourret, J. P., and Le Normand, M. (1971). Les plastes des pétales virescents de la giroflée infectée par des mycoplasmes. *J. Microsc. (Paris)* **12**, 151–156.

Gray, J., Picton, S., Shabbeer, J., Schuch, W., and Grierson, D. (1992). Molecular biology of fruit ripening and its manipulation with antisense genes. *Plant Mol. Biol.* **19**, 69–87.

Green, L. S., Boihon, C., Buchanan, B. B., Kamide, K., Sanada, Y., and Wada, K. (1991). Ferredoxin and ferredoxin-NADP reductase from photosynthetic and nonphotosynthetic tissues of tomato. *Plant Physiol.* **96**, 1207–1213.

Grierson, D., Puton, M. E., Knapp, J. E., and Bathgate, B. (1987). Tomato ripening mutants. *In* "Developmental Mutants in Higher Plants" (H. Tomas and D. Gierson, eds.), pp. 73–94. Cambridge Univ. Press, London and New York.

Grilli, M. (1964). Ultrastructure and involutional phases of some types of chromoplasts. *G. Bot. Ital.* **71**, 574–575.

Grilli, M. (1965a). Origine e sviluppo dei cromoplasti nei frutti di zucca americana (*Cucurbita pepo* L cv. Small Sugar). *Caryologia* **18**, 409–433.

Grilli, M. (1965b). Origine e sviluppo dei cromoplasti nei frutti di zucca americana (*Cucurbita pepo* L. cv. Small Sugar). Origine dei cromoplasti e da proplasti. *Caryologia* **18**, 435–459.

Grilli, M. (1965c). Ultrastrutture e stadi involutivi di alcuni tipi di cromoplasti. *G. Bot. Ital.* **72**, 83–92.

Grob, E. C. (1963). Die pflanzlichen Plastiden im Lichte der Chemie und Biochemie. *Chimia* **17**, 341–348.

Grob, E. C., and Eichenberger, W. (1962). Some aspects of the biogenesis of carotenoids in autumn leaves. *Biochem. J.* **85**, 11P.

Grönegress, P. (1971). The greening of chromoplasts in *Daucus carota* L. *Planta* **98**, 274–278.

Grönegress, P. (1974). The structure of chromoplasts and their conversion to chloroplasts. *J. Microsc. (Paris)* **19**, 183–192.

Gross, A., Brückner, G., Heldt, H. W., and Flügge, U. I. (1990). Comparison of the kinetic properties, inhibition and labelling of the phosphate translocators from maize and spinach mesophyll chloroplasts. *Planta* **180**, 262–271.

Gruissem, W., and Tonkyn, J. C. (1993). Control mechanisms of plastid gene expression. *CRC Crit. Rev. Plant Sci.* **12**, 19–55.

Gruner, S. M. (1985). Intrinsic curvature hypothesis for biomembrane lipid composition: A role for nonbilayer lipids. *Proc. Natl. Acad. Sci. U.S.A.* **82**, 3665–3669.

Guilliermond, A. (1919). Observations vitales sur le chondriome des végétaux et recherches sur l'origine des chromoplastides et le mode de formation des pigments xanthophylliens et carotiniens. *Rev. Gén. Bot.* **31**, 372–413, 446–508, 532–603, 635–770.

Guillot-Salomon, T., Douce, R., and Signol, M. (1967). Relation entre les modifications de l'ultrastructure plastidiale, la teneur en pigments et la composition en lipides polaires des feuilles de maïs traitées par l'aminotriazole. *Bull. Soc. Fr. Physiol. Vég.* **13**, 63–79.

Hadjeb, N., Gounaris, I., and Price, C. A. (1988). Chromoplast-specific proteins in *Capsicum annuum*. *Plant Physiol.* **88**, 42–45.

Halpin, C., Musgrove, J. E., Lord, J. M., and Robinson, C. (1989). Import and processing of proteins by castor bean leucoplasts. *FEBS Lett.* **258**, 32–34.

Hampp, R., and De Filipps, L. F. (1980). Plastid protease activity and prolamellar body transformation during greening. *Plant Physiol.* **65**, 663–668.

Hansmann, P. (1987). Daffodil chromoplast DNA: Comparison with chloroplast DNA, physical map and gene localization. *Z. Naturforsch. C: Biosci.* **42C**, 118–122.

Hansmann, P., and Sitte, P. (1982). Composition and molecular structure of chromoplast globules of *Viola tricolor*. *Plant Cell Rep.* **1**, 111–114.

Hansmann, P., and Sitte, P. (1984). Comparison of the polypeptide complement of different plastid types and mitochondria of *Narcissus pseudonarcissus*. *Z. Naturforsch., C: Biosci.* **39C**, 758–766.

Hansmann, P., Falk, H., Ronai, K., and Sitte, P. (1985). Structure, composition, and distribution of plastid nucleoids in *Narcissus pseudonarcissus*. *Planta* **164**, 459–472.

Hansmann, P., Junker, R., Sauter, H., and Sitte, P. (1987). Chromoplast development in daffodil coronae during anthesis. *Z. Pflanzenphysiol.* **131**, 133–143.

Harada, T., Ishikawa, R., Niizeki, M., and Saito, K. I. (1992). Pollen-derived rice calli that have large deletions in plastid DNA do not require protein synthesis in plastids for growth. *Mol. Gen. Genet.* **233**, 145–150.

Harris, W. M. (1967). Ultrastructural development of chromoplasts in three isogenic lines of *Lycopersicon esculentum* Mill. "Pearson." *Am. J. Bot.* **54**, 642.

Harris, W. M. (1968). Chromoplast development in ripening tomato fruits: Plastid ultrastructure in relation to carotene content of three pigment lines. Ph.D. Thesis, University of California, Davis.

Harris, W. M. (1970). Chromoplasts of tomato fruits. III. The high-delta tomato. *Bot. Gaz. (Chicago)* **131**, 163–166.

Harris, W. M., and Arnott, H. J. (1973). Effects of senescence on chloroplasts of the tobacco leaf. *Tissue Cell* **5**, 527–544.

Harris, W. M., and Spurr, A. R. (1969a). Chromoplasts of tomato fruits. I. Ultrastructure of low-pigment and high-beta mutants. Carotene analyses. *Am. J. Bot.* **56**, 369–379.

Harris, W. M., and Spurr, A. R. (1969b). Chromoplasts of tomato fruits. II. The red tomato. *Am. J. Bot.* **56**, 380–389.

Hauser, I., Debesh, K., and Apel, K. (1984). The proteolytic degradation *in vitro* of the NADPH-protochlorophyllide oxidoreductase of barley (*Hordeum vulgare* L.). *Arch. Biochem. Biophys.* **228**, 577–586.

Heath, M. C. (1974). Chloroplast ultrastructure and ethylene production of senescing and rust-infected cowpea leaves. *Can. J. Bot.* **52**, 2591–2598.

Heim, P. (1946). Sur la localisation des pigments carotiniens chez les phalloidées. *C. R. Hebd. Seances Acad. Sci., Ser. D* **222**, 1354–1355.

Heineke, D., Riens, B., Grosse, H., Hoferichter, P., Peter, U., Flügge, U. I., and Heldt, H. W. (1991). Redox transfer across the inner chloroplast envelope membrane. *Plant Physiol.* **95**, 1131–1137.

Heintze, A., Görlach, J., Leuschner, C., Hoppe, P., Hagelstein, P., Schulze-Siebert, D., and Schultz, G. (1990). Plastidic isoprenoid synthesis during chloroplast development. Change from metabolic autonomy to a division-of-labor stage. *Plant Physiol.* **93**, 1121–1127.

Heintze, A., Riedel, A., Aydogdu, S., and Schultz, G. (1994). Formation of chloroplast isoprenoids from pyruvate and acetate by chloroplasts from young spinach plants. Evidence for a mevalonate pathway in immature chloroplasts. *Plant Physiol. Biochem.* **32**, 791–797.

Heldt, H. W., Flügge, U. I., and Borchert, S. (1991). Diversity of specificity and function of phosphate translocators in various plastids. *Plant Physiol.* **95**, 341–343.

Helm, C. A., Israelchvili, J. N., and McGuiggan, P. M. (1992). Role of hydrophobic forces in bilayer adhesion and fusion. *Biochemistry* **31**, 1794–1805.

Helmann, J. D., and Chamberlin, M. J. (1988). Structure and function of bacterial sigma factors. *Annu. Rev. Biochem.* **57**, 839–872.

Hermann, R. G. (1972). Do chromoplasts contain DNA? II. The isolation and characterisation of DNA from chromoplasts, chloroplasts, mitochondria and nuclei of *Narcissus*. *Protoplasma* **74**, 7–19.

Hess, W. R., Prombona, A., Fieder, B., Subramanian, A. R., and Börner, T. (1993). Chloroplast *rps*15and the *rpo*B/C1/C2 gene cluster are strongly transcribed in ribosome-deficient plastids: Evidence for a functioning non-chloroplast-encoded RNA polymerase. *EMBO J.* **12**, 563–571.

Hill, L. M., and Smith, A. M. (1991). Evidence that glucose 6-phosphate is imported as the substrate for starch synthesis by the plastids of developing pea embryos. *Planta* **185**, 91–96.

Hodgson, R. A. J., Beachy, R. N., and Pakrasi, H. B. (1989). Selective inhibition of photosystem II in spinach by tobacco virus: An effect of the viral coat. *FEBS Lett.* **245**, 267–270.

Hoober, J. K., Marks, D. B., Gabriel, J. L., and Paavola, L. G. (1992). Role of the chloroplast envelope in thylakoid biogenesis. *In* "Regulation of Chloroplast Biogenesis" (J. H. Argyroudi-Akoyunoglou, ed.) pp. 323–330. Plenum, New York.

Hoppe, P., Heintze, A., Riedel, A., Creuzer, C., and Schultz, G. (1993). The plastidic 3-phosphoglycerate-acetylCoA pathway in barley leaves and its involvement in the synthesis of amino acids, plastidic isoprenoids and fatty acids during chloroplast development. *Planta* **190**, 253–262.

Horum, M. A., and Schwartzbach, S. D. (1980). Nutritional regulation of organelle biogenesis in *Euglena*. *Plant Physiol.* **65**, 382–386.

Howes, C. D. (1974a). Nicotine inhibition of carotenoid cyclisation in *Cucurbita ficifolia* cotyledons. *Phytochemistry* **13**, 1469–1472.

Howes, C. D. (1974b). The structural form of nicotine active as an inhibitor of carotenoid cyclization. *Phyton* **32**, 73–76.

Hudak, J. (1981). Plastid senescence 1. Changes of chloroplast structure during natural senescence in cotyledons of *Sinapis alba* L. *Photosynthetica* **15,** 174–178.

Huff, A. (1983). Nutritional control of regreening and degreening in *Citrus* peel segments. *Plant Physiol.* **73,** 243–249.

Huff, A. (1984). Sugar regulation of plastid interconversions in epicarp of *Citrus* fruit. *Plant Physiol.* **76,** 307–312.

Hugueney, P., Römer, S., Kuntz, M., and Camara, B. (1992). Characterization and molecular cloning of a flavoprotein catalyzing the synthesis of phytofluene and ζ-carotene in *Capsicum annuum* chromoplasts. *Eur. J. Biochem.* **209,** 399–407.

Hugueney, P., Badillo, A., Bouvier, F., d'Harlingue, A., Kuntz, A., and Camara, B. (1995a). Developmental and stress regulation of gene expression for plastid and cytosolic isoprenoid pathways. Submitted for publication.

Hugueney, P., Badillo, A., Chen, H. C., Klein, A., Hirschberg, J., Camara, B., and Kuntz, M. (1995b). Metabolism of cyclic carotenoids: A model for the alteration of this biosynthetic pathway in *Capsicum annuum* chromoplasts. *Plant J.* **8** (in press).

Hugueney, P., Bouvier, F., Badillo, A., d'Harlingue, A., Kuntz, M., and Camara, B. (1995c). Identification of a plastid protein involved in vesicle fusion and/or membrane protein translocation. *Proc. Natl. Acad. Sci. U.S.A.* (in press).

Hunt, C. M., Hardison, R. C., and Boyer, C. D. (1986). Restriction enzyme analysis of tomato chloroplast and chromoplast DNA. *Plant Physiol.* **82,** 1145–1147.

Hurtado-Hernandez, H., and Smith, P. G. (1985). Inheritance of mature fruit color in *Capsicum annuum* L. *J. Hered.* **76,** 211–213.

Ida, K. (1981). Eco-physiological studies on the response of taxodiaceous conifers to shading, with special reference to the behaviour of leaf pigments. I. Distribution of carotenoids in green and autumnal reddish brown leaves of gymnosperms. *Bot. Mag.* **94,** 41–54.

Ikeda, T., and Ueda, R. (1964). Light- and electron-microscope studies on the senescence of chloroplasts in *Elodea* leaves. *Bot. Mag.* **77,** 336–341.

Ishida, B. K. (1991). Developmental regulation is altered in the calyx during in vitro ovary culture of tomato. *Plant Cell* **3,** 219–223.

Ismail, M. A., Biggs, R. H., and Oberbacher, M. F. (1967). Effect of gibberellic acid on colour changes in the rind of three sweet orange cultivars (*Citrus sinensis* Blanco). *Proc. Am. Soc. Hortic. Sci.* **91,** 143–149.

Israel, H. W., and Steward, F. C. (1967). The fine structure and development of plastids in cultured cells of *Daucus carota. Ann. Bot. (London)* [N.S.] **31,** 1–18.

Israel, H. W., Mapes, M. O., and Steward, F. C. (1969). Pigments and plastids in cultured cells of *Daucus carota. Am. J. Bot.* **56,** 910–917.

Iwatsuki, N., Hirai, A., and Asahi, T. (1985). A comparison of tomato fruit chloroplast and chromoplast DNA as analyzed with restriction endonucleases. *Plant Cell Physiol.* **26,** 599–602.

James, F. (1994). Changements métaboliques induits par une carence en glucose dans les pointes de racine de maïs. Caractérisation d'une endopeptidase et régulation de son expression par les sucres. Ph.D. Thesis, Université Bordeaux II, Bordeaux.

Jang, J. C., and Sheen, J. (1994). Sugar sensing in higher plants. *Plant Cell* **6,** 1665–1679.

Jen, J. J. (1974a). Influence of spectral quality of light on pigment systems of ripening tomatoes. *J. Food Sci.* **39,** 907–910.

Jen, J. J. (1974b). Spectral quality of light and the ripening characteristics of tomato fruit. *HortScience* **9,** 548–549.

Jordan, E. G., and Chapman, J. M. (1973). Nucleolar and nuclear envelope ultrastructure in relation to cell activity in discs of carrot root (*Daucus carota*). *J. Exp. Bot.* **24,** 197–209.

Journet, E. P., and Douce, R. (1985). Enzymic capacities of purified cauliflower bud plastids for lipid synthesis and carbohydrate metabolism. *Plant Physiol.* **79,** 458–467.

Kaldenhoff, R., and Richter, G. (1990). Light induction of genes preceding chloroplast differentiation in cultured plant cells. *Planta* **181,** 220–228.

Kang, B. G., and Stanley, P. B. (1972). Involvement of ethylene in phytochrome-mediated carotenoid synthesis. *Plant Physiol.* **49,** 631–633.

Kang, F., and Rawsthorne, S. (1994). Starch and fatty acid synthesis in plastids from developing embryos of oilseed rape (*Brassica napus* L.). *Plant J.* **6,** 795–805.

Kayser, H., and Gemmrich, A. R. (1983). Hormone induced changes in carotenoid composition in *Ricinus* cell cultures. I. Identification of rhodoxanthin. *Z. Naturforsch., C: Biosci.* **39C,** 50–54.

Keeling, P. L., Wood, J. R., Tyson, R. H., and Bridges, I. G. (1988). Starch biosynthesis in developing wheat grain. *Plant Physiol.* **87,** 311–319.

Keresztes, A., and Schroth, A. (1979). Light and electron microscopic investigation of *in vitro* starch synthesis in chromoplasts. *Cytobios* **26,** 185–191.

Khudairi, A. K. (1972). The ripening of tomato. *Am. Sci.* **60,** 696–707.

Khudairi, A. K., and Arboleda, O. P. (1971). Phytochrome-mediated carotenoid biosynthesis and its influence by plant hormones. *Physiol. Plant.* **24,** 18–22.

Kieber, J. J., and Ecker, J. R. (1993). Ethylene gas, it's not just for ripening anymore. *Trends Genet.* **9,** 356–362.

Kinzer, S. M., Schwager, S. J., and Mutschler, M. A. (1990). Mapping of ripening-related or -specific cDNA clones of tomato (*Lycopersicon esculentum*). *Theor. Appl. Genet.* **79,** 489–496.

Kirk, J. T. O. (1978). The biochemical basis for plastid autonomy and plastid growth. *In* "The Plastids" (J. T. O. Kirk and R. A. E. Tilney-Bassett, eds.), pp. 525–872. Elsevier/North-Holland, Amsterdam.

Kirk, J. T. O., and Juniper, B. E. (1967). The ultrastructure of the chromoplasts of different colour varieties of *Capsicum*. *In* "Biochemistry of Chloroplasts" (T. W. Goodwin, ed.), pp. 691–701. Academic Press, London and New York.

Kitajima, E. W., and Costa, A. S. (1973). Aggregates of chloroplasts in local lesions induced in *Chenopodium quinoa* Wild. *J. Gen. Virol.* **20,** 413–416.

Kiyosue, T., Yamaguchi-Shinozaki, K., and Shinozaki, K. (1993). Characterization of cDNA for a dehydration-inducible gene that encodes a *Clp* A, B-like proteinin *Arabidopsis thaliana*. *Biochem. Biophys. Res. Commun.* **196,** 1214–1220.

Klein, S., and Ben-Shaul, Y. (1967). "Development of Carotene Bodies in the Carrot," Final Rep. Botany Department, Hebrew University, Jerusalem.

Klein, U., De Camp, J. D., and Bogorad, L. (1992). Two types of chloroplast gene promoters in *Chlamydomonas reinhardtii*. *Proc. Natl. Acad. Sci. U.S.A.* **89,** 3453–3457.

Kleinig, H., and Lievogel, B. (1978). Fatty acid synthesis by isolated chromoplasts from daffodil ^{14}C acetate incorporation and distribution of labelled acids. *Eur. J. Biochem.* **83,** 499–505.

Kleinig, H., and Liedvogel, B. (1980). Fatty acid synthesis by isolated chromoplasts from the daffodil. Energy sources and distribution patterns of the acids. *Planta* **150,** 166–169.

Klösgen, R. B., Saedler, H., and Weil, J. H. (1989). The amyloplast-targeting transit peptide of the waxy protein of maize also mediates protein transport *in vitro* into chloroplasts. *Mol. Gen. Genet.* **217,** 155–161.

Knoth, R. (1981). Ultrastructure of lycopene containing chromoplasts in fruits of *Aglaonema commutatum* Schott (Araceae). *Protoplasma* **106,** 249–259.

Knoth, R., Hansmann, P., and Sitte, P. (1986). Chromoplast of *Palisota barteri*, and the molecular structure of chromoplast tubules. *Planta* **168,** 167–174.

Knypl, J. S. (1969). Accumulation of lycopene in detached cotyledons of pumpkin treated with (2-chlororthyl)-trimethylammonium chloride. *Naturwissenschaften* **56,** 572.

Kobayashi, H., Ngernprasirtsiri, J., and Akazawa, T. (1990). Transcriptional regulation and DNA methylation in plants during transitional conversion of chloroplasts to chromoplasts. *EMBO J.* **9,** 307–313.

Kobmann, J., Müller-Röber, B., Dyer, T. A., Raines, C. A., Sonnewald, U., and Willmitzer, L. (1992). Cloning and expression analysis of the plastidic fructose-1,6-bisphosphate coding sequence from potato: Circumstantial evidence for the import of kexoses into chloroplasts. *Planta* **188,** 7–12.

Kollmann, R. H., Kleinig, H., and Dorr, I. (1969). Fine structure and pigments of plastids in Orobranche. *Cytobiologie.* **1,** 152–158.

Kordan, H. A. (1971). Starch synthesis in quiescent lemon fruits explants. *Z. Pflanzenphysiol.* **65,** 118–123.

Kosegarten, H., and Mengel, K. (1994). Evidence for a glucose1-phosphate translocator in storage tissue amyloplasts of potato (*Solanum tuberosum*) suspension-cultured cells. *Physiol. Plant.* **91,** 111–120.

Kowallik, K. V., and Herrmann, R. G. (1972). Do chromoplasts contain DNA? 1. Electron-microscope investigation of *Narcissus* chromoplasts. *Protoplasma* **74,** 1–6.

Krapp, A., Quick, W. P., and Sitt, M. (1991). Ribulose-1,5-bisphosphate carboxylase-oxygenase, other Calvin-cycle enzymes, and chlorophyll decrease when glucose is supplied to mature spinach leaves *via* the transpiration stream. *Planta* **186,** 58–69.

Kraus, G. (1872) Die Entstehung der Farbtoffkörper in den Beeren von *Solanum pseudocapsicum. Jahrb. Wiss. Bot.* **8,** 131–147.

Kreuz, K., and Kleinig, H. (1981). Chlorophyll synthetase in chlorophyll-free chromoplasts. *Plant Cell Rep.* **1,** 40–42.

Kreuz, K., and Kleinig, H. (1984). Synthesis of prenyllipids in cells of spinach leaf. Compartmentation of enzymes for formation of isopentenyl diphosphate. *Eur. J. Biochem.* **141,** 531–535.

Kreuz, K., Beyer, P., and Kleinig, H. (1982). The sites of carotenogenic enzymes in chromoplasts from *Narcissus pseudonarcissus* L. *Planta* **154,** 66–69.

Kuhn, H. (1970). Chemismus, Struktur und Entstehung der Carotinkriställchen in der Nebenkrone von *Narcissus poeticus* L. var. "La Riante." *J. Ultrastruct. Res.* **33,** 332–355.

Kuhn, H., Frey-Wyssling, A., and Stricker, P. (1969). Über die struktur pflanzlichen Karotinkriställchen. *Experientia* 972–973.

Kumar, A., Bender, L., and Neumann, K. H. (1984). Growth regulation, plastid differentiation and the development of a photosynthetic system in cultured carrot root explants as influenced by exogenous sucrose and various phytohormones. *Plant Cell Tissue Cult.* **3,** 11–28.

Kuntz, M., Evrard, J. L., d'Harlingue, A., Weil, J. H., and Camara, B., (1989). Expression of plastid and nuclear genes during chromoplast differentiation in bell pepper (*Capsicum annuum*) and sunflower (*Helianthus annuus*). *Mol. Gen. Genet.* **216,** 156–163.

Kuntz, M., Camara, B., Weil, J. H., and Schantz, R. (1992). The *psbL* gene from bell pepper (*Capsicum annuum*): Plastid RNA editing also occurs in non-photosynthetic chromoplasts. *Plant Mol. Biol.* **20,** 1185–1188.

Kuroiwa, T., Suzuki, T., Ogawa, K., and Kawano, S. (1981). The chloroplast nucleus: Distribution, number, size, and shape, a model for the multiplication of the chloroplast genome during chloroplast development. *Plant Cell Physiol.* **22,** 381–396.

Laborde, J. A. (1969). Effect of two fruit colour genes in *Capsicum annuum* (*Capsicum frutescens*) (red pepper) on chlorophyll carotenoids and ultrastructure of the chromoplast. Ph.D. Thesis, University of California, Davis.

Laborde, J. A., and Spurr, A. R. (1973). Chromoplast ultrastructure as affected by genes controlling grana retention and carotenoids in fruits of *Capsicum annuum. Am. J. Bot.* **60,** 736–744.

Lanahan, M. B., Yen, H. C., Giovannoni, J. J., and Klee, H. J. (1994). The *Never ripe* mutation blocks ethylene perception in tomato. *Plant Cell* **6,** 521–530.

Lance-Nougarède, A. (1960). Développement infra-microscopique des chromoplastes au cours de l'ontogénèse des fleurs de ligules de *Chrysanthemum segetum* L. *C. R. Hebd. Seances Acad. Sci.* **250,** 173–175.

Lang, N. J. (1968). Electron microscopic studies of extraplastidic astaxanthin in *Haematococcus*. *J. Phycol.* **4**, 12–19.

Laudi, G., and Albertini, A. (1967). Richerche infrastrutturali sui plastidi dell piante parassite III. *Orobanche ramosa*. *Caryologia* **20**, 207–216.

Laval-Martin, D. (1969). Evolution des pigments et des plastes au cours de la maturation de la tomate "cerise." *Bull. Soc. Fr. Physiol. Vég.* **15**, 77–79.

Laval-Martin, D. (1974). La maturation du fruit de tomate cerise: Mise en évidence, par cryodécapage, de l'évolution des chloroplastes en deux types de chromoplastes. *Protoplasma* **82**, 33–60.

Laval-Martin, D., Quennemet, J., and Monéger, R. (1975). Pigment evolution in *Lycopersicon esculentum* fruits during growth and ripening. *Phytochemistry* **14**, 2357–2362.

Ledbetter, M. C., and Porter, K. R. (1970). Chromoplasts. *In* "Introduction to the Fine Structure of Plant Cells" (M. C. Ledbetter, ed.), pp. 146–147. Springer-Verlag, Berlin and New York.

Lee, T. H., Erickson, L. C., and Chichester, C. O. (1971). An evaluation of the use of grafted fruit for study of carotenoid and chlorophyll changes in regreening "Valencia orange peel." *HortScience* **6**, 231–232.

Leech, R. M., and Pyke, K. A. (1988). Chloroplast division in higher plant with particular reference to wheat. *In* "The Division and Segregation of Organelles" (S. A. Boffey and D. Lloyd, eds.), pp. 39–62. Cambridge Univ. Press, Cambridge, UK and New York.

Lendzian, K., and Ziegler, H. (1970). Uber die Regulation der Glucose-6-phosphat-Dehydrogenase in Spinatchloroplasten durch Licht. *Planta* **94**, 27–36.

Lewis, L. N., and Coggins, J. C. W. (1964). Chlorophyll concentration in the navel orange rind as related to potassium gibberellate, light intensity, and time. *Proc. Am. Soc. Hortic. Sci.* **84**, 177–180.

Lichtenthaler, H. K. (1966). Plastoglobuli und Plastidenstruktur. *Ber. Dtsch. Bot. Ges.* **79**, 82–88.

Lichtenthaler, H. K. (1968). Plastoglobuli and the fine structure of plastids. *Endeavour* **27**, 144–149.

Lichtenthaler, H. K. (1969). Zur synthese der lipophilen Plastidenchinone und Sekundärcarotinnoide wahrene der Chromoplastenentwicklung. *Ber. Dtsch. Bot. Ges.* **82**, 483–497.

Lichtenthaler, H. K. (1970a). Die Lokalisation der Plastidenchinone und Carotinoide in den Chromoplasten der Petalen von *Sarathamunus scoparius* L. Wimm ex Koch. *Planta* **90**, 142–152.

Lichtenthaler, H. K. (1970b). Die Feinstruktur der Chromoplasten in plasmochromen Pergon-Blättern von *Tulipa*. *Planta* **93**, 143–151.

Lichtenthaler, H. K. (1971). Formation and function of plastoglobuli in plastids. *Congr. Int. Microsc. Electron., 7th,* Grenoble, *1970,* pp. 205–206.

Lichtenthaler, H. K. (1977). Zur Synthese der lipophilen Plastidenchinone und Sekundärcarotinoide während der Chromoplastenen-twicklung. *Ber. Dtsch. Bot. Ges.* **82**, 483–497.

Liedvogel, B., and Falk, H. (1980). Leucoplasts mimicking membranous chromoplasts. *Z. Pflanzenphysiol.* **98**, 371–375.

Liedvogel, B., and Kleinig, H. (1976). Galactolipids synthesis in chromoplast internal membranes of the daffodil. *Planta* **129**, 19–21.

Liedvogel, B., and Kleinig, H. (1980). Phosphate translocator and adenylate translocator in chromoplast membranes. *Planta* **150**, 170–173.

Liedvogel, B., Sitte, P., and Falk, H. (1976). Chromoplasts in daffodil: Fine structure and chemistry. *Cytobiologie* **12**, 155–174.

Lim, H. T., Gounaris, I., Hardison, R. C., and Boyer, C. D. (1989). Organization of the plastid DNA of *Cucurbita pepo* L. *In* "Physiology, Biochemistry, and Genetics of Nongreen Plastids" (C. D. Boyer, J. C. Shannon, and R. C. Hardison, eds.), pp. 269–273. Am. Soc. Plant Physiol., Rockville, MD.

Lindbeck, A. G. C., Dawson, W. O., and Thomson, W. W. (1991). Coat protein-related polypeptides from *in vitro* tobacco mosaic virus coat protein mutants do not accumulate in the chloroplasts of directly inoculated leaves. *Mol. Plant-Microbe Interact.* **4,** 89–94.

Link, G. (1984). DNA sequence requirements for the accurate transcription of a protein coding plastid *in vitro* system from mustard (*Sinapis alba*). *EMBO J.* **3,** 1697–1704.

Lippmaa, T. (1926a). Sur la formation des chromoplastes chez les phanérogames. *C. R. Hebd. Seances Acad. Sci.* **182,** 1040–1042.

Lippmaa, T. (1926b). Sur les hématocarotinoïdes et les xanthocarotinodes. *C. R. Hebd. Seances Acad. Sci.* **182,** 1350–1352.

Liu, X. Q., and Jagendorf, A. T. (1984). ATP-dependent proteolysis in pea chloroplasts. *FEBS Lett.* **166,** 248–252.

Livne, A., and Gepstein, S. (1988). Abundance of the major chloroplast polypeptides during development and ripening of tomato fruits—An immunological study. *Plant Physiol.* **87,** 239–243.

Ljubesic, N. (1970). Fine structure of developing chromoplasts in outer yellow fruit parts of *Cucurbita pepo* var. pyriformis. *Acta Bot. Croat.* **29,** 51–56.

Ljubesic, N. (1977). The formation of chromoplasts in fruits of *Cucurbita maxima* Duch. "Turbaniformis." *Bot. Gaz. (Chicago)* **138,** 286–290.

Ljubesic, N. (1979). Chromoplasts in the petals of *Liriodendron tulipefera* L. *Z. Pflanzenphysiol.* **91,** 49–52.

Mackinney, G., Rick, C. M., and Jenkins, J. A. (1956). The phytoene content of tomatoes. *Proc. Natl. Acad. Sci. U.S.A.* **42,** 404–408.

MacNicol, P. K. (1976). Rapid metabolic changes in the wounding response of leaf discs following excision. *Plant Physiol.* **57,** 80–84.

Maillot-Vernier, P., Gondet, L., Schaller, H., Benveniste, P., and Belliard, G. (1991). Genetic study and further biochemical characterization of a tobacco mutant that overproduces sterols. *Mol. Gen. Genet.* **231,** 33–40.

Malek, L., Bogorad, L., Ayers, A. R., and Goldberg, A. L. (1984). Newly synthesized proteins are degraded by an ATP-stimulated proteolytic process in isolated pea chloroplasts. *FEBS Lett.* **166,** 253–257.

Manzara, T., Carrasco, P., and Gruissem, W. (1991). Developmental and organ-specific changes in promoter DNA-protein interactions in the tomato *rbcS* gene family. *Plant Cell* **3,** 1305–1316.

Marano, M. R., and Carrillo, N. (1991). Chromoplast formation during tomato fruit ripening. No evidence for plastid DNA methylation. *Plant Mol. Biol.* **16,** 11–20.

Marano, M. R., and Carrillo, N. (1992). Constitutive transcription and stable RNA accumulation in plastids during the conversion of chloroplasts to chromoplasts in ripening tomato fruits. *Plant Physiol.* **100,** 1103–1113.

Marano, M. R., Serra, E. C., Orellano, E. G., and Carrillo, N. (1993). The path of chromoplast development in fruits and flowers. *Plant Sci.* **94,** 1–17.

Martineau, B., C. M., H., Sheehy, R. M., and Hiatt, W. R. (1994). Fruit-specific expression of the *A. tumefaciens* isopentenyl transferase gene in tomato: Effects on fruit ripening and defense-related gene expression in leaves. *Plant J.* **5,** 11–19.

Marty, F. (1978). Cytochemical studies on GERL, provacuoles, and vacuoles in root meristematic cells of *Euphorbia. Proc. Natl. Acad. Sci. U.S.A.* **75,** 852–856.

Masterson, C., Wood, C., and Thomas, D. R. (1990). L-acetylcarnitine, a substrate for chloroplast fatty acid synthesis. *Plant, Cell Environ.* **13,** 755–765.

Matienco, B. T. (1965). Organisation inframicroscopique des chromoplastes des Cucurbitacées et classification morphologique des chromoplastes. *Electron Microsc., Proc. Eur. Reg. Conf., 3rd,* Prague, *1964,* pp. 153–154.

Matienco, B. T. (1967). Chromoplasts of the red-fleshed fruits of watermelon. *Bot. Zh. (Leningrad)* **52,** 229–239.

Matienco, B. T., Salinsky, S. M., and Solovey, V. K. (1968). Etude électronmicroscopique des caroténoidoplastes (chromoplastes) sans fixation. *Eur. Reg Congr. Electron Microsc., 4th, 1968,* Vol. 2, pp. 387–388.

Matile, P. (1990). The toxic compartment of plant cell. *In* "Progress in Plant Cellular and Molecular Biology" (H. J. J. Nijkamp, I. H. W. van der Plas, and U. van Aartrijk, eds.), pp. 557–566. Kluwer Academic Publishers, Dordrecht, Boston, and London.

Matile, P., and Winkenbach, F. (1971). Function of lysosomes and lysosomal enzymes in the senescing corolla of the morning glory (*Ipomea purpurea*). *J. Exp. Bot.* **22,** 759–771.

Mayfield, S. P., and Huff, A. (1986). Accumulation of chlorophyll, chloroplastic proteins and thylakoid membranes during reversion of chromoplasts to chloroplasts in *Citrus sinensis* epicarp. *Plant Physiol.* **80,** 30–35.

Menke, W. (1962). Structure and chemistry of plastids. *Annu. Rev. Plant Physiol.* **13,** 27–44.

Mesquita, J. F. (1976). La différenciation des plastes dans les fleurs de *Narcissus* L. I. Modifications ultrastructurales et pigmentaires pendant la morphogénèse des chromoplastes chez *N. bulbocodium* L. *Rev. Biol. (Lisbon)* **10,** 127–150.

Mesquita, J. F., and Dias, J. D. S. (1990). Electron microscopic study of the plastids of variegated leaves of *Croton* species. 2. Utrastructural and cytochemical aspects of induced starch synthesis in chromoplasts. *Cytobios* **64,** 121–128.

Mesquita, J. F., Santos Dia, J. D., and Martino, A. P. (1980). Chromoplast differentiation in fruits of *Musa sp.* (banana-tree). *Biol. Cell* **39,** 343–348.

Mikulska, E., Zolnierowicz, H., and Narolewska, B. (1973). Ultrastructure of chromoplasts in detached cotyledons of cucumber treated with growth retardant (2-chloroethyl-tri-methylammonium chloride). *Biochem. Physiol. Pflanz.* **164,** 514–522.

Miller, E. V., Winston, J. R., and Schomer, A. (1940). Physiological studies of plastid pigments in rinds of maturing oranges. *J. Agric. Res.* **60,,** 259–267.

Miller, O. L. J., Hamkalo, B. A., and Thomas, C. A. J. (1970). Visualization of bacterial genes in action. *Science* **169,** 392–395.

Mollenhauer, H. H., and Kogut, C. (1968). Chromoplast development in daffodil. *J. Microsc. (Paris)* **7,** 1045–1050.

Moore, T., and Keegstra, K. (1993). Characterization of a cDNA clone encoding a chloroplast-targeted Clp homologue. *Plant Mol. Biol.* **21,** 525–537.

Morden, C. W., Wolfe, K. H., dePamphilis, C. W., and Palmer, J. D. (1991). Plastid translation and transcription genes in a non–photosynthetic plant: Intact, missing and pseudo genes. *EMBO J.* **10,** 3281–3288.

Morigasaki, S., Talorta, S., Suzuki, T., and Wada, K. (1990). Purification and characterisation of ferredoxin-NADP$^+$ oxidoreductase-like enzyme from radish root tissue. *Plant Physiol.* **93,** 896–901.

Morré, D. J., Morré, J. T., Morré, S. R., Sundqvist, C., and Sandelius, A. S. (1991). Chloroplast biogenesis: Cell-free transfer of envelope monogalactosylglycerides to thylakoids. *Biochim. Biophys. Acta* **1070,** 437–445.

Mühlethaler, K. (1971). The ultrastructure of plastids. *In* "Structure and Function of Plastids" (M. Gibbs, ed.), pp. 7–34. Springer-Verlag, New York.

Mühlethaler, K., and Frey-Wyssling, A. (1959). Entwicklung und Struktur der Proplastiden. *J. Biophys. Biochem. Cytol.* **6,** 501–517.

Mullet, J. E., and Klein, R. R. (1987). Transcription and RNA stability are important determinants of higher plant chloroplast RNA levels. *EMBO J.* **6,** 1571–1579.

Murata, N., Sato, N., Omata, T., and Kuwabara, T. (1981). Separation and characterization of thylakoid and cell envelope of the blue-green alga (*Cyanobacterium*) *Anacystis nidulans*. *Plant Cell Physiol.* **22,** 855–866.

Murphy, D. J., and Stumpf, P. K. (1981). The origin of chloroplastic acetyl coenzyme A. *Arch. Biochem. Biophys.* **212,** 730–739.

Nath, I., and Laal, S. (1990). Nucleotide sequence and deduced amino acid sequence of *Mycobacterium leprae* gene showing homology to bacterial *atp* operon. *Nucleic Acids Res.* **18**, 4935.

Nedukha, E. M. (1969). Ultrastructure of parenchyma cells of *Daucus carota* explantat in culture *in vitro*. *Dokl. Akad. Nauk Ukr.,* pp. 844–846.

Nelson, M. A., Morelli, G., Carattoli, A., Romano, N., and Macino, G. (1989). Molecular cloning of *Neurospora crassa* carotenoid biosynthetic gene (*albino-3*) regulated by blue light and product of the white collar genes. *Mol. Cell. Biol.* **9**, 1271–1276.

Nemo, Y., Kawano, S., Kondoh, K., Nagata, T., and Kuroiwa, T. (1990). Studies on plastid-nuclei (nucleoids) in *Nicotiana tabacum* L. III. Isolation of chloroplast-nuclei from mesophyll protoplasts and identification of chloroplast DNA-binding proteins. *Plant Cell Physiol.* **31**, 767–776.

Neuhaus, H. E., Batz, O., Thom, E., and Scheibe, R. (1993a). Purification of highly intact plastids from various heterotrophic plant tissues: Analysis of enzymic equipment and precursor dependency for starch biosynthesis. *Biochem. J.* **296**, 395–401.

Neuhaus, H. E., Henrichs, G., and Scheibe, R. (1993b). Characterization of glucose-6-phosphate incorporation into starch by isolated intact cauliflower-bud plastids. *Plant Physiol.* **101**, 573–578.

Ngernprasirtsiri, J., Kobayashi, H., and Akazawa, T. (1988). DNA methylation occurred around lowly expressed genes of plastid DNA during tomato fruit development. *Plant Physiol.* **88**, 16–20.

Nielsen, O. F. (1974a). Macromolecular physiology of plastids. XI. Carotenes in etiolated *tigrina* and *xantha* mutants of barley. *Physiol. Plant.* **30**, 246–254.

Nielsen, O. F. (1974b). Macromolecular physiology of plastids. XII. *Tigrina* mutants in barley: Genetic, spectroscopic and structural characterization. *Hereditas* **76**, 269–306.

Nougarède, A. (1964). Evolution infrastructural des chromoplastes au cours de l'ontogénèse des pétales chez *Spartium junceum* L(Papilinacées). *C. R. Hebd. Seances Acad. Sci.* **258**, 683–685.

Ogura, T., Tomoyasu, T., Yuki, T., Morimura, S., Begg, K. J., Donachie, W. D., Mori, N., Niki, H., and Hiraga, S. (1991). Structure and function of the *ftsH* gene in *Escherichia coli*. *Res. Microbiol.* **142**, 279–282.

Oliveira, L. (1982). The development of chloroplasts in root meristematic tissue of *Secale cereale* L. seedlings. *New Phytol.* **91**, 263–275.

Orcival, J. (1968). Sur les modifications particulières provoquées par une Urédinale sur la structure des chloroplastes de *Tussilago farfara*. *C. R. Hebd. Seances Acad. Sci., Ser. D* **266**, 1272–1274.

Osumi, M. (1961). Electron-microscopical studies on chromoplasts. 1. The ultrastructure of chromoplast in oranges. *Bot. Mag.* **74**, 165–168.

Pacini, E., Taylor, P. E., Singh, M. B., and Knox, R. B. (1992). Development of plastids in pollen and tapetum of reye-grass, *Lolium perenne* L. *Ann. Bot. (London)* **70**, 179–188.

Parthier, B. (1979). The roles of phytohormones (cytokinins) in chloroplast development. *Biochem. Physiol. Pflanz.* **174**, 173–214.

Pasqua, G., Vecchia, F. D., Rascio, N., and Casadoro, G. (1989). Influence of exogenous sucrose on the greening of oat. *J. Ultrastruct. Mol. Struct. Res.* **102**, 249–254.

Pecker, I., Chamovitz, D., Linden, H., Sandmann, G., and Hirschberg, J. (1992). A single polypeptide catalyzing the conversion of phytoene to γ-carotene is transcriptionally regulated during tomato fruit ripening. *Proc. Natl. Acad. Sci. U.S.A.* **89**, 4962–4966.

Pérez, A. G., Sanz, C., Richardson, D. G., and Olias, J. M. (1993). Methyl jasmonate vapor promotes β-carotene synthesis and chlorophyll degradation in Golden Delicious apple peel. *J. Plant Growth Regul.* **12**, 163–167.

Piazza, G. J., Smith, M. G., and Gibbs, M. (1982). Characterization of the formation and distribution of photosynthetic products by *Sedum praeltum*chloroplasts. *Plant Physiol.* **70**, 1748–1758.

Piechulla, B., Cholnoles-Imlay, K. R., and Gruissem, W. (1985). Plastid gene expression during fruit ripening in tomato. *Plant Mol. Biol.* **5**, 373–384.

Piechulla, B., Pichersky, E., Cashmore, A. R., and Gruissem, W. (1986). Expression of nuclear and plastid genes for photosynthesis specific proteins during tomato fruit development and ripening. *Plant Mol. Biol.* **7**, 367–376.

Plumpley, F. G., and Schmidt, G. W. (1989). Nitrogen-dependent regulation of photosynthetic gene expression. *Proc. Natl. Acad. Sci. U.S.A.* **70**, 2678–2682.

Poling, S. M., Hsu, W. J., and Yokoyama, H. (1980). Chemical induction of poly-*cis* carotenoid biosynthesis. *Phytochemistry* **19**, 1677–1680.

Porter, J. W., and Spurgeon, S. L. (1979). Enzymatic synthesis of carotenes. *Pure Appl. Chem.* **51**, 609–622.

Possingham, J. V., Hashimoto, H., and Oross, J. (1988). Factors that influence plastid division in higher plants. *In* "The Division and Segregation of organelles" (S. A. Boffey and D. Lloyd, eds.), pp. 1–20. Cambridge Univ. Press, Cambridge, UK and New York.

Prat, S. (1924). Die Farstoffe der Potamogetonblätter. *Biochem. Z.* **152**, 495–497.

Price, C. A., Hadjeb, N., and Newman, L. A. (1989). A search for chromoplast-specific genes and proteins in *Capsicum annuum*. *In* "Physiology, Biochemistry, and Genetics of Nongreen Plastids" (C. J. Boyer, J. C. Shannon and R. C. Hardison, eds.), pp.215–226. Am. Soc. Plant Physiol., Rockville, MD.

Price, C. A., Cervantes-Cervantes, M., Hadjeb, N., Newman, L. A., and Oren-Shamir, M. (1993). Molecular biology of chromoplast development. *In* "Pigment-Protein Complexes in Plastids: Synthesis and Assembly" (C. Sundqvist and M. Ryberg, eds.), pp. 485–505. Academic Press, San Diego, CA.

Pryer, N. K., Wuestehube, L. J., and Schekman, R. (1992). Vesicle-mediated protein sorting. *Annu. Rev. Biochem.* **61**, 471–516.

Purvis, A. C., and Barmore, C. R. (1981). Involvement of ethylene in chlorophyll degradation in peel of *Citrus* fruits. *Plant Physiol.* **68**, 854–856.

Purvis, A. C. (1980). Sequence of chloroplast degreening in calamondin fruit as influenced by ethylene. *Plant Physiol.* **66**, 624–627.

Rabinowitch, H. D., and Rudich, J. (1972). Effects of ethephon and CPTA on color development of tomato fruits at high temperatures. *HortScience* **7**, 76–77.

Radin, D. N. (1986). A model cell culture system to study isoprenoid regulation in plants. *Curr. Top. Plant Biochem. Physiol.* **5**, 153–163.

Ramirez, D. A., and Tomes, M. L. (1964). Relationship between chlorophyll and carotenoid biosynthesis in dirty-red (*green-flesh*) mutant in tomato. *Bot. Gaz.* (*Chicago*) **125**, 221–226.

Rand, R. P., Fuller, N., Pasegian, V. A., and Rau, D. C. (1988). Variation in hydration forces between neutral phospholipid bilayers: Evidence for hydration attraction. *Biochemistry* **27**, 7711–7722.

Rau, W. (1985). Mechanism of photoregulation of carotenoid biosynthesis in plants. *Pure Appl. Chem.* **57**, 777–784.

Raymundo, L. C. (1971). The biosynthesis of carotenoids in the tomato fruit. Ph.D. Thesis, University of Rhode Island, Kingston.

Raymundo, L. C., Griffiths, A. E., and Simpson, K. L. (1967). Effect of dimethyl sulfoxide (DMSO) on the biosynthesis of carotenoids in detached tomatoes. *Phytochemistry* **6**, 1527–1532.

Raymundo, L. C., Chichester, C. O., and Simpson, K. L. (1976). Light-dependent carotenoid synthesis in the tomato fruit. *J. Agric. Food Chem.* **24**, 59–64.

Raymundo, L. C., Chichester, C. O., and Simpson, K. L. (1976). Light-dependent carotenoid synthesis in the tomato fruit. *J. Agric. Food Chem.* **24**, 59–64.

Reiss, T., and Link, G. (1985). Characterization of transcriptionally active dna-protein complexes from chloroplasts and etioplasts of mustard (*Sinapis alba* L.). *Eur. J. Biochem.* **148**, 207–212.

Reith, M., and Munholland, J. (1993). A high-resolution gene map of the chloroplast genome of the red alga *Porphyra purpurea*. *Plant Cell* **5,** 465–475.

Rezende-Pinto, M. C., and Salema, R. (1959). Concerning the genesis of the chromoplasts of some fruit. *Rev. Biol.* **2,** 51–54.

Roberts, E. A. (1946). Electron microscope studies of plant cells and their contents showing structural and functional units of less than 100 angströms. *Am. J. Bot.* **33,** 231–232.

Roberts, E. A., and Southwick, M. D. (1948). Contribution of studies with electron microscope to studies of the relationship of chromoplasts to carotene bodies and carotene bodies to vitamin A. *Plant Physiol.* **23,** 621–633.

Robertson, D. S. (1975). Survey of the albino and white-endosperm mutant of maize. *J. Hered.* **66,** 67–74.

Robertson, I. S., Anderson, I. C., and Bachmann, M. D. (1978). Pigment-deficient mutants: Genetic, biochemical, and developmental studies. *In* "In Maize Breeding and Genetics" (D. B. Walden, ed.), pp. 461–494. New York, John Wiley and Sons.

Robinson, R. W., and Tomes, M. L. (1968). Ripening inhibitor; a gene with multiple effect on ripening. *Rep. Tomato Genet. Coop.* **18,** 36–37.

Roby, D., Marco, Y., and Esquerre-Tugaye, M. T. (1988). Ribulose1,5-bisphosphate carboxylase/oxygenase gene expression in melon plants infected with *Colletotrichum lagenarium*. Activity level and rate of synthesis of mRNAs coding for the large and small subunits. *Physiol. Mol. Pathol.* **32,** 411–424.

Rohmer, M., Knani, M., Simonin, P., Sutter, B., and Sham, H. (1993). Isoprenoid synthesis in bacteria: A novel pathway for the early steps leading to isopentenyl diphosphate. *Biochem. J.* **295,** 517–524.

Römer, S., d'Harlingue, A., Camara, B., Schantz, R., and Kuntz, M. (1992). Cysteine synthase from *Capsicum annuum* chromoplasts. Characterization and cDNA cloning of an up-regulated enzyme during fruit development. *J. Biol. Chem.* **267,** 17966–17970.

Römer, S., Hugueney, P., Bouvier, F., Camara, B., and Kuntz, M. (1993). Expression of the genes encoding the early carotenoid biosynthetic enzymes in *Capsicum annum*. *Biochem. Biophys. Res. Commun.* **196,** 1414–1421.

Rosso, S. W. (1967a). Studies on tomato fruit plastids. *Am. J. Bot.* **54,** 637.

Rosso, S. W. (1967b). An ultrastructural study of the mature chromoplasts of the tangerine tomato (*Lycopersicum esculentum* var. "Golden Jubilee"). *J. Ultrastruct. Res.* **20,** 179–189.

Rosso, W. (1968). The ultrastructure of chromoplast development in red tomatoes. *J. Ultrastruct. Res.* **25,** 307–322.

Rothman, J. E., and Orci, L. (1992). Molecular dissection of the secretory pathway. *Nature* (*London*) **355,** 409–415.

Salema, R. (1968). Amido: Estudio ultrastructural da sua biogenese em plantas superiores. *Broteria, Sèr. Trimest.: Cienc. Nat.* **38,** 1–126.

Salunkhe, D. K., Jadhav, S. J., and Yu, M. H. (1974). Quality and nutritional composition of tomato fruit as influenced by certain biochemical and physiological changes. *Qual. Plant.— Plant Foods Hum. Nutr.* **24,** 85–113.

Saniewski, M., and Czapski, J. (1983). The effect of methyl jasmonate on lycopene and β-carotene accumulation in ripening red tomatoes. *Expeirentia* **39,** 1373–1374.

Santos, M. F., and Mesquita, J. F. (1984). Ultrastructural study of *Haematococcus lacustris* (Girod.) Rostafinski (volvocales) I. Some aspects of carotenogenesis. *Cytologia* **49,** 215–228.

Sato, N., Albrieux, C., Joyard, J., Douce, R., and Kuroiwa, T. (1993). Detection and characterization of a plastid envelope DNA-binding protein which may anchor plastid nucleoids. *EMBO J.* **12,** 555–561.

Saunders, C. A. (1991). Regulation of isoprenoid biosynthesis. *Am. Soc. Symp.*, Atlanta, Abstr., p. 108.

Saunders, M. J., and Helper, P. K. (1982). Calcium ionophore A23187 stimulates cytokinin-like mitosis in *Funaria*. *Science* **217,** 943–945.

Saunders, M. J., and Helper, P. K. (1983). Calcium antagonists and calmoludin inhibitors block cytokinin-induced bud formation in *Funaria. Dev. Biol.* **99,** 41–49.

Savelli, R. (1938a). Remarques optiques sur les plastes.V. Nouvelles remarques sur les substances bir—fringentes des chromoplastes. *Protoplasma* **29,** 1–8.

Savelli, R. (1938b). Sur la distribution du caroténoïde rouge dans les chlorochromoplastes. *Protoplasma* **29,** 601–607.

Savelli, R. (1960). I chlorocromoplasti. *Atti. Accad. Naz. Lincei, Cl. Sci. FiS., Mat. Nat. Rend.* [8] **29,** 162–169.

Scannerini, S., and Bonfante-Fasolo, P. (1977) Unusual plastids in an endomycorrhizal root. *Can. J. Bot.* **55,** 2471–2474.

Scaramuzzi, C. D., Hiller, R. G., and Stokes, H. W. (1992). Identification of a chloroplast-encoded *secA* gene homologue in a chromophytic alga: Possible role in chloroplast protein translocation. *Curr. Genet.* **22,** 421–427.

Schäfer, G., Heber, U., and Heldt, H. W. (1977). Glucose transport into spinach chloroplasts. *Plant Physiol.* **60,** 286–289.

Schimper, A. F. W. (1885). Untersuchungen über die Chlorophyllkörper und die ihnen homologen Gebilde. *Jahrb. Wiss. Bot.* **16,** 1–247.

Schnepf, E., and Czygan, F. C. (1966). Feinbau und Carotinoide von Chromoplasten im Spadix-Appendix von *Typhonium* und *Arum. Z. Planzenphysiol.* **54,** 345–355.

Schötz, F. (1961). Untersuchungen an den Plastiden von *Oenothera. Ber. Dtsch. Bot. Ges.* **74,** 215–216.

Schötz, F., and Senser, F. (1961). Reversible Plastidenumwandlung bei der Mutante "Weissherz" von *Onothera suaveolens* Desf. *Planta* **57,** 235–238.

Schünemann, D., and Borchert, S. (1994). Specific transport of inorganic phosphate and C3- and C6-sugar-phosphates across the envelope membranes of tomato (*Lycopersicon esculentum*) leaf-chloroplasts, tomato fruit-chromoplasts and fruit-chromoplasts. *Acta Botanica* **107,** 461–467.

Schünemann, D., Borchert, S., Flügge, U.-I., and Heldt, H. W. (1993). ADP/ATP translocator from pea root plastids. Comparison with translocators from spinach chloroplasts and pea leaf mitochondria. *Plant Physiol.* **103,** 131–137.

Schweitz, F. D., Cisneros, P. L., Jagield, J. A., Comero, J. L., and Butterfield, K. A. (1978). The relationship of fixed carbon and nitrogen sources on the greening process in *Euglena gracilis* strain. Z. *J. Protozool.* **25,** 257–261.

Scolnik, P. A., Hinton, P., Greenblatt, I. M., Giuliano, G., Delanoy, M. R., Spector, D. L., and Pollock, D. (1987). Somatic instability of carotenoid biosynthesis in the tomato *ghost* mutant and its effect on plastid development. *Planta* **171,** 11–18.

Seol, J. H., Yoo, S. J., Kim, H. I., Kang, M. S., Ha, D. B., and Chung, C. H. (1994). The 65-kDa protein derived from the internal translational initiation site of the *ClpA* gene inhibits the ATP-dependent protease Ti in *Escherichia coli. J. Biol. Chem.* **269,** 29468–29473.

Seyama, N., and Splittstoesser, W. E. (1975). Pigment synthesis in *Cucurbita moschata* cotyledons as influenced by CPTA and several inhibitors. *Plant Cell Physiol.* **16,** 13–19.

Sheen, J. (1990). Metabolic repression of transcription in higher plants. *Plant Cell* **2,** 1027–1038.

Shewfelt, A. L., and Halpin, J. E. (1967). The effect of light quality on the rate of tomato color development. *Proc. Am. Soc. Hortic. Sci.* **91,** 561–565.

Shifriss, O. (1981). Origin, expression, and significance of gene B in *Cucurbita pepo* L. *J. Am. Soc. Hortic. Sci.* **106,** 220–232.

Shifriss, O., and Paris, H. S. (1981). Identification of modifier genes affecting the extent of precocious fruit pigmentation in *Cucurbita pepo* L. *J. Am. Soc. Hortic. Sci.* **106,** 653–660.

Shimokawa, K., and Horiba, K. (1975). Ultrastructural observations of the yellowish green peels of *Citrus unshiu* Marc. *Bull. Fac. Agric., Miyazaki Univ.* **22,** 1–19.

Shimokawa, K., Sakanoshita, A., and Horiba, K. (1978). Ethylene-induced changes of chloroplast structure in *Satsuma mandarin* (*Citrus unshiu* Marc.). *Plant Cell Physiol.* **19,** 229–236.

Shinozaki, K., Ohme, M., Tanaka, M., Wakasugi, T., Hayashida, N., Matssubayashi, T., Zaita, N., Chunwongse, J., Obokata, J., Yamaguchi-Shinozaki, K., Ohto, C., Torazawa, K., Meng, B., Sugita, M., Deno, H., Kamogashira, T., Yamada, K., Kusuda, J., Takaiwa, F., Kato, A., Tohdoh, N., Shimada, H., and Sugiura, M. (1986). The complete nucleotide sequence of tobacco chloroplast genome: Its gene organization and expression. *EMBO J.* **5,** 2043–2049.

Simcox, P. D., Reid, E. E., Canvin, D. T., and Dennis, D. T. (1977). Enzymes of the glycolytic and pentose phosphate pathways in proplastids from the developing endosperm of *Ricinus communis* L. *Plant Physiol.* **59,** 1128–1132.

Simpson, D. J. (1974). Aspects of plastid development. Ph.D. Thesis, University of New South Wales, Kensington.

Simpson, D. J., and Lee, T. H. (1976a). The fine structure and formation of fibrils of *Capsicum annuum* L. Chromoplasts. *Z. Pflanzenphysiol.* **77,** 127–138.

Simpson, D. J., and Lee, T. H. (1976b). Plastoglobules of leaf chloroplasts of two cultivars of *Capsicum annuum. Cytobios* **15,** 139–147.

Simpson, D. J., Chichester, C. O., and Lee, T. H. (1974a). Chemical regulation of plastid development. I. The inhibition of chlorophyll biosynthesis in detached pumpkin cotyledons by CPTA. A pigment and ultrastructural study. *Aust. J. Plant Physiol.* **1,** 119–133.

Simpson, D. J., Rahman, F. M. M., Buckle, K. A., and Lee, T. H. (1974b). Chemical regulation of plastid development. II. Effect of CPTA on the ultrastructure and carotenoid composition of chromoplasts of *Capsicum annuum* cultivars. *Aust. J. Plant Physiol.* **1,** 135–147.

Simpson, D. J., Baqar, M. R., and Lee, T. H. (1975a). Ultrastructure and carotenoid composition of chromoplasts of the sepals of *Strelitzia reginae* Aiton during floral development. *Ann. Bot. (London)* [N.S.] **39,** 175–183.

Simpson, D. J., Baqar, M. R., and Lee, T. H. (1975b). Unusual ultrastructural features of the chloroplast-chromoplast transformation in *Solanum luteum* fruit. *Aust. J. Plant Physiol.* **2,** 235–245.

Simpson, D. J., Baqar, M. R., and Lee, T. H. (1977a). Chemical regulation of plastid development. III. Effect of light and CPTA on chromoplast ultrastructure and carotenoids of *Capsicum annuum. Z. Pflanzenphysiol.* **82,** 189–209.

Simpson, D. J., Baqar, M. R., and Lee, T. H. (1977b). Chromoplast ultrastructure of *Capsicum annuum* carotenoid mutants. I. Ultrastructure and carotenoid composition of a new mutant. *Z. Pflanzenphysiol.* **83,** 293–308.

Simpson, D. J., Baqar, M. R., and Lee, T. H. (1977c). Fine structure and carotenoid composition of the fibrillar chromoplasts of *Asparagus officinalis* L. *Ann. Bot. (London)* [N.S.] **41,** 1101–1108.

Sinden, S. L., and Durbin, R. D. (1988). Glutamine synthetase inhibition: The possible mode of action of wildfire toxin from *Pseudomonas tabaci. Nature (London)* **219,** 379–380.

Sink, K. C., Herner, R. C., and Knowlton, L. L. (1974). Chlorophyll and carotenoids of the *rin* tomato mutant. *Can. J. Bot.* **52,** 1657–1560.

Sitte, P. (1974). Plastid metamorphosis and chromoplasts in *Chrysosplenium. Z. Pflanzenphysiol.* **73,** 243–265.

Sitte, P. (1977). Functional organization of biomembranes. *In* "Lipids and Lipid Polymers in Higher Plants" (M. Tevini and H. K. Lichtenthaler, eds.), pp. 1–28. Springer-Verlag, Berlin.

Sitte, P. (1987). Development and division of chromoplasts in petals of *Forsythia. Cellule* **74,** 59–76.

Sitte, P., Falk, H., and Lievogel, B. (1980). Chromoplasts. *In* "Plant Pigments" (F. C. Czygan, ed.), pp. 117–148. Fischer Verlag, Stuttgart.

Slabas, A. R., and Fawcett, T. (1992). The biochemistry and molecular biology of plant lipid biosynthesis. *Plant Mol. Biol.* **19,** 169–191.

Smirra, I., Halevy, A. H., and Vainstein, A. (1993). Isolation and characterization of a chromoplast-specific carotenoid-associated protein from *Cucumis sativus. Plant Physiol.* **102,** 491–496.

Smith, L. L. W., and Smith, S. (1931). Light and the carotenoid content of certain fruits and vegetables. *Plant Physiol.* **6**, 265–275.

Smith, M., and Butler, R. D. (1971). Ultrastructural aspects of petal in *Cucumis sativus* with particular reference to chromoplasts. *Protoplasma* **73**, 1–13.

Smith, P. G. (1950). Inheritance of brown and green mature peppers. *J. Hered.* **41**, 138–140.

Smith, O. (1936). Effects of light on carotenoid formation in tomato fruits. *Mem.–N.Y., Agric. Exp. Stn. (Ithaca)* **187**, 3–26.

Smith, R. G., Gauthier, D. A., Dennis, D. T., and Turpin, D. H. (1992). Malate- and pyruvate-dependent fatty acid synthesis in leucoplasts from developing castor bean endosperm. *Plant Physiol.* **98**, 1233–1238.

Soler, E., Clastre, M., Bantignies, B., Marigo, G., and Ambid, C. (1993). Uptake of isopentenyl diphosphate into plastids isolated from *Vitis vinifera* L. suspension. *Plant Mol. Biol.* **26**, 1867–1873.

Soll, J., Schultz, G., Joyard, J., Douce, R., and Block, M. A. (1985). Localization and synthesis of prenylquinones in isolated outer and inner envelope membrane from spinach chloroplasts. *Arch. Biochem. Biophys.* **238**, 290–299.

Sprey, B. (1970). Die Lokalisierung von Skundäcarotinoïden von *Hematococcus pluvialis* Flotow cm. Wille. *Protoplasma* **71**, 235–250.

Spurr, A. E. (1969). The thylakoid plexus in chloroplast of tomato fruit. *Proc. Int. Bot. Cong., 11th,* Seattle, p. 207.

Spurr, A. R. (1970). Morphological changes in ripening fruit. *HortScience* **5**, 5–7.

Spurr, A. R., and Harris, W. M. (1968). Ultrastructure of chloroplasts and chromoplasts in *Capsicum annuum* I. Thylakoid membrane changes during fruit ripening. *Am. J. Bot.* **55**, 1210–1224.

Stabel, P., Sundas, A., and Engström, P. (1991). Cytokinin treatment of embryos inhibits the synthesis of chloroplast proteins in Norway spruce. *Planta* **183**, 520–527.

Stahl, R. J., and Sparace, S. A. (1991). Characterization of fatty acid biosynthesis in isolated pea root plastids. *Plant Physiol.* **96**, 602–608.

Stead, A. D., and Duckett, J. G. (1980). Plastid ontogeny in the corolla cells of *Digitalis purpurea* L. cv. Foxy. *Ann. Bot. (London)* [N.S.] **46**, 549–555.

Steffen, K. (1955). Die submikroscopische Struktur der chromoplasten. *Ber. Dtsch. Bot. Ges.* **68**, 23.

Steffen, K. (1964a). Chromoplastenstudien 11. Die ontogenese der Chromoplasten in den Beeren von *Convallaria majalis* L. *Protoplasma* **58**, 579–588.

Steffen, K. (1964b). Chromoplastenstudien. I. Der amöboide Chromoplastentyp. *Planta* **60**, 506–522.

Steffen, K., and Walter, F. (1955). Die Submikroskopische Struktur der Chromoplasten. *Naturwissenchaften* **42**, 395–396.

Steffen, K., and Walter, F. (1958). Die Chromoplasten von *Solanum capsicastrum* L. und ihre Genese. *Planta* **50**, 640–670.

Steginck, S. J., and Vaughn, K. C. (1988). Norflurazon (SAN-9789) reduces abscisic acid levels in cotton seedlings: A glandless isoline is more sensitive than its glanded conterpart. *Pestic. Biochem. Physiol.* **31**, 269–275.

Steinmüller, D., and Tevini, M. (1985). Composition and function of plastoglobuli I. Isolation and purification from chloroplasts and chromoplasts. *Planta* **163**, 201–207.

Stern, A. I., Schiff, J. A., and Epstein, H. T. (1964a). Studies of chloroplast development in *Euglena.* V. Pigment biosynthesis, photosynthetic oxygen evolution and carbon dioxide fixation during chloroplast development. *Plant Physiol.* **39**, 220–226.

Stern, A. I., Epstein, H. T., and Schiff, J. A. (1964b). Studies of chloroplast development in *Euglena.* VI. Light intensity as controlling factor in development. *Plant Physiol.* **39**, 226–231.

Stewart, I., and Wheaton, T. A. (1971). Effects of ethylene and temperature on carotenoid pigmentation of *Citrus* peel. *Proc. Fa. State Hortic. Soc.,* pp. 264–266.

Stitt, M., von Schaewenn, A., and Willmitzer, L. (1990). "Sink" regulation of photosynthetic metabolism in transgenic tobacco expressing yeast invertase in their cell wall involves a decrease of the Calvin-cycle enzymes and increase of glycolytic enzymes. *Planta* **183**, 40–50.

Straus, W. (1954). Properties of isolated carrot chromoplasts. *Exp. Cell Res.* **6**, 392–402.

Strzalka, K., Ngernprasirtsiri, J., Watanabe, A., and Akazawa, T. (1987). Sycamore amyloplasts can import and process precursors of nuclear encoded chloroplast proteins. *Biochem. Biophys. Res. Commun.* **149**, 799–806.

Suzuki, S. (1974). Ultrastructural development of plastids in cherry peppers during fruit ripening. *Bot. Mag. (Tokyo)* **87**, 165–178.

Tagaya, M., Wilson, D. W., Brunner, M., Arango, N., and Rothman, J. E. (1993). Protein structure of an N-ethylmaleimide-sensitive fusion protein involved in vesicular transport. *J. Biol. Chem.* **268**, 2662–2666.

Theologis, A., Oeller, P. W., Wong, L. M., Rottmann, W. H., and Gantz, D. M. (1993). Use of a tomato mutant constructed with the reverse genetics to study fruit ripening, a complex developmental process. *Dev. Genet.* **14**, 282–295.

Thomas, R. L., and Jen, J. J. (1975a). Red light intensity and carotenoid biosynthesis in ripening tomatoes. *J. Food Sci.* **40**, 566–568.

Thomas, R. L., and Jen, J. J. (1975b). Phytochrome-mediated carotenoids biosynthesis in ripening tomatoes. *Plant Physiol.* **56**, 452–453.

Thompson, A. E., Tomes, M. L., Erickson, H. T., Wann, E. V., and Armstrong, R. J. (1967). Inheritance of crimson fruit colour in tomatoes. *Proc. Am. Soc. Hortic. Sci.* **91**, 495–504.

Thompson, J. A. (1980). Apparent identity of chromoplast and chloroplast DNA in the daffodil, *Narcissus pseudonarcissus. Z. Naturforsch. C: Biosci.* **35C**, 1101–1103.

Thomson, W. W. (1965). The ultrastructural development of chromoplasts in Navel oranges. *Am. J. Bot.* **62**, 622.

Thomson, W. W. (1966a). Observations on the ultrastructure of the plasmalemma in oranges. *J. Ultrastruct. Res.* **16**, 640–650.

Thomson, W. W. (1966b). Ultrastructural development of chromoplasts in Valencia oranges. *Bot. Gaz. (Chicago)* **127**, 133–139.

Thomson, W. W. (1969). Ultrastructural studies on the epicarp of ripening oranges, *Proc. Int. Citrus Symp., 1st,* University of California, *1968,* pp. 1163–1169.

Thomson, W. W., and Whatley, J. M. (1980). Development of nongreen plastids. *Annu. Rev. Plant Physiol.* **31**, 375–394.

Thomson, W. W., Weier, T. E., and Drever, H. (1964). Electron microscope studies on chloroplasts from phosphorus-deficient plants. *Am. J. Bot.* **51**, 933–938.

Thomson, W. W., Lewis, L. N., and Coggins, C. W. (1967). The reversion of chromoplasts to chloroplasts in Valencia oranges. *Cytologia* **32**, 117–124.

Tiller, K., and Link, G. (1993). Sigma-like transcription factors from mustard (*Sinapis alba* L.) etioplast are similar in size to, but functionally distinct from their chloroplast counterparts. *Plant Mol. Biol.* **21**, 503–513.

Tolbert, N. E. (1960). (2-Chloroethyl)trimethylammonium chloride and related compounds as plant growth substances. 1. Chemical structure and bioassay. *J. Biol. Chem.* **235**, 475–479.

Tomes, M. L. (1963). Temperature inhibition of carotene synthesis in tomato. *Bot. Gaz. (Chicago)* **124**, 180–185.

Tomes, M. L. (1969). *Delta*-carotene in the tomato. *Genetics* **56**, 227–232.

Tomes, M. L., and Verkerk, K. (1965). Pigment analyses of three flesh color mutants. *Rep. Tomato Genet. Coop.* **16**, 61–62.

Tôyama, S. (1967). Electron microscope studies on the morphogenesis of plastids. IV. Fine structure development of chromoplasts in the root of *Daucus carota* L. *Sci. Rep. Tokyo Kyoiku Daigaku, Sect. B* **12**, 245–253.

Tôyama, S., and Ueda, K. (1965). Electron microscope studies on the morphogenesis of plastids. II. Changes in fine structure and pigment composition of the plastids in autumn leaves of *Gingko biloba. Sci. Rep. Tokyo Kyoiku Daigaku, Sect. B* **12**, 31–37.

Trabucci, B. (1964). Richerche al microscopio elettronico sulla sviluppo e sulla struttura dei cromoplasti di *carota*. *Ann. Fac. Agrar.* (*Univ. Cattol. Saero. Cuore.*) [2] **1**, 135–147.

Trebitsh, T., Goldschmidt, E. E., and Riov, J. (1993). Ethylene induces *de novo* synthesis of chlorophyllase, a chlorophyll degrading enzyme, in *Citrus* fruit peel. *Proc. Natl. Acad. Sci. U.S.A.* **90**, 9441–9445.

Trethewey, R. N., and ap Rees, T. (1994). The role of the hexose transporter in the chloroplasts of *Arabidopsis thaliana* L. *Planta* **195**, 168–174.

Trimming, B. A., and Emes, M. J. (1993). Glycolytic enzymes in non-photosynthetic plastids of pea (*Pisum sativum* L.) roots. *Planta* **190**, 439–445.

Tyson, R. H., and ap Rees, T. (1988). Starch synthesis by isolated amyloplasts from wheat endosperm. *Planta* **175**, 33–38.

Tzen, J. T. C., and Huang, A. H. C. (1990). Surface structure and properties of plant seed oil bodies. *J. Cell Biol.* **117**, 327–335.

Ueda, R., and Momose, T. (1968). Observations on the autumnal reddening of leaves in Gymnospermae and Pteridophyta in reference to their phylogenetic relationship. *Sci. Rep. Tokyo Kyoiku Daigaku, Sect. B* **13**, 199–205.

Ursin, V. M., and Bradford, K. J. (1989). Auxin and ethylene regulation of petiole epinasty in two developmental mutants of tomato, *diageotropica*, and *epinastic*. *Plant Physiol.* **90**, 1341–1346.

Valentin, K. (1993). SecA is a plastid-encoded in a red alga: Implications for the evolution of plastid genomes and the thylakoid protein import apparatus. *Mol. Gen. Genet.* **236**, 245–250.

Van der Wilden, W., Herman, E. M., and Chrispeels, M. J. (1980). Protein bodies of mung bean cotyledons as autophagic organelles. *Proc. Natl. Acad. Sci. U.S.A* **77**, 428–432.

Varshavsky, A. (1992). The N-end rule. *Cell* (*Cambridge, Mass.*) **69**, 725–735.

Vaughn, K. C., and Duke, S. O. (1981). Evaginations from the plastid envelope: A method for transfer of substances from plastid to vacuole. *Cytobios* **32**, 89–95.

Vaughn, K. C., and Wilson, K. G. (1981). Improved visualisation of plastid fine structure: plastid microtubules. *Protoplasma* **108**, 21–27.

von Schaewen, A., Stitt, M., Schmidt, R., Sonnewald, U., and Willmitzer, L. (1990). Expression of a yeast-derived invertase in the cell wall of tobacco and *Arabidopsis* plants leads to accumulation of carbohydrate and inhibition of photosynthesis and strongly influences growth and phenotype of transgenic tobacco plants. *EMBO J.* **9**, 3033–3044.

von Wettstein, D. (1958). The formation of plastid structures. *Broohaven Symp. Biol.* **11**, 138–159.

von Wettstein, D., Henningsen, K. W., Boynton, J. E., Kannangara, G. C., and Nielsen, O. F. (1971). The genetic control of chloroplast development in barley. *In* "Autonomy and Biogenesis of Mitochondria and Chloroplasts" (N. K. Boardman, A. W. Linnane, and R. M. Smillie, eds.), pp. 205–223. North-Holland Publ., Amsterdam.

Walles, B. (1971). Chromoplast development in a carotenoid mutant maize. *Protoplasma* **73**, 159–175.

Walles, B. (1972). Chromoplast development in a maize mutant. *J. Ultrastruct. Res.* **38**, 210.

Walters, D. R., and Ayres, P. G. (1984). Ribulose bisphosphate caboxylase protein and enzymes of CO_2 assimilation in barley infected by powdery mildew (*Erysiphe graminis hordei*). *Phytopathol. Z.* **109**, 208–218.

Washitani, I., and Sato, S. (1977). Studies on the function of proplastids in the metabolism of *in vitro*-cultured tobacco cells. III. Source of reducing power for amino acid synthesis from nitrite. *Plant Cell Physiol.* **18**, 1235–1241.

Washitani, I., and Sato, S. (1981) Sudies on the function of proplastids in metabolism of *in vitro* cultured tobacco cells. VI. ATP generation by 3-phosphoglycerate kinase. *Plant Cell Physiol.* **22**, 145–148.

Weissenborn, D. L., Denbow, C. J., Laine, M., Lang, S. S., Zhenbiao, Y., Yu, X., and Cramer, C. L. (1995). HMG-CoA reductase and terpenoid phytoalexins: Molecular specialization within a complex pathway. *Physiol. Plant.* **93**, 393–400.

Went, F. W., LeRosen, A. L., and Zechmeister, L. (1942). Effect of external factors on tomato pigments as studied by chromatographic methods. *Plant Physiol.* **17**, 91–100.

Whatley, J. M. (1974). Chloroplast development in primary leaves of *Phaseolus vulgaris. New Phytol.* **73**, 1097–1110.

Whatley, J. M. (1978). A suggested cycle of plastid developmental interrelationships. *New Phytol.* **80**, 489–502.

Whatley, J. M. (1983). Plastids-past, present, and future. *Int. Rev. Cytol., Suppl.* **14**, 329–373.

Whatley, J. M., and Whatley, F. R. (1987). When is a chromoplast? *New Phytol.* **106**, 667–678.

Whatley, J. M., McLean, B., and Juniper, B. E. (1991). Continuity of chloroplast and endoplasmic reticulum membranes in *Phaseolus vulgaris. New Phytol.* **117**, 209–217.

Whitaker, B. D. (1986). Fatty-acid composition of polar lipids in fruit and leaf chloroplasts of "16:3"- and "18:3"-plant species. *Planta* **169**, 313–319.

Whitaker, B. D. (1992). Glycerolipid-fatty-acid desaturase deficiencies in chloroplasts from fruits of *Capsicum annuum* L. *Planta* **187**, 261–265.

Whiteheart, S. W., Rossnagel, K., Buhrow, S. A., Brunner, M., Jaenicke, R., and Rothman, J. E. (1994). N-ethylmaleimide-sensitive fusion protein: A trimeric ATPase whose hydrolysis of ATP is required for membrane fusion. *J. Cell Biol.* **126**, 945–954.

Whithers, N. W., and Haxo, F. T. (1978). Isolation and characterization of carotenoid-rich lipid globules from *Peridinium foliaceum. Plant Physiol.* **62**, 36–39.

Wiley, R. C., Bueso, C. E., and Angell, F. F. (1972). Surface color and pigment development in tomatoes treated by ethephon and CPTA. *HortScience* **7**, 325.

Williams, M., and Randall, D. D. (1979). Pyruvate dehydrogenase complex from chloroplasts of *Pisum sativum. Plant Physiol.* **64**, 1099–1103.

Wilson, D. W., Wilcox, C. A., Flynn, C. A., Ellson, C., Kuang, W. J., Henzel, W. J., Block, M. R., Ullrich, A., and Rothman, J. E. (1989). A fusion protein required for vesicle-mediated transport in both mammalian cells and yeast. *Nature (London)* **339**, 355–359.

Winkenbach, F., Falk, H., Liedvogel, B., and Sitte, P. (1976). Chromoplast of *Tropaeolum majus* L.: Isolation and characterization of lipoprotein elements. *Planta* **128**, 23–28.

Wittenbach, V. A., Lin, W., and Hebert, R. R. (1982). Vacuolar localization of proteases and degradation of chloroplasts in mesophyll protoplasts from senescing primary wheat leaves. *Plant Physiol.* **69**, 98–102.

Wooding, F. B. P., and Northcote, D. H. (1965). The fine structure of the mature resin canal cells of *Pinus pinea. J. Ultrastruct. Res.* **13**, 233–244.

Wozny, A., and Szweykowska, A. (1975). Effect of cytokinins and antibiotics on chloroplast development in cotyledons of *Cucumis sativus. Biochem. Physiol. Pflanz.* **168**, 195–209.

Wrischer, M. (1972). Transformation of plastids in young carrot callus. *Acta Bot. Croat.* **31**, 41–46.

Wrischer, M. (1978). Ultrastructural localization of diaminobenzidine photooxidation in etiochloroplasts. *Protoplasma* **97**, 85–92.

Wrischer, M., and Ljubesic, N. (1984). Plastid differentiation in *Claceolaria* petals. *Acta Bot. Croat.* **43**, 19–24.

Wu, M., and Salunkhe, D. K. (1971). Influence of soil fumigation of telone and nemagon on the ultrastructure of chromoplasts in carrot roots (*Daucus carota*). *Experientia* **27**, 712–713.

Wuttke, H. G. (1976). Chromoplasts in *Rosa rugosa*: Development and chemical characterization of tubular elements. *Z. Naturforsch. C: Biosci.* **31C**, 456–460.

Young, R., and Jahn, O. (1972). Ethylene induced carotenoid accumulation in *Citrus* fruit rinds. *Proc. Am. Soc. Hortic. Sci.* **97**, 258–261.

Yu, H. S., and Russell, S. D. (1994). Populations of plastids and mitochondria during male reproductive cell maturation in *Nicotiana tabacum* L.: A cytological basis for occasional biparental. *Planta* **193**, 115–122.

Yu, J., and Woo, K. C. (1988). Glutamine transport and the role of glutamine translocator in chloroplasts. *Plant Physiol.* **88**, 1048–1054.

Yuasa, A. (1960). The green-spiral theory of the plastid and its elucidation. *Kromosomo* **46–47,** 1549–1553.

Ziegler, H., Schafer, E., and Schneider, M. M. (1983). Some metabolic changes during chloroplast-chromoplast transition in *Capsicum annuum. Physiol. Vég.* **21,** 485–494.

Zscheile, F. P., and Lesley, J. W. (1967). Pigment analysis of sherry flesh colour mutation resembling yellow in the tomato. *J. Hered.* **58,** 193–194.

Zurzycki, J. (1962). Studies on chromoplasts. I. Morphological changes of plastids in the ripening fruit. *Acta Soc. Bot. Pol.* **23,** 161–164.

Sperm-Binding Proteins

Kathleen R. Foltz

Division of Molecular, Cell, and Developmental Biology and the Marine Science Institute, University of California at Santa Barbara, Santa Barbara, California 93106

Gamete recognition and binding are mediated by specific proteins on the surface of the sperm and egg. Identification and characterization of some of these proteins from several model systems, particularly mouse and sea urchin, have focused interest on the general properties and functions of gamete recognition proteins. Sperm-binding proteins located in egg extracellular coats as well as sperm-binding proteins that are localized to the egg plasma membrane are presented in the context of their structure and function in gamete binding. Unifying and disparate characteristics are discussed in light of the diverse biology of fertilization among species. Outstanding questions, alternative mechanisms and models, and strategies for future work are presented.

KEY WORDS: Fertilization, Gamete recognition, Sperm-binding proteins, Sperm receptors.

I. Introduction

The purpose of this review is to present and discuss proteins serving as "sperm receptors" on eggs. Although their presence has been assumed for many years, relatively few sperm receptors have been positively identified, purified, and characterized. While there have been many outstanding reviews (cited throughout) regarding gamete interactions and sperm receptors for particular species, this review aims to take a comprehensive, comparative approach to these molecules. This hopefully will serve to integrate information from distant sources and will provide a resource for students entering the field and for colleagues in other disciplines who wish to become acquainted with the molecular mechanisms of gamete recognition and binding.

When comparing receptor molecules from species as disparate as echinoderms, ascidians, amphibians, and mammals, it is important to consider the various biological parameters of fertilization (e.g., external versus internal fertilization and differences in egg coat structures), which necessarily introduce problems into any sort of comprehensive treatment of the structural basis of sperm receptor function. Indeed, merely trying to define "sperm receptor" often initiates heated debate. Nonetheless, I hope to present a synthesis of the data pertaining to egg proteins which bind to sperm and to point out some of the outstanding questions in this field. One major area of emphasis will be a comparison of the structural basis of gamete interactions and the assessment of the molecules' recognition and adhesive properties. Following a brief overview of fertilization and gamete recognition (Sections II and III), the various sperm-binding proteins will be discussed in turn, beginning with those localized to the egg outer vestments (Section IV) followed by those few proteins that have been identified as oolemmal proteins that bind to sperm (Section V). Structural and functional similarities as well as disparities will be discussed and major problems and questions pointed out in the last two sections (VI and VII). In as much as it is possible, historical perspective will be included; however, because of space constraints, not all references can be cited. Wherever appropriate, I have tried to include references to the most recent, as well as historically relevant, and most comprehensive reviews and papers.

II. An Overview of Fertilization

A. The Significance of Sperm–Egg Interactions

The long-standing interest in sperm–egg interactions stems from the logical view that this event represents the first step in development. Further, the meeting of the two gametes encompasses several basic biological phenomena, including cell–cell recognition and adhesion, membrane fusion, and activation of the cell cycle. Therefore, while fertilization is worthy of understanding in and of itself, the potential insight into additional fundamental biological processes has made the study of gamete interaction and activation particularly attractive to biologists from many fields of interest (Lallier, 1977). An understanding of fertilization has its practical side as well. Interest in sperm–egg interactions has led to new ideas and strategies for improving contraceptive and *in vitro* fertilization (IVF) methods. In the last decade or so, much emphasis has been placed upon identifying target gamete surface molecules for the development of contraceptive antibodies, with some success (Primakoff *et al.*, 1988a; Primakoff and Myles, 1990; Alexander and Bialy, 1994; Epifano and Dean, 1994; Primakoff, 1994).

B. The Major Steps of Fertilization

As the first step in the development of an organism, successful fertilization has several key prerequisites that must be met, resulting in the meeting of the two gametes (Fig. 1). First, the sperm must find the egg. In broadcast spawners such as sea urchins, small peptides are released from the egg which increase sperm motility and respiration (Garbers, 1989; Ward and Kopf, 1993; Hardy *et al.*, 1994). Whether or not chemoattraction, in its strictest definition, is involved has not been established (Miller, 1985; Ward and Kopf, 1993). A true sperm chemotaxis, whereby the sperm changes direction and moves up a concentration gradient of the chemoattractant, has not been documented for any species except *Arbacia punctulata* (Ward *et al.*, 1985, Hardy *et al.*, 1994). The egg peptides produced by the egg are called sperm-activating peptides (SAPs) and are not strictly species specific in their ability to stimulate sperm respiration and motility (Suzuki and Yoshino, 1992). No vertebrate SAP-like molecule has been discovered to date, but there has been a report of a sperm-stimulating activity in human

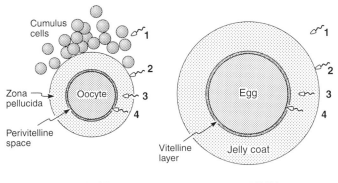

MOUSE   SEA URCHIN

FIG. 1 Comparison of sperm–egg interactions in that mouse and the sea urchin. Sperm must first find and contact the oocyte/egg (1). In mammals, this involves penetration through the cumulus cells and capacitation. In sea urchins, the sperm swim to the egg through the sweater in response to peptides released by the egg. The sperm acrosome reaction (2) occurs in response to binding to the zona pellucida (mouse) or the egg jelly coat (sea urchin). This exposes new sperm membranes and proteins and releases enzymes (glycosidases and proteases) which are required for penetration of the extracellular vestment (3). Successful fertilization occurs only when the sperm has penetrated to the plasma membrane where binding and fusion occur (4). Each step is mediated by specific proteins on the surface of each gamete.

follicular fluid (Ralt *et al.,* 1991). Overall, precisely how the sperm finds the egg remains a mystery.

Once a spermatozoon reaches the egg, it is faced with a physical barrier. Eggs have a variety of extracellular coats, depending on the species, and the sperm must recognize, bind to, and then penetrate these vestments to reach the egg surface proper (Fig. 1). Some species, most notably insects and fish, have micropyles through which the sperm must travel (cf. Nakano, 1969; Sander, 1985). In many other organisms, the sperm must penetrate directly through the extracellular coats. This generally involves an activation, per se, of the sperm; the sperm undergoes an "acrosomal reaction" that results in morphological changes and the release of glycosidases and proteases which facilitate sperm penetration and which expose the acrosomal membrane for fusion (Dan, 1967; Yanagimachi, 1988; Myles, 1993; Ward and Kopf, 1993). Specific egg coat proteins mediate the acrosomal reaction (Trimmer and Vacquier, 1986; Vacquier, 1989; Wassarman, 1992). In addition, the acrosome reaction often results in the exposure of proteins which mediate secondary sperm–egg binding events. Following the contact with the extracellular coat, the sperm then penetrates to the egg plasma membrane (the oolemma). Binding and fusion of the sperm and egg plasma membranes results in an array of physiological changes within the egg, collectively referred to as egg activation (Whitaker and Swann, 1993; Miyazaki *et al.,* 1993; Jaffe, 1995). Although the details of egg activation vary among species, there are a few general features. These include calcium release, a polyspermy block, and resumption of the cell cycle (Jaffe, 1995). In most species, there are immediate blocks to polyspermy in response to sperm binding. In echinoderms and most other invertebrates, as well as in amphibians, there is a fast, transient electrical block at the plasma membrane (Jaffe and Gould, 1985; Jaffe and Cross, 1986; Jaffe, 1995) followed by a slower, physical block, the cortical reaction, which modifies the egg surface (Kay and Shapiro, 1985; Schuel, 1985; Larabell and Chandler, 1991; Shapiro, 1991). The cortical reaction involves, in most species, Ca^{2+}-stimulated exocytosis of cortical granules (L. F. Jaffe, 1985; L. A. Jaffe, 1995).

Depending on the organism, the extent of species specificity in these interactions between the sperm and the egg surfaces varies. For example, in the free spawners, such as sea urchins and many other marine invertebrates, there is some degree of specificity at each step, particularly the final interaction between the gametes at the egg surface proper (Summers and Hylander, 1975, 1976; Yanagimachi, 1988). Lillie (1914, 1919) first noted the necessity of the specificity component; the gametes must be highly selective as a barrier to cross-fertilization in organisms that release their gametes into the environment. This is in contrast to internal fertilizers such as mammals, where adult behaviors help ensure that cross-fertilizations are very rare (O'Rand, 1988). Indeed, hamster oocytes that have had their extracellular coats removed are easily fertilized by human sperm which have been capacitated (see below) and acrosome-reacted *in vitro* (Yanagimachi, 1988). Zoma-free oocytes of some mammalian species (most notably the hamster) can be recognized nonspecifically by the capacitated and acrosome-reacted sperm of other mammalian species (O'Rand, 1988). However, internal fertilization does not always correlate with reduced species specificity. For example, mouse oocytes are extremely species specific in their ability to bind sperm (Yanagimachi, 1978, 1988). Several excellent reviews address the issue of specificity in mammalian gamete interactions (O'Rand, 1986, 1988; Yanagimachi, 1981, 1988).

C. The Role of the Gamete Surface Components

A goal of fertilization biologists for almost 100 years has been the identification of the gamete surface components which mediate each step of sperm–egg recognition and binding. Lillie (1914, 1919) proposed that any theory of fertilization would have to account for the specificity observed in gamete interactions. In particular, he postulated that specific receptors on both the sperm and the egg were required.

For each step in the process of sperm–egg contact (Fig. 1), recognition seems to be a critical aspect. With regard to the binding of the sperm to the egg, adhesion then becomes a critical component as well. The binding of the sperm to the egg extracellular coat(s) triggers changes in the sperm. This results in a different sperm surface now being presented to the egg plasma membrane. Thus, the sperm component(s) mediating the initial binding to the egg coat may be involved in transducing a signal. Binding of the sperm to the egg plasma membrane is followed quickly by membrane fusion and then a variety of physiological changes in the egg. Thus, one must consider that the gamete surface component(s) involved in this final interaction may participate in fusogenic and signaling events as well. In fact, a central question has been whether egg activation is due to receptor-mediated signal transduction by an egg transmembrane protein (a sperm receptor) or the delivery of an "activating factor" by the sperm which then engages the egg's internal signaling machinery (Epel, 1989; Turner and Jaffe, 1989; Nuccitelli, 1991; Swann, 1993; Whitaker and Swann, 1993). It is apparent that identification and characterization of the sperm and egg components that mediate gamete interactions will be necessary before a complete molecular understanding of fertilization and egg activation can be reached. It is also worth noting that none of these interactions need be mediated by a single sperm component or a single egg component, but may in fact be the result of complex interactions.

III. Understanding Gamete Interaction Molecules

A. A Brief History of the Search for Gamete Interaction Molecules

Over the years, several basic strategies have been employed in attempts to identify and characterize gamete surface proteins involved in recognition and adhesion. Careful observation of the sperm–egg interaction suggested that specific surface components were involved (Lillie, 1914, 1919). Ultrastructural investigation of the acrosome reaction and bound gametes indicated morphological changes in the gametes' surfaces, raising the possibility that the biochemical components could be pursued (Dan, 1967; Vacquier, 1979). Thus, biochemical fractionation of gamete surface proteins has been employed in conjunction with assays in which fractions are assessed for their ability to perturb the interaction via binding competition (cf. Ruiz-Bravo and Lennarz, 1989). Heightened interest in identifying egg and sperm proteins developed following the advent of antibody techniques in the 1940s and 1950s (Metz, 1967). A common approach was to prepare a crude preparation of the gamete surface (or the whole gamete itself), inject this into a rabbit, and then assess the resulting polyclonal antisera for its ability to bind to the gamete surface and block a particular interaction (Tyler and O'Melveny, 1941; Metz et al., 1964; Perlmann, 1954, 1956; Perlmann and Perlmann, 1957). Unfortunately, these studies had only limited success because of the complexity of the antigens; it was not until the development of monoclonal antibodies that immunological strategies succeeded (Myles et al., 1981; Saling et al., 1986; Primakoff et al., 1987, 1988a,b; Alexander and Bialy, 1994).

The biochemical intractability of the gamete surface proteins has been a continual problem. Generally speaking, the surface proteins are heavily glycosylated and insoluble. Many are labile in the sense that they are modified by proteases as part of the biology of fertilization (Vacquier, 1979). Many eggs have multiple vestments which are difficult to separate, thereby further complicating the biochemical fractionation (Hedrick and Nishihara, 1991; Larabell and Chandler, 1991). Finally, their relatively low abundance, particularly in the mammals, has made the brute force biochemical approach to identifying individual proteins a daunting challenge to even the most ambitious investigator. Nonetheless, major advances in the identification, purification, and characterization of several recognition and binding proteins involved in gamete interaction have been achieved in the past few years (Foltz and Lennarz, 1993; Myles, 1993; Ward and Kopf, 1993; Ramarao et al., 1994; Shaw et al., 1994). These advances have been made possible by the advent of more sensitive biochemical techniques, highly specific proteases, and the development of batteries of mono-clonal antibodies.

Somewhat surprising is the lack of a genetic approach to the problem of fertilization. While a great deal of effort has gone into studying gameto-genesis in systems such as *Drosophila* (Spradling, 1993; Mahajan-Miklos and Cooley, 1994) and *Caenorhabditis elegans* (Hirsh *et al.*, 1985; Li and Chalfie, 1990; Stern and DeVore, 1994), virtually nothing is known about the molecular basis of gamete interaction in these species. One reason for the lack of a genetic model is that the study of fertilization historically has had a large observational component. The species that tend to be the model systems of choice are those with relatively large eggs or oocytes, gametes that are easily obtainable, and those in which the fertilization event can be observed readily under the microscope (Monroy, 1986). Both *Drosophila* and *C. elegans* have internal fertilization and no standard methods for external fertilization have been worked out. In addition, mutational analy-ses of gamete recognition proteins would be inherently difficult; loss of function mutations would preclude the production of a zygote. Although the power of *Drosophila* genetics probably would provide a route to circum-vent this problem, no obvious approach or screen has been put forth that would result in the identification of genes that encode the gamete interaction proteins. It is likely that once such a gene is found in another system, *Drosophila* would then become a model system for assessing the function of the protein once the fruit fly homolog was identified.

Despite the similarities in the fertilization process across species and the temptation to draw comparative conclusions, the obvious diversity in the basic biology of fertilization requires that any comparative approach take into account numerous caveats (Tyler, 1967; Monroy and Rosati, 1979). Different species have different means of optimizing sperm–egg interac-tions. Even among closely related species there can be great differences. For example, depending upon which species of anuran (frogs and toads) one investigates, the oocyte can have a variety of different jelly layers, each with their own unique biochemical properties (Hedrick and Nishihara, 1991; Larabell and Chandler, 1991). Nonetheless, criteria first suggested by Lillie (1914, 1919) and expanded in recent years (Schmell *et al.*, 1977; Wassarman *et al.*, 1989) have been established for identifying putative sperm-binding proteins. These necessarily imply that the physiochemical proteins of the gamete recognition protein are compatible with the biological observations. These criteria include: (1) Proper location on the gamete surface at the time of interaction; (2) Species-specific binding properties (if the mediated interaction is species specific); and (3) Ability to competitively inhibit sperm–egg interactions.

Despite the formidable technical obstacles involved in working with egg proteins (discussed above), a number of egg proteins satisfying these criteria have been purified and characterized. The remainder of this review will focus on these molecules, following a brief discussion of some of the special problems and considerations (Section III.B).

B. Special Problems in Identifying Egg Proteins That Bind to Sperm

1. Terminology

One frustration to any student of fertilization biology, including the most established researcher, is the historical terminology which has been applied to gamete interaction proteins. The term "receptor" sometimes has been applied to any protein on the sperm or egg surface that is involved in the cell–cell interaction (Monroy, 1985; Wassarman, 1990). Hence, proteins located in the extracellular coat of the egg have been termed "sperm receptors" by many researchers. The most restrictive definition of a functional sperm receptor would place that protein in the egg plasma membrane and require specific properties such as signal transduction, endocytosis, or channel gating. For the purposes of this review, sperm-binding proteins will be defined as egg proteins which have been shown to be involved in sperm binding. These sperm-binding proteins will be classified into two groups; those located in the extracellular coats and layers and those located in the egg plasma membrane. The sperm-binding proteins of the latter group are referred to as "receptors."

2. Defining the Egg Surface

Different species naturally have different egg surface structures. At the same time, close examination of how the gamete surfaces interact reveals functional similarities (Fig. 1). In the sea urchin, as in many other organisms such as amphibians, the jelly layer serves to induce the acrosome reaction of the sperm (Trimmer and Vacquier, 1986; Vacquier, 1989). In the mouse, the zona pellucida has this function (Wassarman, 1987a, 1992). Only acrosome-reacted sperm are capable of binding to and fusing with the egg plasma membrane; this functional comparison is a useful guide when considering the egg component(s) which mediate this event.

Sea urchins (and many other species) have an additional extracellular coat called the vitelline layer (VL). The VL is composed of complexed glycoproteins and lies in close juxtaposition to the plasma membrane (Chandler, 1991; Larabell and Chandler, 1991). Binding to an penetration through this layer is poorly understood as it is difficult to obtain pure preparations that remain free from contaminating plasma membrane and egg jelly (Kinsey, 1986). Depending on the species, the vitelline layer or coat (sometimes called the chorion) may have different thickness. Once sperm penetrate, bind to, and fuse with the plasma membrane, the cortical granule contents are released in an exocytic response. This results in the rapid biochemical modification of the VL components (via proteolysis, glycosidase activity, dityrosine crosslinking, and hydration) such that it elevates away from the plasma membrane and forms a tough envelope around the zygote (Chandler, 1991; Larabell and Chandler, 1991). This "slow block" to polyspermy is also seen in mammals, but in this case, the cortical reaction results in the modification of the zona proteins such that sperm can no longer bind. Regardless, modification of the extracellular coat seems to be a common mechanism to prevent polyspermy.

3. Carbohydrates as Binding Sites

As mentioned above, most gamete surface proteins are heavily glycosylated. A major point of contention over the years has been the role of these carbohydrate moieties in recognition, binding, and adhesion (Litscher and Wassarman, 1993). Many cell adhesion molecules recognize and bind to carbohydrates on cell surfaces (Section VI.A). There are several examples of carbohydrate-mediated binding of the sperm to the egg extracellular coat (Wassarman, 1992; Litscher and Wassarman, 1993; Hoshi *et al.*, 1994). This complicates matters with regard to structural comparisons and identification of the individual proteins engaged in a particular function. Carbohydrate side chains cannot be cloned and the same carbohydrate moiety can be present on different proteins. In addition, glycoproteins are inherently difficult to work with from a biochemical point of view. The characterization of the oligosaccharides and their contributions to the functions of gamete interaction proteins will be discussed in greater detail in Section IV.G.

IV. Proteins on the Outer Egg Surface That Bind to Sperm

A. Overview

The outer egg surface contains the protein(s) responsible for initial sperm recognition and binding as well as for the induction of the sperm acrosome reaction. Elegant ultrastructural work in a number of laboratories paved the way for the more recent biochemical and molecular analyses. For example, work by Dan (1952, 1956, 1967), Colwin and Colwin (1963, 1967), and others (Dan and Wada, 1955; Tegner and Epel, 1973); Vacquier and Moy,

1977; Tilney *et al.*, 1978; Moy and Vacquier, 1979; Tilney and Jaffe, 1980; Tilney, 1985) indicated that contact of the sea urchin sperm with the egg resulted in the extension of the acrosomal tip and the exposure of an "electron dense substance" on the tip of which seemed to be involved in the next binding step (Moy and Vacquier, 1979). This assay for the acrosome reaction served as a means to fractionate and characterize the jelly components responsible for the induction, particularly in the sea urchin (SeGall and Lennarz, 1979, 1981; T. Shimizu *et al.*, 1990; Mikami-Takei *et al.*, 1991; Keller and Vacquier, 1994) and the mouse (Bleil and Wassarman, 1980c; Wassarman *et al.*, 1989). Ultrastructural analyses of gamete interactions continues to augment the biochemical work. Some of the more pronounced examples are in the penaeoidean shrimp (Clark *et al.*, 1990), the polychaete (Sato and Osanai, 1986), the frog (Hedrick and Nishihara, 1991; Larabell and Chandler, 1988a,b, 1989, 1990, 1991), and the sea urchin (Mozingo and Chandler, 1991, 1993; Bonnell *et al.*, 1993, 1994).

Table I provides a list of those proteins which have been at least partially purified and which have been demonstrated to bind to sperm. A quick glance at the table reveals that the sea urchin and the mouse are the best characterized. Many of the proteins on the list trigger the acrosome reaction, based on the ultrastructural assay, and all appear to be heavily glycosylated. While characterization of the mouse zona pellucida proteins has resulted in the identification by homology of the corresponding cDNAs or by immunological cross reactivities in other mammals, biochemical characterization in these other species is notably absent.

B. Sperm-Binding Proteins in the Vitelline Coat of the Ascidians

The ascidians (Ascidacea) are interesting marine invertebrate members of the Phylum Chordata. Their eggs are encased in a system of complicated vestments including follicle cells, the vitelline coat (chorion), and the test cells (De Santis *et al.*, 1980; Rosati, 1985; Hoshi *et al.*, 1994). Sperm bind to the vitelline coat, undergo the acrosome reaction, penetrate the extracellular layer(s), and finally fuse with the egg plasma membrane. It is thought that sperm glycosidases and proteases play a major role in the initial binding and penetration steps (Hoshi *et al.*, 1994). In *Ciona intestinalis,* a high M_r glycoprotein fraction of the vitelline coat (consisting of five major glycoproteins) has the ability to bind sperm and induce the acrosome reaction (De Santis *et al.*, 1983a,b; De Santis and Pinto, 1987). Fucosyl residues, which are a major constituent of the glycoprotein complex, appear to be the binding component (Rosati and De Santis, 1980; Pinto *et al.*, 1981), probably via a sperm α-L-fucosidase, which has been identified and shown to be necessary for sperm–egg binding (Hoshi *et al.*, 1983). There is some evidence that the polypeptide backbone of the vitelline coat glycoprotein(s) is responsible for the acrosome-inducing activity (De Santis and Pinto, 1987), but this aspect requires further characterization.

A similar situation exists for another ascidian, *Phallusia mammillata*. In this case, a different oligosaccharide moiety appears to be important—*N*-acetylglucosamine (Honnegger, 1982, 1986). In this species, two proteins have been isolated and characterized from the vitelline coat (ca. M_r 450K and 180K); they are heavily glycosylated and bind to wheat germ agglutinin (Litscher and Honnegger, 1991). The major sugar components of these glycoproteins are glcNAc and galNAc. Sperm from *Phallusia* have a β-D-*N*-acetylglucosaminidase which presumably mediates recognition and binding via the glcNAc. The partially purified proteins inhibit sperm–egg interactions, but it is not known if they trigger the sperm acrosome reaction (Litscher and Honnegger, 1991).

Thus, ascidians employ a common mechanism of using sperm glycosidases as binding proteins for vitelline coat proteins which carry the complementary saccharide chains (Hoshi *et al.*, 1994). A similar mechanism may also occur in mammals (Shur, 1993; Miller and Shur, 1994; see also Section IV.F.1). Further, it appears that the egg may itself release glycosidases and proteases at fertilization which then serve as blocks to polyspermy by destroying the binding sites (Hoshi *et al.*, 1994).

C. Sperm-Binding Proteins of the Echinoderm Egg Jelly

Perhaps one of the most well-characterized gamete interactions is the acrosome reaction of sea urchin sperm in response to egg jelly (Dan, 1967; Trimmer and Vacquier, 1986). Sea urchin gametes are readily available and the jelly coat is easily isolated (Suzuki, 1990), making this an attractive system to study the interaction of the sperm with the egg. The sperm–jelly interaction is a somewhat species-specific event, although not as restricted as plasma membrane binding and fusion (Summers and Hylander, 1975). Ishihara and Dan (1970) originally provided evidence for the role of a jelly coat glycoprotein, but later studies suggested that a fucose sulfate polymer (FSP) of the jelly layer was responsible for inducing the acrosome reaction and that this was species specific (SeGall and Lennarz, 1979, 1981). No detectable protein was found in the active FSP fraction. Subsequently, several other groups found evidence for the existence of the jelly coat glycoprotein that induced the acrosome reaction (Garbers *et al.*, 1983; DeAngelis and Glabe, 1985). However, they did not demonstrate directly the ability of the partially pure glycoprotein fraction to induce the acrosome reaction. More recently, improved purification and characterization suggest

TABLE I

Sperm-Binding Proteins Located in the Egg Extracellular Coat

Species	ca. M_r ($\times 1000$)	Function/properties	References[a]
Echinoderms			
Strongylocentrotus purpuratus	82 and 138	Major constituents of egg jelly which induce the sperm acrosome reaction; heavily glycosylated; complexed with a "fucose polymer superstructure"; may trigger calcium channel opening in sperm; specificity not assessed; not clear if both components necessary	Keller and Vacquier (1994)
Hemicentrotus pulcherrimus	135 and 140	Major components of egg jelly; induce the acrosome reaction; reduction of disulfide-bonded complex yields these heavily fucosylated proteins; specificity not assessed	T. Shimizu *et al.* (1990)
Asterius amurensis	High M_r	High M_r glycoconjugate isolated from the jelly coat that induces the acrosome reaction; specificity not assessed; sulfated and fucose-rich; probably requires a saponin for activity	Hoshi *et al.* (1991)
Ascidians			
Ciona intestinalis	$>10^7$	Glycoprotein complex (five proteins) from the vitelline coat; fucosyl residues mediate binding; protein may trigger the acrosome reaction	De Santis *et al.* (1983a,b), De Santis and Pinto (1987)
Phallusia mammillata	450, 180	Major constituents of soluble fraction of vitelline coat; both bind wheat germ agglutinin and inhibit sperm–egg interactions; specificity not assessed	Litscher and Honegger (1991)

260

	Mass (kDa)	Comments	References
Fish			
Acipenser transmontanus (sturgeon)	66	Purified from layer 3 of the egg envelope; proteolytic product of a 70-kDa protein; triggers the acrosome reaction; species specific	Cherr and Clark (1985)
Mammals			
Zona pellucida protein 3 (ZP3)			
Mouse	83	One of three major zona pellucida glycoproteins; O-linked sugars responsible for binding activity purified and recombinant protein binds sperm and induces the acrosome reaction; sugars alone do not induce the acrosome reaction	Bleil and Wassarman (1980a,b,c), Florman and Wassarman (1985), Ringuette et al. (1988)
Hamster	56	Same size mRNA as mouse ZP3; differential glycosylation?	Moller et al. (1990), Kinloch et al. (1990)
Human	65	Same size coding region as mouse ZP3; partial biochemical characterization; 74% identity with mouse	Chamberlin and Dean (1990), Shabanowitz and O'Rand (1988)
Porcine	55	Heavily glycosylated, mostly acidic lactosaminoglycans	Yurewicz et al. (1987, 1992, 1993)
Marmoset	?	92% homology with huZP3; no biochemical characterization	Thillai-Koothan et al. (1993)
Rat	115	Partial characterization; oligosaccharides important for sperm binding	Araki et al. (1992), Shalgi et al. (1986)
Rabbit	?	Sequence similarity varies with respect to other species although similarity to porcine is substantial	Dunbar et al. (1981)
Zona pellucida protein 2 (ZP2)			
Mouse	120	Binds only to acrosome-reacted sperm; may be the secondary sperm binding protein of mouse zona pellucida; glycosylated	Bleil et al. (1988)

[a] References given are primary; see text for further references and explanation.

that a fucosylated glycoprotein probably is responsible, although a single, functional polypeptide has not been isolated (T. Shimizu *et al.,* 1990; Mikami-Takei *et al.,* 1991; Keller and Vacquier, 1994).

In *Hemicentrotus pulcherrimus,* a complex of heavily glycosylated proteins was isolated from the jelly coat (T. Shimizu *et al.,* 1990). Two proteins, of ca. M_r 135K and 140K, and associated by disulfide bonds, were found to induce the acrosome reaction, and these proteins contained fucose residues. Using a sensitive assay for acrosome reaction induction, Keller and Vacquier (1994) isolated two glycoproteins from the jelly coat of *Strongylocentrotus purpuratus.* These proteins, of ca. M_r 82K and 138K, also are heavily fucosylated. The current hypothesis is that one or both of these proteins binds to a receptor on the sperm and that this signals a calcium channel to open, triggering the acrosome reaction (Keller and Vacquier, 1994; Trimmer *et al.,* 1987; Sendai and Aketa, 1989). Further, ultrastructural analysis of the two purified proteins indicates that they are small (~ 8 nm in diameter), spherical molecules bound to a fucose polymer superstructure to compose the jelly layer of the sea urchin egg (Bonnell *et al.,* 1994).

In another echinoderm, the starfish *Asterias amurensis,* a high M_r, sulfated, fucose-rich glycoprotein fraction of the jelly layer appears to be the acrosome reaction-inducing substance (Hoshi *et al.,* 1991). Interestingly, a second factor, a saponin, appears to be necessary for the proper functioning of the substance. Neither of these factors has yet been purified to homogeneity and characterized in detail.

In summary, it appears that the echinoderms have a common mechanism of inducing the acrosomal reaction: fucosylated proteins, as part of a polymeric superstructure, bind to sperm and trigger the reaction. The precise role of the fucosyl moieties—as binding sites and/or inducers—as well as the species specificity of the glycoproteins, remains to be determined.

D. Sperm-Binding Proteins in the Extracellular Coats of Other Marine Invertebrates

There are a number of other species, particularly marine invertebrates, for which there are presumed sperm-binding proteins located in the egg vestments. In the abalone, the thick vitelline layer presumably contains a species-specific receptor for the sperm protein lysin (Hoshi, 1985; Shaw *et al.,* 1994). Lysin has been characterized, cloned, and sequenced from at least seven different abalone species (Vacquier *et al.,* 1990; Lee and Vacquier, 1992; Vacquier and Lee, 1993), and its three-dimensional structure has been determined (Shaw *et al.,* 1993). It is a protein (M_r 16K) that allows for sperm penetration via a nonenzymatic dissolution of the vitelline coat proteins (Haino and Kigawa, 1966; Haino-Fukushima, 1974; Lewis *et al.,*

1982). A hypervariable region, which confers species-specific binding, presumably is recognized by and binds to an as yet unidentified vitelline component (Shaw *et al.,* 1993, 1994). Because the biochemistry and molecular biology of lysin structure and function are so well known, elucidation of its cognate receptor will be extremely useful toward developing a comprehensive view of the mechanism of gamete interactions.

The penaeoidean shrimp, *Sicyonia ingentis,* also provides an interesting system in which to study gamete interactions. Unlike most of the other model systems, penaeoid oocytes self-activate when they are shed into the sea water and the sperm (which is stored in the female seminal receptacles, where it becomes capacitated) is important only for cleavage and subsequent development (Clark *et al.,* 1994). The actual gamete contact occurs in the sea water, where the swimming action of the female is necessary to propel the nonmotile sperm toward the eggs. Primary binding of the sperm to the vitelline envelope appears to be carbohydrate dependent; i.e. a carbohydrate component of this extracellular coat binds to the sperm (Wikramanayake and Clark, 1992, 1993). Once the sperm binds, the acrosomal reaction is induced and penetration through the VE occurs. The sperm then bind to the oocyte's surface coat, which is juxtaposed to the plasma membrane. The molecular basis for this specific binding, however, is not known at this time.

E. Sperm-Binding Proteins in Fishes and Amphibians

The structure and function of the extracellular matrices of anuran (frog and toad) and urodele (salamander) eggs have been studied for well over 100 years (Elinson, 1986; Hedrick and Nishihara, 1991). There are various layers, depending on the species, and they are composed largely of glycoproteins which must function in sperm binding and the induction of the acrosome reaction (Dumont and Brummett, 1985; Katagiri, 1987; Larabell and Chandler, 1991). Although several proteins have been identified, particularly with regard to the fact that they are modified at fertilization, no purified protein has been demonstrated to be unequivocally responsible for sperm binding or induction of the acrosome reaction (Hedrick and Nishihara, 1991).

In most fish, sperm penetrate to the egg surface through a micropyle and often do not undergo a true acrosome reaction (Nakano, 1969). However, a protein of M_r 66K has been purified from the third layer of the egg of the sturgeon, *Acipenser transmontanus,* which appears to trigger the acrosome reaction in this species (Cherr and Clark, 1985). This glycoprotein, which is the mature, proteolytic product of a larger, 70-kDa protein, induces the acrosome reaction species specifically (it has no effect on *A. fulvescens* sperm). Exposure of the egg to fresh water triggered the proteolytic processing of the M_r 70K protein, while trypsin inhibitors blocked the proteolysis and thus the acrosome reaction. The protein was isolated by fractionating egg proteins on a gel filtration column followed by gel purification. Interestingly, Cherr and Clark (1985) also showed that unlike the situation in echinoids, the egg jelly had no effect on the sperm in the sturgeon.

F. Sperm-Binding Proteins of the Mouse Egg Zona Pellucida

There are many excellent reviews that provide a thorough description and discussion of the mammalian zona pellucida proteins (Wassarman *et al.,* 1985; Wassarman, 1987b, 1988, 1991; Dunbar *et al.,* 1994; Sidhu and Guraya, 1991; Litscher and Wassarman, 1993) and our overview here is not meant to be exhaustive. The zona proteins of the mouse oocyte were first described by Bleil and Wassarman (1980a,b,c) in a series of important papers. They are integral structural proteins of the zona pellucida, probably existing as ZP2–ZP3 heterodimer filaments crosslinked by ZP1 around the egg to form the 2- to 25-μm-thick (depending on the species) zona pellucida (Greve and Wassarman, 1985; Wassarman *et al.,* 1989; Sidhu and Guraya, 1991; see also Tong *et al.,* 1995). All three ZP proteins are glycosylated via both asparagine and serine/threonine linkages. As in most of the gamete interactions described above, particular carbohydrate moieties appear to be critical for binding (Wassarman, 1989; Miller and Ax, 1990; Litscher and Wassarman, 1993).

1. ZP3 as a Primary Sperm-Binding Protein

Mouse oocyte ZP3 is a glycoprotein of ca. M_r 83K which is thought to be responsible for the initial binding of capacitated sperm to the oocyte. First described in 1980 (Bleil and Wassarman, 1980a,b,c), ZP3 from a variety of mammals appear to be immunologically related (Maresh and Dunbar, 1987; C. C. Miller *et al.,* 1992) and homologs of mZP3 have been identified and cloned in human, guinea pig, rat, rabbit, hamster, pig, and marmoset (see Table I for references). It is heavily glycosylated with both O- and N-linked oligosaccharyl moieties (Florman and Wassarman, 1985; Bleil and Wassarman, 1988). Approximately 53% of its M_r is due to the polypeptide backbone, and there is some degree of heterogeneity in the glycosylation (Litscher and Wassarman, 1993). Purified mouse ZP3 protein binds to sperm and competes for their binding to isolated zonae or intact oocytes (Bleil and Wassarman, 1980c). Purified ZP3 or recombinant ZP3 expressed in mammalian tissue culture cells also induces the acrosome reaction (Kinloch *et al.*, 1991; Beebe *et al.*, 1992). Further, ZP3 binding is localized to the acrosomal cap region of acrosome-intact sperm, which is where primary binding is predicted to occur (Bleil and Wassarman, 1986).

Binding of ZP3 to and activation of the sperm are species specific to some degree, although this has not been tested extensively (Litscher and Wassarman, 1993). It has been demonstrated that the O-linked oligosaccharides, primarily the mucin-like *N*-acetyl lactosamine with terminally linked galactose residues, are required for the recognition and binding event (Florman *et al.*, 1984; Florman and Wassarman, 1985; Bleil and Wassarman, 1988). Glycopeptides derived from ZP3 bind to sperm (based on their ability to compete for sperm–zona binding) but they do not activate the acrosome reaction; the inference is that the polypeptide backbone confers the ability to trigger the acrosomal reaction (Florman and Wassarman, 1985; Beebe *et al.*, 1992). Thus, these two activities of ZP3, recognition/binding and acrosomal reaction induction, are separable on a structural basis. The degree to which the oligosaccharides contribute to species specificity is not yet determined unequivocally (Moller *et al.*, 1990; see Section VI.A.1). For example, mouse ZP3 expressed in tissue culture cells derived from hamster is glycosylated and binds to both mouse and hamster sperm (Kinloch *et al.*, 1991). Similarly, recombinant mouse ZP3 synthesized in mouse L-929 cells as well as green monkey CV-1 cells competes for sperm binding and induces the acrosome reaction in mouse sperm (Beebe *et al.*, 1992). No careful examination of the ZP3 oligosaccharide components has been conducted across species as a means to assess the role of the carbohydrates in species specificity, although hamster ZP3 has been investigated (Moller *et al.*, 1990). Clonin and molecular analysis of the cDNA encoding ZP3 in other species has indicated that the coding regions are very similar in size and sequence, yet analyses of the proteins indicated a wide range of heterogeneity (see Table I). This suggests a great deal of differential glycosylation across species; the signifiance of this with regard to recognition remains to be seen. It is a complex problem; for example, fucoidan acts as a competitive inhibitor of sperm–zona binding in almost all mammalian species tested. Perhaps there are two aspects to the interaction: a nonspecific lectin-like adhesion and a more specific interaction relying on a more precise carbohydrate structure (cf. O'Rand, 1988; Litscher and Wassarman, 1993).

What sperm protein binds to ZP3? There are a number of candidate proteins that have been identified and characterized, and the jury is still out as to which proteins are truly serving as the ZP3 sperm cognate(s). It is possible that more than one protein is involved (Sidhu and Guraya, 1991; Litscher and Wassarman, 1993). It is generally accepted that aggregation of sperm ZP3 binding protein is a requirement for the acrosome reaction (O'Rand, 1988; O'Rand *et al.*, 1988; Leyton and Saling, 1989a). Perhaps one protein is responsible for initial binding and another is responsible for adhesion through aggregation and the acrosome reaction. Three of the candidate sperm proteins that have been shown to bind to mouse ZP3 directly are discussed briefly. For a thorough listing of candidate sperm proteins, see Litscher and Wassarman (1993). In addition, the rabbit sperm autoantigen Sp17 can be crosslinked to rabbit ZP1 (also called R55) and rabbit ZP3 (also called R45) proteins and may be responsible for sperm–zona binding in rabbits and other mammals (Richardson et al., 1994; Yamasaki et al., 1995).

One of the better-characterized candidate ZP3 receptors is β-1,4-galactosyltransferase (GalTase), which is presented on all mammalian sperm (D. J. Miller et al., 1992; Shur, 1993; Miller and Shur, 1994). Treatment of zona-intact oocytes with N-acetylglucosaminidase renders them incapable of binding to sperm, and it may in fact be released from cortical granules as a polyspermy block (Miller et al., 1993). ZP3, which contains O-linked N-acetylglucosamine moieties, binds to purified GalTase unless those glcNAc residues are removed (D. J. Miller et al., 1992). Surprisingly, transgenic mice which overexpress GalTase under the control of a heterologous promoter produce sperm which undergo precocious acrosomal reactions, rendering them less capable of successful fertilization (Youakim et al., 1994). The possible role of GalTase as a gamete adhesion molecule has been presented in at least two recent reviews (Shur, 1993; Miller and Shur, 1994).

A second sperm protein which has been implicated in ZP3-mediated binding to mammalian oocytes is referred to as Sp56. This protein (of ca. M_r 56K) was purified from sperm on the basis of its specific affinity for ZP3 (Bleil and Wassarman, 1990). Recent work (Cheng et al., 1994) using Sp56 purified by the ZP3 affinity method and by standard ion exchange chromatography has revealed that it is a homomultimer and that specific anti-Sp56 IgGs localize the protein to the outer surface of the sperm head plasma membrane. Further, cross-linking experiments indicate that Sp56 and ZP3 interact specifically via ZP3 O-linked oligosaccharides, and suggest strongly that Sp56 is at least involved in, if not responsible for, murine sperm–zona binding (Cheng et al., 1994). However, since Sp56 is a peripheral membrane protein, it is unlikely that it is directly responsible for triggering the acrosome reaction upon binding to the ZP.

Finally, a third candidate for a specific mouse ZP3 binding protein is referred to as p95. This protein was first identified as a plasma membrane component which increased in phosphotyrosine content in response to zona pellucida and ZP3 binding (Leyton and Saling, 1989b; Leyton et al., 1992). The p95 protein and other tyrosyl-phosphorylated proteins localize to the acrosomal region of the sperm and serve as a substrate for the tyrosine kinase activity which appears to be required for the acrosomal reaction to occur (Saling, 1991; Leyton et al., 1992). However, there is no direct evidence that p95 binds to purified ZP3. Recent analyses of p95 indicate that it is a sperm-specific hexokinase (Kalab *et al.*, 1994). Amino acid sequence analysis of a subset of peptide fragments of the purified protein indicated 100% identity to a known mouse hexokinase and an antibody against the enzyme specifically immunoprecipitated p95 (Kalab *et al.*, 1994). Interestingly, no other known hexokinases are tyrosine phosphorylated; however, the significance of this in regard to the function of p95 in ZP3 interactions and the acrosomal reaction is at present unknown (Kalab *et al.*, 1994).

2. ZP2 as a Secondary Sperm-Binding Protein

After the mouse sperm undergoes the acrosomal reaction, it can no longer bind to the ZP3 protein; however, it does bind to a second zona protein, designated ZP2 (Bleil and Wassarman, 1986; Bleil *et al.*, 1988; Mortillo and Wassarman, 1991). ZP2 binds only to acrosome reacted sperm and does not induce the acrosome reaction. It, too, is a glycoprotein, ca. M_r 120K (Bleil and Wassarman, 1986; Bleil *et al.*, 1988, Wassarman, 1992). However, the necessity of the glycosylation for ZP2 function has not been assessed. The current model of how ZP2 and ZP3 function as sperm receptors of the zona pellucida is that the ZP3 glycoprotein is responsible for primary binding and triggering the acrosome reaction. The acrosome-reacted sperm then maintain their interaction with the egg through the ZP2 glycoproteins (Wassarman, 1992). If this is accurate, then it would seem likely that the sperm protein that is responsible for the ZP2 interaction is either exposed or modified at the time of the acrosome reaction.

There are at least two sperm protein candidates that may serve as secondary binding proteins and as ZP2 cognates. The first is the PH-20 antigen which has been characterized in guinea pig (Ramarao *et al.*, 1994). In response to the acrosome reaction, PH-20 alters its localization to the inner acrosomal membrane, increasing the surface density of the protein in this area (Myles and Primakoff, 1984; Cowan *et al.*, 1991) and is proteolytically cleaved (Primakoff *et al.*, 1988b). In addition, antibodies against the PH-20 antigen inhibit binding of acrosome-reacted (but not acrosome-intact) sperm to the zona of guinea pig oocytes (Primakoff *et al.*, 1985; Myles *et al.*, 1987). Another twist is that the deduced amino acid sequence of PH-20 (Lathrop *et al.*, 1990) is similar to that of bee venom hyaluronidase (Gmachl and Kreil, 1993) and both the native and recombinant PH-20 proteins have hyaluronidase activity (Lin *et al.*, 1993a, 1994; Hunnicutt *et al.*, 1993; Gmachl *et al.*, 1993). Whether or not this activity is critical for sperm–zona interactions and the identity of the PH-20 binding protein on the guinea pig oocyte remain to be determined (Ramarao *et al.*, 1994). Interestingly, rat and mouse sperm hyaluronidase activity was described as a necessary enzymatic activity for fertilization in the early part of the century, and after detailed characterizations it was suggested that it was necessary for sperm penetration through the surrounding follicle cells (reviewed in Tyler, 1948). Guinea pig PH-20 appears to serve this function (Lin *et al.*, 1994). PH-20 also has been cloned from mouse (Lathrop *et al.*, 1991) and from human and monkey (Gmachl *et al.*, 1993; Lin *et al.*, 1993b), but no extensive functional studies have been conducted in these species.

Another mammalian sperm protein that may function in some sort of secondary binding event is acrosin (Urch, 1991). Acrosin (which also exists in a "pre" form called proacrosin) is a serine protease. Notably, there is some evidence that serine protease inhibitors block secondary sperm binding (Bleil *et al.*, 1988). Proacrosin will bind to fucoidan (Jones, 1991; Urch and Patel, 1991), which blocks sperm–egg interactions (Boldt *et al.*, 1989; Jones *et al.*, 1988), and a subpopulation of mature acrosin has been localized to the sperm plasma membrane (Tesarik *et al.*, 1990). Nonetheless, very little is known about the role of proacrosin and acrosin in mammalian gamete interactions.

G. Common Themes—Recognition and Activation of the Sperm

Based on the available data, primary binding of sperm to egg or oocyte outer coats generally involves recognition between carbohydrate moieties of egg proteins and sperm surface proteins. While in many cases the identities of the egg glycoprotein have not been determined precisely, this requirement for an oligosaccharyl component is clearly evident (Litscher and Wassarman, 1993). Because of the nature of the protein–carbohydrate interaction, the binding events have often been referred to as "lectin-like" (Miller and Ax, 1990). In the strictest sense, a lectin is a carbohydrate binding protein other than an antibody or an enzyme (Barondes, 1988; Gabius, 1994). If one accepts this definition, then the ascidian sperm glycosidases which seem to mediate sperm–egg interactions by binding to egg coat oligosaccharides (Hoshi *et al.*, 1994) are not true lectins. The mouse has provided the most information for mammalian fertilization, but an issue to consider is whether the murine model system is a paradigm for all mammalian fertilization. For example, in other mammals, there is evidence that acrosome-reacted as well as acrosome-intact sperm can bind to the zona (Huang *et al.*, 1981; Myles *et al.*, 1987; Morales *et al.*, 1989). As the molecules involved in gamete interactions for more species are discerned, the issues of carbohydrate-mediated specificity and the differences across species will be resolved one way or another. Regardless, one common aspect among all of the species presented here is that the recognition of egg protein carbohydrate is critical, and from that perspective, a knowledge of lectin-based interactions is applicable. Vasta *et al.* (1994) have reviewed lectins as "self/non-self recognition molecules," particularly with regard to invertebrates, and it certainly is possible that carbohydrates can serve as recognition structures (Feizi, 1989; Wassarman, 1989). This aspect of gamete recognition will be discussed in greater detail in Section VI.A.

V. Proteins Located in the Egg Plasma Membrane That Bind to Sperm

A. Overview

There are very few examples of egg/oocyte membrane proteins which have been shown clearly to be involved in gamete interactions (Table II). Most examples of these putative sperm-binding proteins involve indirect evidence, usually based upon antibody studies Nonetheless, in both the sea urchin and the mouse, progress has been made (Foltz, 1994; Ramarao *et al.*, 1994) regarding proteins that may be referred to as "receptors" in that they span the plasma membrane and "respond" to ligand binding. Keep in mind that the "ligand" in this case is bound to another cell (the sperm) and the context here is very much a situation of cell–cell adhesion (discussed further in Section VI.A.2).

B. The Mammalian Egg Membrane Proteins That Bind Sperm

Although the binding and fusion of the sperm to the plasma membrane of the mammalian oocyte are rather nonspecific events (Yanagimachi, 1988), the evidence indicates that a plasma membrane protein(s) is necessary for proper binding and fusion (Monroy, 1985; O'Rand, 1988; Myles, 1993). Protease treatment of zona-free mouse oocytes renders them incapable of binding to or fusing with sperm (Boldt *et al.*, 1988, 1989; Kellom *et al.*, 1992). However, the identity of any mammalian egg plasma membrane proteins involved in gamete interaction has yet to be ascertained directly. The examples provided here are those for which at least some positive identification has occurred.

1. FcR and CD4-like Proteins

The link between immune recognition and gamete recognition was forged by Lillie (1914, 1919) and revisited by Tyler (1948; Tyler and Tyler, 1966)

TABLE II
Sperm-Binding Proteins Located on the Egg Plasma Membrane

Species	M_r (×1000)	Function/properties	References[a]
Brown alga			
Fucus	30	Partially purified glycoprotein from plasma membrane which inhibits sperm binding to egg; α-fucosidase and α-mannosidase destroy activity; specificity unknown	Bolwell *et al.* (1980)
Echinoderms			
Hemicentrotus pulcherrimus	66	May be vitelline layer or plasma membrane protein; initially part of a high M_r complex; specific antibody used to identify p66; anti-p66 blocks sperm binding; sperm bind to beads coated with SBF; species specific; presumably glycosylated	Tsuzuki *et al.* (1977), Yoshida and Aketa (1978, 1979, 1987), Aketa *et al.* (1979)
Anthocidaris crassispina	80, 87, 225	Components of high M_r plasma membrane/vitelline layer preparation; antibody against 225-kDa protein inhibits sperm–sperm binding; species specific	Yoshida and Aketa (1982, 1983)
Strongylocentrotus purpuratus	High M_r	High M_r glycoprotein fraction from egg membrane/vitelline layer preparation; binds to purified bindin; species specific	Glabe and Vacquier (1978)
S. purpuratus	305	Preparation of fertilization envelope proteins, isolated in the presence of protease inhibitors; antibody against impure 305K preparation blocked sperm binding species specifically (may be a vitelline layer component)	Acevedo-Duncan and Carroll (1986)

270

	M_r		References[a]
S. purpuratus	350	Type I transmembrane glycoprotein; binds to acrosome-reacted sperm and to purified bindin species specifically; antibody blocks sperm binding; recombinant protein has same properties as native protein; is tyrosine phosphorylated at fertilization; probably exists as an oligomer (tetramer?); probably the active component of previously characterized high M_r fractions (see text)	Foltz and Lennarz (1990, 1992), Foltz et al. (1993), Ohlendieck et al. (1993, 1994), Abassi and Foltz (1994)
Mammals			
Hamster FcγR	nd	Indirect evidence; antibody against IgGFc receptor blocked sperm binding and stained eggs; IgGFc competed for sperm binding; beads coated with IgGFc rosetted on oocytes	Bronson et al. (1990)
Mouse CD4-like protein	53	Antibody against CD4 stains oocytes; 53-kDa protein can be immunoprecipitated from eggs with CD4 antibody; may associate with a p56^lck-like kinase: indirect evidence	Mori et al. (1990, 1991, 1992)
Mouse integrin	?	Presumed cognate of sperm PH30/fertilin protein (a disintegrin); a -TDE- peptide inhibits sperm–egg fusion	Primakoff et al. (1987), Blobel et al. (1990, 1992), Myles et al. (1994)

[a] References given are primary; see text for further references and explanation.

and has since received sporadic, intense attention (Monroy and Rosati, 1979). Bronson *et al.* (1990) found that polyclonal antibodies raised against the human IgG Fc receptor (FcγR) stained hamster oocytes, suggesting the presence of the receptor on the gamete. When zona-free oocytes were preincubated with IgGFcγ, sperm binding to the oocytes were inhibited. And finally, beads coated with IgGFcγ formed rosettes on the zona-free oocytes (Bronson *et al.*, 1990). However, no FcγR-like protein has been isolated and characterized from the oocytes of any species, and it remains to be seen if indeed this protein truly is involved in sperm binding.

Another immune recognition molecule which has been implicated in sperm–egg interactions is CD4-like. An anti-mouse CD4 antibody stained mouse oocytes, and the same antibody immunoprecipitated a protein of ca. M_r 53K from the oocytes. In addition, the antibody appeared to block sperm binding in a concentration-dependent manner (Mori *et al.*, 1991), although the controls for this experiment were not extensive. Interestingly, when the antibody was used to crosslink the CD4-like proteins on the oocytes, the CD4 coimmunoprecipitated with a kinase activity presumed to be related to p56^{lck} (Mori *et al.*, 1991). Using reverse transcriptase PCR, mRNA for both CD4 and p56^{lck} have been detected in the murine egg (Mori *et al.*, 1992). The supposed presence of these thymocyte-signaling proteins in oocytes has led to the speculation that fertilization is similar to the type of recognition and signaling that occurs between T cells and antigen-presenting cells (Mori *et al.*, 1990, 1991, 1992). A protein complex with cross-reactivity to Major Histocompatibility Complex Class II (MHCII) proteins has been identified on the surface of murine sperm and could serve as the CD4 recognition protein on the sperm (Mori *et al.*, 1990), although all of the evidence is indirect.

While it is tempting to make dramatic correlations between immune cell recognition and gamete recognition, it is difficult to ascertain the specificity of any gamete interaction using an antibody raised against an immune cell protein. If the similarity does exist, it will be best determined by identifying the important gamete interaction proteins based on the criteria of their mediating species-specific gamete interactions and their proper location on both the sperm and the egg and then assessing their relationship to known immune recognition proteins. Tyler (1948) cautioned against the assumption that the molecules mediating gamete recognition and immune cell recognition would be structurally similar; rather, he advocated an open-minded approach that sought functional similarity. Historically, antibodies have bound to components of the egg surface in rather nonspecific manners and have been able to interfere with fertilization (Metz, 1961; Baxandall *et al.*, 1964; Tyler and Tyler, 1966; Ackerman and Metz, 1972). Thus, the approach of screening gametes with antibodies against known immune recognition proteins must be complemented by rigorous biochemical (or perhaps eventually genetic) analyses before any conclusions may be drawn.

2. The PH-30/Fertilin Cognate (Integrins)

With the advent of monoclonal antibody (mAb) technology, the strategy of fractionating gametes, injecting these fractions into mice, and then assaying the resulting mAbs for the ability to perturb gamete interactions became a popular and productive strategy (Alexander and Bialy, 1994; Primakoff, 1994). In the guinea pig, this strategy has proven to be extremely successful (Myles *et al.*, 1981; Primakoff and Myles, 1983; Primakoff, 1994). Two sperm antigens, PH-20 (Section IV.F.2) and fertilin (formerly called PH-30), have been characterized as a result of the initial success with mAbs (Ramarao *et al.*, 1994). In particular, the mAb PH-30 was found to inhibit sperm–egg fusion (but not sperm–zona binding) species specifically (Primakoff *et al.*, 1987). The protein was purified using the PH-30 antibody (Primakoff *et al.*, 1987; Blobel *et al.*, 1990) and eventually the cDNA was isolated (Blobel *et al.*, 1992; Wolfsberg *et al.*, 1993). Fertilin, the protein recognized by the PH-30 mAb, is a heterodimer (Blobel *et al.*, 1990) with a disintegrin-like domain in the β subunit (Blobel and White, 1992). Disintegrins encompass a family of snake venom proteins which inhibit the binding of fibrinogen and von Willebrand factor to the GPIIb–IIIa ($\alpha_{IIb}\beta_3$) integrin of activated platelets (Gould *et al.*, 1990; Weskamp and Blobel, 1994). Thus, it has been hypothesized that fertilin recognizes and binds to the oocyte at the plasma membrane via an integrin and that the fertilin protein α subunit and its cognate integrin then mediate membrane fusion (Ramarao *et al.*, 1994).

It had been shown previously that peptides containing the -RGD- sequence (which is recognized by some integrins) blocked hamster fertilization in a concentration-dependent manner (Bronson and Fusi, 1990a,b). Recently, it was demonstrated that several integrins are present on mouse oocytes (Taron *et al.*, 1993; Evans *et al.*, 1995). The $\alpha_6\beta_1$ integrin is differentially distributed on the oocyte microvilli (Tarone *et al.*, 1993), as is the vitronectin receptor ($\alpha_5\beta_3$, $\alpha_5\beta_5$) (Evans *et al.*, 1995). Fertilin contains the sequence -TDE- in the β subunit disintegrin domain and -TDE- peptides are potent inhibitors of guinea pig sperm–egg fusion (Myles *et al.*, 1994). It is reasonable to propose that an interaction between a sperm disintegrin (PH30/fertilin) and an egg integrin mediates sperm–egg binding at the plasma membrane and perhaps fusion as well (Ramarao *et al.*, 1994). The goal now is to identify and purify an oocyte integrin and demonstrate directly that it plays a role in gamete interactions. This issue is discussed further in Sections VI.A.2 and VII.

C. Sperm-Binding Factors of the Sea Urchin Egg Plasma Membrane

A long history of investigation into the basis of gamete interactions in the sea urchin exists in the literature. Here we will present some of the more relevant work that has been carried out with regard to sperm-binding proteins that reside in the sea urchin egg plasma membrane. Much of the earlier work may refer to proteins that are a component of the vitelline layer as opposed to the plasma membrane. To date, it has not been determined if these "sperm-binding activities" or "sperm-binding factors" are localized to the plasma membrane or the vitelline layer. They are presented in this section because the methodology and characterization served as preludes to the identification of a transmembrane glycoprotein that has been identified as a gamete recognition molecule. For more thorough reviews of the history of such studies, see Metz (1967, 1978) and Vacquier (1979).

1. The Sperm-Binding Factor

Aketa (1967) described a "bonding substance" isolated from the *Hemicentrotus pulcherrimus* egg surface, presumably the vitelline layer (which is referred to in his paper as the "vitelline membrane"; see Fig. 1). The isolation procedure involved first fertilizing jelly-free eggs and then suspending them in 1 M urea to prevent the fertilization envelope from hardening. The envelopes were stripped from the eggs and then dissolved. Precipitates from the dissolved, dialyzed preparation were then tested for the ability to bind sperm (Aketa, 1967). A series of papers followed, in which the sperm-binding factor was used as an immunogen and the resulting antisera were assessed for the ability to perturb sperm–egg interactions species specifically (Aketa and Onitake, 1969; Aketa, 1973). The caveat to the approach was the fact that the starting material was derived from fertilized eggs, which had undergone the cortical reaction. It has been shown that the release of the cortical granule contents results in the modification of surface proteins and certainly can affect sperm binding (Vacquier *et al.,* 1972a,b, 1973). Thus, it is possible that sperm-binding proteins would be modified prior to their isolation. Nonetheless, this approach set the stage for the next decade of investigation.

Later work on the same sperm-binding factor fractions of vitelline layer resulted in further purification of the activity (Tsuzuki *et al.,* 1977). Size fractionation and ion exchange chromatography yielded a very high M_r glycoprotein fraction which retained the ability to interfere with fertilization species specifically (Tsuzuki *et al.,* 1977). A polyclonal antibody raised against this glycoprotein complex stained egg surfaces and also blocked sperm binding (Yoshida and Aketa, 1978). Interestingly, agarose beads that were coated with the fraction containing the sperm-binding factor (SBF) triggered the acrosome reaction species specifically and the sperm then bound to the beads (Aketa *et al.,* 1979). This has led to the speculation that perhaps the SBF fraction contained some components of egg jelly. An improved purification method finally resulted in the identification of a glycoprotein of ca. M_r 61K which inhibited sperm–egg binding. Further, an antibody directed against this protein blocked fertilization (Yoshida and Aketa, 1987). It has been suggested that this 61-kDa SBF is at least functionally related to that of another urchin species, *Anthocidaris crassispina*.

Using antisera raised against the *H. pulcherrimus* SBF, Yoshida and Aketa (1979) suggested that a similar SBF also existed in *A. crassispina*, a hypothesis which was supported by partial purification of the activity from this species (Yoshida and Aketa, 1982). Three major components subsequently were identified, of ca. M_r 80K, 87K, and 225K (Yoshida and Aketa, 1983). Antibodies against the 225-kDa glycoprotein revealed that it was likely to be the active core component; these antibodies (and not those against the other two proteins) inhibited fertilization (Yoshida and Aketa, 1983). The authors also suggest that this may be the *A. crassispina* homolog to the bindin receptor isolated by Glabe and Vacquier (1978).

2. The Sperm Receptor

The presence of a proteinaceous sperm receptor on the membrane of sea urchin eggs was further confirmed by the fact that trypsin treatment of *Arbacia punctulata* eggs rendered them incapable of binding sperm (Schmell et al., 1977). Membrane preparations from these eggs inhibited. *A. punctulata* sperm–egg binding but not *S. purpuratus* fertilization. This work established criteria for any sperm receptor, which included: (i) that it be a surface component from the egg distinct from the jelly coat; (ii) that it should inhibit fertilization of eggs by binding to sperm; and (iii) that this activity should be species specific.

As discussed above, the sea urchin acrosomal reaction exposes bindin on the tip of the sperm (Moy and Vacquier, 1979; Hofmann and Glabe, 1994). Bindin is believed to be the sperm component which determines specific interaction with the egg surface once the sperm penetrates the jelly layer (Vacquier and Moy, 1977; Glabe and Vacquier, 1977; Hofmann and Glabe, 1994; Vacquier et al., 1995). Purified bindin agglutinates eggs species preferentially (Glabe and Vacquier, 1977; Glabe and Lennarz, 1979), and molecular analysis of the cDNA from multiple species indicates a species-specific structural domain which may account for its binding properties (Minor et al., 1991, 1993; Lopez et al., 1993; Hofmann and Glabe, 1994; Vacquier et al.,1995). Taken together with the work of Aketa and his colleagues, these data led to the reasonable hypothesis that a protein exists on the egg surface capable of recognizing bindin. Candidate sperm receptors thus had yet another criteria to satisfy.

Glabe and Vacquier (1978) isolated a high M_r glycoconjugate from *S. purpuratus* eggs that had the ability to bind to bindin species specifically (it did not recognize *S. franciscanus* bindin). Peptides derived from this glycoconjugate had less binding ability, and this led to the conclusion that perhaps multivalency was a component of the interaction (Glabe and Vacquier, 1978; Glabe, 1979). However, no single protein component of the activity could be purified to homogeneity. Despite the difficulty in working with high M_r, insoluble glycoconjugates, a great deal of progress was made in the next decade or so with regard to characterizing the sperm receptor activity (cf. Ruiz-Bravo and Lennarz, 1989; Foltz and Lennary, 1993; Foltz, 1994).

In particular, it became apparent that the sperm receptor activity was species specific (Kinsey *et al.*, 1980; Glabe and Lennarz, 1981; Rossignol *et al.*, 1981, 1984; Kinsey and Lennarz, 1981; Ruiz-Bravo *et al.*, 1986). The $>10^6$ M_r glycoconjugate appeared to have multivalent binding properties and was able to bind only to sperm that had been acrosome-reacted, thereby exposing bindin (Rossignol *et al.*, 1981). Interestingly, pronase-derived glycopeptides also retained the ability to bind to sperm, but the species-specific component was lost (Rossignol *et al.*, 1981; Kinsey and Lennarz, 1981). Characterization of the carbohydrate moieties present indicated that the sperm receptor fraction ws glycosaminoglycan-like, including the presence of sulfate and mannose (Glabe and Lennarz, 1981; Rossignol *et al.*, 1981, 1984). Some degree of dityrosine cross-linking was also suggested (Rossignol *et al.*, 1981, 1984). Acevedo-Duncan and Carroll (1986) isolated a preparation from *S. purpuratus* egg envelopes that was reported to have sperm receptor activity, and this preparation has proteins of ca. M_r 305K, 225K, and 180K. Antibodies raised against the individual proteins, isolated from a gel, localized to the vitelline envelope and indicated that the 305K and 225K proteins were immunologically related; the antiserum blocked sperm binding in *S. purpuratus* but had no effect on *L. variegatus* fertilization (Acevedo-Duncan and Carroll, 1986). Unfortunately, the 305K protein, which was presumed to be analogous to Aketa and co-workers' sperm-binding factor, was not characterized further, particularly in terms of its ability to recognize bindin or its carbohydrate composition.

The idea that perhaps the carbohydrate portion of the receptor was responsible for initial binding and that the polypeptide backbone was somehow conferring specificity was investigated further (Ruiz-Bravo *et al.*, 1986; Ruiz-Bravo and Lennarz, 1986). Trypsin treatment of dejellied eggs resulted in the generation of proteolytic peptides that retained the ability to bind sperm species specifically, but only if the protein content was at least 30% by weight; glycopeptides less than 30% protein by weight would bind but were not specifc (Ruiz-Bravo and Lennarz, 1986). Unfortunately, no glycopeptide or tryptic fragment was purified to homogeneity so that a positive identification of the intact receptor could be made. Antibodies against sperm receptor fractions, although raised against impure preparations, suggested that the sperm receptor was localized to the egg surface (Ruiz-Bravo *et al.*, 1986, 1989) and one of the antibody preparations was found to inhibit fertilization species specifically (Ruiz-Bravo *et al.*, 1989). However, these antibodies were not useful in terms of identifying the intact sperm receptor.

Based on the trypsinization strategy (Ruiz-Bravo and Lennarz, 1986), various other proteases were used to remove proteins from the egg surface in the hopes of isolating a stable, sperm-binding peptide. One protease, lysylendoproteinase C (LysC), resulted in reproducible digests that were stable and amenable to further biochemical purification. A glycosylated proteolytic fragment of ca. M_r 70K was purified to homogeneity from these LysC digests and eventually led to the identification of the intact receptor (Foltz and Lennarz, 1993). The 70-kDa fragment bound only to acrosome-reacted sperm and prevented fertilization species specifically. The fragment also bound to purified bindin (Foltz and Lennarz, 1990). An antibody raised against the purified 70-kDa fragment localized to egg plasma membranes and inhibited fertilization species specifically (Foltz and Lennarz, 1992). The intact receptor was identified in plasma membrane fractions and migrated on SDS polyacrylamide gels as a ca. M_r 350K protein (Foltz and Lennarz, 1992). Pronase-derived glycopeptides of the 70-kDa fragment lose the species-specific component of their sperm-binding activity, as was found consistently for the high M_r glycoconjugate preparations discussed previously (Foltz and Lennarz, 1990).

Sequence analysis of the cDNA encoding the receptor supported the hypothesis that the receptor was an integral membrane protein (Foltz *et al.*, 1993) and subsequent topology studies have confirmed that the receptor is a Type I transmembrane protein (Hoang and Foltz, 1995). An extracellular domain of approximately 400 residues binds to sperm and to purified bindin species specifically. Recombinant (nonglycosylated) protein purified from *Escherichia coli* binds to sperm and to bindin species specifically, suggesting that the recognition and binding is a protein–protein interaction (Foltz *et al.*, 1993). Likewise, the recombinant protein competes for sperm–egg binding to the same extent as the native protein and the 70-kDa fragment (Foltz *et al.*, 1993; Ohlendieck *et al.*, 1993; Ohlendieck and Lennarz, 1995). The primary structure of this domain is species specific, while the cytoplasmic domain is highly conserved among species (Foltz *et al.*, 1993). Although the sperm receptor bears no homology to any proteins present in the current databases, there are conserved tyrosine motifs in the cytoplasmic domain and the receptor is tyrosine phosphorylated rapidly in response to binding of conspecific sperm or purified bindin (Abassi and Foltz, 1994). Biochemical characterization of the intact, native receptor has reaffirmed that the carbohydrate component (which is 70% by weight) is predominantly sulfated fucose, mannose, galNAc, and glcNAc (Ohlendieck *et al.*, 1993). Further, the intact receptor probably exists as a homomultimer in the egg plasma membrane (Ohlendieck *et al.*, 1994), which has implications for its function and binding properties (see Section VI.A). See Foltz (1994) and Ohlendieck and Lennarz (1995) for more complete discussions of the structure and function of this interesting cell recognition molecule.

D. Other Receptors, Known and Inferred

1. *Fucus*

The brown alga *Fucus* has been a model system for studies of fertilization and egg activation for a number of years (cf. Callow *et al.*, 1985; Brawley, 1987, 1991, 1992). The plasma membrane of *Fucus* eggs can be isolated quite readily and these preparations have been used as immunogens to obtain specific antibodies (Bolwell *et al.*, 1980; Stafford *et al.*, 1993). Some of these antibodies interfere with sperm–egg interactions including binding and fusion (Stafford *et al.*, 1993). A partially purified plasma membrane glycoprotein has demonstrated sperm-binding activity, although the specificity of this has not been determined (Bolwell *et al.*, 1980; Stafford *et al.*, 1993). In this case, the glycosylation appears to play a critical role; α-fucosidase or α-mannosidase treatment destroys the activity (Bolwell *et al.*, 1980). Purification and further characterization of this glycoprotein will be necessary in order to assess its role in gamete interaction.

2. *Urechis*

The echiurian worm *Urechis caupo* is the only species known in which external binding of a purified sperm protein to the oocyte triggers complete egg activation (Gould *et al.*, 1986; Gould and Stephano, 1987, 1991). Proteins isolated from acrosome-reacted sperm were fractionated and assayed based upon their ability to bind to and agglutinate eggs and then trigger activation as defined by the induction of the polyspermy block, completion of meiosis, and DNA synthesis (Gould *et al.*, 1986). Presumably, the sperm factor identified (ca. M_r 25–30K) binds to a proteinaceous receptor on the egg plasma membrane (Gould and Stephano, 1987, 1991). It is known that trypsin treatment of the eggs stimulates fertilization responses (Paul, 1975; Jaffe *et al.*, 1979) and it is possible that this sperm protein is interacting with a trypsin-sensitive receptor. A peptide (Val–Ala–Lys–Lys–Pro–Lys) derived from this highly basic protein was found to bind to eggs and, in so doing, activated development completely; the eggs underwent the fertilization potential, completed meiosis, and synthesized DNA. If a centrosome was injected, the fertilized eggs divided and the embryos reached the trochophore larvae stage in parallel with control embryos derived from normal fertilizations using sperm (Gould and Stephano, 1991).

There are several very intriguing aspects of this sperm protein. It does not shown a high degree of species specificity, which actually is in agreement with the observed specificity of *Urechis* sperm. For example, *Urechis* sperm will acrosome react and bind to sea urchin eggs, causing partial fertilization envelope elevation (Jaffe *et al.*, 1982). The intact *Urechis* sperm protein does induce partial elevation of sea urchin egg and oyster envelopes (Gould *et al.*, 1986), but the peptide triggers a response only in oyster while having no effect on sea urchin eggs (Gould and Stephano, 1991). Another note of interest is that the sperm protein and the peptide induce electrical responses in the egg (Gould and Stephano, 1987, 1991). Very little is known about the mechanism of these responses in *Urechis* other than that there are channels that exist in the egg plasma membrane which are opened in response to sperm but not to voltage (for overviews in *Urechis* and other species, see Steinhardt *et al.*, 1971; Hagiwara and Jaffe, 1979; Jaffe *et al.*, 1979, 1982; Cross, 1981). It is possible that the receptor could be a channel protein or another protein that directly influences channel opening. Regardless of its function, the identification and characterization of the *Urechis* egg surface protein cognate for this sperm protein would add substantial knowledge to the data base.

3. Others

There are several examples of species in which proteolytic modification of the egg surface results in egg activation, implying the existence of receptors on the egg surface. In addition to *Urechis,* mentioned above, *Dendraster* (sand dollar) eggs are also stimulated by trypsin to round out and activate as defined by elevation of the fertilization envelope, entry into the cell cycle, and increase in protein synthesis (Moore, 1951; Hand, 1971). Sand dollar eggs also produce a fertilization potential in response to trypsin (Steinhardt *et al.*, 1971). The eggs of the Japanese palolo worm *Tylorrhynchus heterochaetus* also have been reported to be activated by Pronase P treatment (Osanai, 1976). Iwao *et al.* (1994) have shown that an extract of *Cynops* (newt) sperm has been shown to activate *Xenopus* eggs (as measured by the changes in membrane potential and internal calcium levels), and activation by the extract appears to require a protease activity (Iwao *et al.*, 1994).

Perhaps the best-studied example of egg activation by external protease treatment is that of *Asterina miniata,* tha bat star. Application of trypsin, chymotrypsin, or pronase to the oocytes caused increases in intracellular free calcium, triggered exocytosis, and stimulated of DNA synthesis (Carroll and Jaffe, 1995). Importantly, the authors established that the increase in intracellular calcium and resulting exocytosis was due to proteolysis of transmembrane proteins and not to proteolytic damage to the egg surface that triggered an external calcium leak. Thus, it now seems possible to isolate the specific membrane protein responsible. Whether or not a sperm protease binds to and hydrolyzes a specific egg protein to trigger activation remains unclear but should be actively pursued based on the available data.

The situation in sea urchins is less clear. Hand (1971) demonstrated that trypsin treatment resulted in increased protein synthesis, but no other parameters of activation were noted. Trypsin (Ruiz-Bravo and Lennarz, 1986) or lysylendoproteinase C (Foltz and Lennarz, 1990) treatment of *S. purpuratus* eggs does not result in cortical reaction, envelope elevation, or cleavage, and in fact renders the egg unable to bind sperm. Zalokar (1980) reported that the lectins WGA and ConA caused a patch-cap response and partial activation in ascidian (*Phallusia mammillata*) eggs, but the specific egg surface proteins have not been identified. The observations that cross-linking of surface proteins and protease treatment of many species of eggs can trigger activation strongly suggest that egg membrane proteins are involved in this step of fertilization (see discussion in Section VI.C).

VI. Functional Roles of Egg Receptors for Sperm

A. Recognition

1. Sperm Receptors of Egg Outer Coats

The proteins located in egg extracellular coats which are responsible for sperm recognition and binding, as discussed in Section IV, generally are thought to mediate the primary binding of sperm and egg and often trigger the sperm acrosome reaction. Although only a few of these egg proteins have been well characterized (namely the mammalian ZP3 proteins; reviewed in Litscher and Wassarman, 1993), it seems clear that the presence of specific, O-linked oligosaccharides are the critical components in terms of recognition and binding (Section IV.G). Whether or not this recognition is lectin-like remains unclear, and relatively little work has been done to clarify how the oligosaccharyl moieties confer species-specific binding. In mammals, primary binding of the sperm to the zona pellucida is much more species specific than binding and fusion of the sperm to the egg plasma membrane (Gwatkin, 1977; Yanagimachi, 1988; Litscher and Wassarman, 1993). For organisms such as sea urchins, just the opposite is true; plasma membrane binding and fusion generally are more species specific than jelly coat–sperm interactions (Summers and Hylander, 1975, 1976; Vacquier, 1979).

In terms of characterizing the recognition and binding events, progress has been hampered by conflicting data concerning the sperm proteins involved (see Section IV.F) and a complete dearth in the knowledge regarding kinetic and equilibrium constants. If binding of sperm to the egg outer vestment is lectin-like, it probably has relatively low-affinity binding events that are strengthened by avidity (multivalent binding) mechanisms (Simmons, 1993). Sperm protein(s) may cluster as a part of this interaction (Leyton and Saling, 1989a; Macek et al., 1991). Along these same lines, if the O-linked carbohydrates are mediating the primary binding events between mammalian sperm and egg, then the issue of species specificity comes to the fore. Likewise, the role of the fucosylated glycoproteins in the echinoderm egg jelly, which seem to mediate the binding of sperm and the acrosome reaction, remains unclear. Are the carbohydrates responsible for the specific recognition as well as adhesion? This is of fundamental importance; until the precise carbohydrate structures that mediate the binding can be identified and until these can be compared across at least a few species, this will remain a key unresolved issue. For murine ZP3, an N-acetylgalactosamine at the reducing terminus and a galactose moiety in an α linkage at the penultimate sugar at the nonreducing terminus seem critical for binding (Wassarman, 1989; Litscher and Wassarman, 1993; see Section IV.F); whether or not these specific oligosaccharyl structures are responsible for specific recognition remains unknown. There is precedence for oligosaccharides being able to determine species-specific recognition, most notably in the interactions between nitrogen-fixing rhizobacteria and their various legume hosts (Denarie and Cullimore, 1993; Kannenberg and Brewin, 1994; see Feizi, 1989, for discussion of oligosaccharides as recognition structures).

The role of the oligosaccharyl moieties of the zona proteins in mediating specific sperm binding has been addressed somewhat indirectly. ZP3 purified from both hamster and mouse eggs will bind to sperm from either species, and the cross-species binding of hamster and mouse gametes is common (Cherr et al., 1986; Moller et al., 1990). Murine and hamster ZP3 expressed in murine embryonic carcinomal (EC) cells are differentially glycosylated and murine sperm are unable to bind to the EC hamster ZP3 (Kinloch et al., 1991). These data would seem to support the hypothesis that sperm can discriminate among species of sperm based on the glycosylation pattern. However, the hZP3 was not tested for its ability to bind hamster sperm, an experiment which would allow a more rigorous interpretation of the result. Also, the differential glycosylation patterns may be a reflection more of the EC cell context rather than the species. In a similar experiment, murine ZP3 expressed in mouse L-929 cells and in green monkey CV-1 cells, although glycosylated differently, had the same binding activity and specificity as purified murine ZP3, suggesting that the recombinant ZP3 from both mouse and primate cells contained the necessary structures for function (Beebe *et al.*, 1992). Because ZP3 can be purified and has been cloned from a variety of species (see Table II), it should be possible to determine the role of the O-linked oligosaccharides in species-specific binding by comparisons among species that normally exhibit strict specificity.

Another question is the nature of the mechanism of the acrosome reaction. In mammals, this seems to be dependent upon the polypeptide portion of the ZP3 protein (Wassarman, 1989; Litscher and Wassarman, 1993). Gabius (1994) provides a review of animal lectins from the point of view that both carbohydrate–protein interactions and protein–protein interactions are often important. However, the evidence for this being the case for ZP3 is indirect; the fact that glycopeptides and oligosaccharyl fractions of mZP3 will bind to sperm and compete for zona binding but will not trigger the acrosome reaction (Florman *et al.*, 1984; Florman and Wassarman, 1985) has been interpreted to mean that the polypeptide is responsible (Wassarman, 1989). Chemical deglycosylation of native ZP3 suggests that N-linked sugars play no role in either binding or the acrosome reaction; however, removal of the O-linked oligosaccharides destroys both activities (Florman and Wassarman, 1985). It is possible that chemical deglycosylation can damage the polypeptide, thereby rendering it nonfunctional; recombinant, bacterially produced (and therefore nonglycosylated) protein or portions thereof could be tested for the ability to bind to sperm and induce the acrosome reaction.

One explanation for why glycopeptides bind sperm but do not trigger the acrosome reaction may be that the smaller glycopeptides are unable to mediate the clustering of the sperm proteins to which they bind. Although the actual mechanism of the mammalian acrosome reaction has not received a great deal of attention, it is possible that ZP3 is involved in crosslinking sperm proteins to which it binds as a part of the induction. Leyton and Saling (1989a) found that mZP3 glycopeptides bound to sperm but did not cause the acrosome reaction. However, if they subsequently applied anti-ZP3 IgG, the sperm then displayed the acrosome reaction (Leyton and Saling, 1989a). Interestingly, aggregation of GalTase on sperm, presumably via ZP3 crosslinking, also triggers the acrosome reaction (Macek *et al.*, 1993). An adhesive event involving aggregation to increase avidity followed by cellular activation is a common mechanism for cell–cell adhesion (Simmons, 1993) and certainly could be operating in this situation. Clearly, comparisons across mammalian species will be critical in resolving the issue of the role of carbohydrates in recognition and sperm binding as well as the mechanism of the acrosome reaction.

2. Sperm-Binding Proteins of the Egg Plasma Membrane

The interaction of the sperm with the egg plasma membrane certainly can be considered a cell adhesion event in which the egg provides a cell adhesion molecule that interacts with another molecule tethered to the sperm (see Simmons, 1993, for a review of cell adhesion mechanisms). In the case of both the mammalian egg integrin and the sea urchin egg receptor for sperm, it is likely that cytoskeletal attachment would strengthen the adhesive interaction; integrins are known cell adhesion molecules which are linked to the actin-based cytoskeleton (Hynes, 1994) and the urchin sperm receptor likely interacts with the cortical actin cytoskeleton (Foltz and Lennarz, 1993; Foltz, 1994). There are several recent reviews concerning "low affinity" adhesive interactions, and the reader is directed to these for in-depth discussions of adhesive mechanisms (Simmons, 1993; Hynes, 1994; Loftus *et al.,* 1994; van der Merwe and Barclay, 1994).

Cell adhesion molecules are a large class of cell surface proteins defined as having dissociation constants (K_d) in the range of $10^{-5}-10^{-8}$ M; for frame of reference, cytokine and growth factor receptors have ligand dissociation constants of $10^{-9}-10^{-12}$ M (Simmons, 1993). Integrins overcome the weakness of the interaction—and the very real risk of being pulled from the membrane—by strengthening the interaction via crosslinks to the actin-based cytoskeleton (Simmons, 1993; Hynes, 1994). Another mechanism to strengthen cell adhesion is to cluster the receptor molecules (Simmons, 1993; van der Merwe and Barclay, 1994). The urchin sperm receptor undergoes a patching response when live eggs are incubated with antibody to the receptor and then crosslinked with a secondary antibody (Foltz and Lennarz, 1992). Further, the sperm receptors are localized primarily to the tips of the microvilli, where sperm bind primarily (Ohlendieck *et al.,* 1994; Ohlendieck and Lennarz, 1995; see also Summers and Hylander, 1974), promoting localized binding and perhaps multivalency in binding. K. Rood and E. Davidson (unpublished) have measured the K_d for the bindin–receptor interaction, using recombinant proteins, at approximately 10^{-8} M. Thus, it seems likely that the sea urchin sperm receptor would require an increase in avidity to compensate for the relatively low affinity.

Although the K_d values are useful in terms of classifying or categorizing an interaction, the measurement usually is not biologically relevant in that it defines the interaction at equilibrium; K_{on} and K_{off} rates therefore would be more useful. Cell adhesion molecules that mediate transient cell–cell interactions tend to have relatively low affinities (because of the transient nature of the interaction) and extremely fast dissociation rates (van der Merwe and Barclay, 1994). On the other hand, adhesion molecules that mediate more stable interactions have slightly higher affinities and tend to have a recognition event that is then followed by an adhesion event, most often linking the protein to the cytoskeleton and increasing the number of receptors involved in the binding (Simmons, 1993). For example, integrins have K_ds in the 10^{-7} M range, and the key elements of their binding sites are relatively short stretches of amino acids. Assuming that the mammalian egg uses an integrin to mediate sperm binding and that the sea urchin egg uses the sperm receptor (see above, Sections V.B and V.C), it would seem that a common mode of adhesion is to tether the transmembrane receptors to the actin-based cytoskeleton.

In the case of mammals, it remains to be established definitively that an integrin is responsible for sperm binding (see Section V.B). The integrins have a variety of specificities (Hynes, 1992; Hogg and Landis, 1993; Haas and Plow, 1994; Loftus *et al.*, 1994) and could contribute to recognition as well as adhesion. In the case of the sea urchin, further characterization of the receptor and its cytoskeletal association is necessary.

Finally, the available data indicate that glycosylation is not a critical parameter of gamete recognition and binding at the plasma membrane. Integrins, although glycosylated, are not thought to require the oligosaccharyl modifications to carry out their functions (Hynes, 1992, 1994). Likewise, the sea urchin egg receptor for sperm is heavily glycosylated (Foltz, 1994; Ohlendieck and Lennarz, 1995), but the carbohydrates do not seem to play a critical role in recognition and binding. Although pronase-derived glycopeptides bind to sperm and compete for egg binding, the species specificity is greatly reduced (see Ruiz-Bravo and Lennarz, 1989). This was originally interpreted to mean that the carbohydrate moieties had some role in binding (Foltz and Lennarz, 1993). However, bacterially expressed, and therefore nonglycosylated, sea urchin sperm receptor protein recognizes sperm species specifically and has the same ability to compete for sperm binding as does the native protein (Foltz *et al.*, 1993). Still, careful measurements of the dissociation constants and comparisons of the native, deglycosylated, and aglycosylated receptor would be valuable and unequivocally resolve the question of whether or not the oligosaccharyl moieties play a role in binding and adhesion.

B. Fusion

One of the obvious consequences of sperm–egg interactions is fusion of the plasma membranes. This event happens quickly (Myles, 1993; Jaffe, 1995) and has been compared to viral–host membrane fusions (White, 1990; Myles *et al.*, 1994; Ramarao *et al.*, 1994). Relatively little is known about the molecules that mediate the fusion event in any species, but several excellent review articles address the issue from a cell biological perspective

(Monroy, 1985; Yanagimachi, 1988; Glabe *et al.*, 1991; Myles, 1993). The recent work on the PH30/fertilin protein of mammalian sperm (Section V.B) suggests that fertilin has a fusogenic domain that could mediate membrane fusion (Ramarao *et al.*, 1994). An antibody against fertilin blocks sperm–egg fusion (Primakoff *et al.*, 1987), and modeling studies suggest that this fusogenic peptide sequence, located in the α subunit of the fertilin protein, could indeed function in this capacity (Blobel *et al.*, 1992; Ramarao *et al.*, 1994). Likewise, there is some evidence that sea urchin sperm bindin has fusogenic properties; bindin induces fusion of mixed phase vesicles (Glabe, 1985a,b). However, no protein from any system has yet been shown to directly mediate gamete fusion.

C. Signaling

One question which consistently arises is whether or not the gamete recognition molecules engage in intracellular signaling (Nuccitelli, 1991; Epel, 1989; Whitaker and Swann, 1993). Integrins have been shown to integrate signaling through nonreceptor tyrosine kinases that result in cytoskeletal rearrangements and the activation of the ras/MAP kinase pathway (Damsky and Werb, 1992; Hynes, 1992, 1994; Schwartz, 1992). Could this also occur at fertilization? It will be necessary to identify the integrins involved and then to assess their possible role in signaling. The sea urchin sperm receptor becomes tyrosine phosphorylated within seconds of sperm binding or even the application of purified bindin to eggs (Abassi and Foltz, 1994), but whether or not this phosphorylation event is responsible for cytoskeletal rearrangements or signaling of some aspect of egg activation, or for that matter is even specific, remains to be established. Identification and characterization of the responsible kinase and of other interacting proteins will be necessary before these questions can be answered definitively.

It is also valuable to consider recent evidence that indicates that optimal intracellular signaling often involves not only a specific receptor such as a growth factor receptor tyrosine kinase, but also proper adhesion mediated through integrins (McNamee *et al.*, 1993; Hynes, 1992). Costimulatory signaling through these "dual pathways" has been shown to result in changes in phospholipid metabolism, tyrosine phosphorylation, calcium fluxes, and pH changes (Hynes, 1992; McNamee *et al.*, 1993; Chong *et al.*, 1994). Section VII provides a more thorough consideration of this point. Neither the sea urchin egg receptor for sperm nor the presumed egg integrin have been shown to be involved directly in signaling; nonetheless, by identifying the egg membrane receptors for sperm, the first steps toward discerning the mechanism(s) of egg activation have been taken.

VII. Concluding Remarks

Considering the many ways in which organisms achieve gamete recognition and adhesion, is there a coherent way in which to view the molecular mechanisms? Can we expect the mouse egg to have a homolog of the sea urchin egg receptor for sperm? Can we expect the sea urchin egg to express an integrin that recognizes a sperm fertilin homolog? The answers are largely dependent not only on continued progress in identifying the sperm receptors for various species, but, more importantly, on establishing their role in recognition, adhesion, fusion, and signaling. It is possible, and even likely, that more than one egg surface protein is involved in this multistep process (see Ohlendieck and Lennarz, 1995). Some components could be common among species, others more restricted.

Now that sperm receptors are beginning to be identified, cloned, and sequenced in several model systems, it will be possible to identify the homologs, if they exist. Until the cross-species surveys are conducted, it is difficult to draw firm conclusions about similarities or differences among the sperm receptors from different species. It is possible that proteins that serve similar functions may not be structurally related. Clearly, the sea urchin egg should be assessed for the presence of surface integrins that may be involved in fertilization. Likewise, the presence of any sea urchin egg receptor for sperm homologs should be determined in other systems, particularly the mouse.

In order to gain a complete understanding of fertilization, the molecular players must be identified (Green, 1993; see also Buxbaum, 1995). There are at least two basic strategies that would lend themselves well to this effort. The first, as mentioned in the previous section, is a concerted effort to find homologs of the already-identified proteins. A PCR-based approach using degenerate primers to the conserved cytoplasmic domain of the sea urchin egg receptor for sperm is already underway in our laboratory. If indeed a homolog is identified in another species, there would then be a need to establish that the homolog did function as a gamete recognition protein. Integrins surely are present on the surfaces of oocytes and eggs (Tarone et al., 1993; Whittaker and DeSimone, 1993; Evans et al., 1995), but their precise role in gamete recognition, binding, and fusion has yet to be determined.

Perhaps one of the most satisfying and productive lines of investigation would be to establish a genetic approach to the problem of fertilization in general and the sperm receptors in particular (see Section III.A). If mutants could be isolated that were disrupted in their ability to undergo successful gamete interactions, new gene products necessary for fertilization and egg activation could be discovered. Similarly, if an integrin necessary for fertilization or a homolog of the sea urchin sperm receptor could be identified and studied in a system amenable to genetic analyses (e.g., *Drosophila*), functional analyses and identification of interacting components would become much more tractable. The mouse, which already is a well-established model for studying fertilization, has recently emerged as a system that allows for at least reverse genetic analysis. Despite the sometimes disappointing results with gene knockouts in mice (Melton, 1994), it may be possible to take this approach. For example, mice that overexpress the sperm β-1,4-galactosyltransferase have sperm which exhibit precocious acrosome reactions and a decreased binding to oocytes (Youakim *et al.*, 1994). Whether or not application of this reverse genetics approach to the problems of fertilization will be informative remains to be determined.

Advances in other broad areas of research, such as immune recognition, adhesion mechanisms, and cellular activation (signal transduction), should continue to provide insight into the mechanism of fertilization as well. In fact, recent evidence concerning the mechanisms of T cell activation and adhesion suggest that a costimulatory recognition event is necessary for full activation (Hynes, 1992, 1994). For example, although the specific interaction between the T cell and the antigen-presenting cell is mediated by the T cell receptor (TCR) and the MHC, this initial binding event triggers an increase in the avidity of the T cell integrins, increasing the adhesion event (Y. Shimizu *et al.*, 1990; Chan *et al.*, 1991; Wilkins *et al.*, 1991; Miyake *et al.*, 1994). Whether or not integrin signaling is also a part of the T cell activation pathway is still under investigation, but seems likely (Matsuyama *et al.*, 1989; Davis *et al.*, 1990; Nojima *et al.*, 1990; van Seventer *et al.*, 1990; Burkly *et al.*, 1991; Yamada *et al.*, 1991; Sánchez-Mateos *et al.*, 1993). Interestingly, T lymphocytes express only a subset of integrins, including $\alpha_6\beta_1$ and $\alpha_5\beta_1$ (Hynes, 1992), which are also found on mouse (Tarone *et al.*, 1993; Evans *et al.*, 1995). Frog oocytes also contain mRNA for several integrins (Whittaker and DeSimone, 1993).

It is possible that oocytes and eggs, like T lymphocytes, also require a two-step interaction with sperm to achieve full activation; perhaps a specific receptor (e.g., the sea urchin sperm receptor) mediates the initial binding event and then the adhesive interaction is strengthened by the activation of the integrins (the PH30/fertilin cognate?). This is a testable hypothesis limited only by the availability of reagents that work across species. To date, no integrin antibodies are available that recognize sea urchin egg proteins and the sea urchin sperm receptor antibodies are echinoderm-specific. However, because of the progress in cloning and sequencing of the egg proteins that bind sperm, a molecular approach to identifying any homologs should be possible. As discussed in Section VI.C, application of bindin to sea urchin eggs results in partial activation (induction of a tyrosine kinase activity). If an integrin needs to be engaged also, this could be accomplished by antibody crosslinking or application of the integrin ligand. Regardless, it is imperative that the question of whether sea urchin eggs have integrins that are also involved in sperm binding and whether mouse oocytes have an urchin sperm receptor homolog be resolved as soon as possible. An answer to this question will necessarily provide either a point of convergence or a departure in the way in which the molecular mechanisms of gamete recognition and egg activation are viewed. This information is critical if we are to pursue a unifying theory of the mechanism of fertilization.

The study of gamete interactions was once described as "obscure and hypothetical" due to the "lack of specific information" by Albert Tyler (1948) as he vented his frustration and echoed that of others in their collective efforts to elucidate the molecular mechanisms that mediate sperm–egg recognition and adhesion. Nearly 50 years later, perhaps Tyler would be gratified to learn that the chase was still on and that the "obscure" field was beginning to blossom with new information about the sperm receptors that he and his predecessors and then students so painstakingly endeavored to describe. Tyler's students and contemporaries (Metz and Monroy, 1985) pushed the field on, and today's students of fertilization biology hopefully will continue to shed light on the beautiful mystery that is fertilization.

Acknowledgments

I thank the members of my lab (Yama Abassi, Robert Belton, and Kenneth Hoang) and Stu Feinstein, Rindy Jaffe, Dave Carroll, Gary Hunnicutt, Nick Cross, Cathy Thaler, and Rich Cardullo for helpful comments and suggestions. Special thanks go to Vic Vacquier for access to his extensive library and to Nick Cross for urging me to stop writing and get back into the lab. Thanks also to Michael O'Rand, Bill Lennarz, Vic Vacquier, Doug Chandler, and Eric Davidson for sharing unpublished results and providing preprints of manuscripts in press. Work in my lab is supported by an American Cancer Society Junior Faculty Award, the Searle Scholars Trust Fund, and NIH R01 HD30698.

References

Abassi, Y. A., and Foltz, K. R. (1994). Tyrosine phosphorylation of the egg receptor for sperm at fertilization. *Dev. Biol.* **164,** 430–443.

Acevedo-Duncan, M., and Carroll, E. J. (1986). Immunological evidence that a 305-kilodalton vitelline envelope polypeptide isolated from sea urchin eggs is a sperm receptor. *Gamete Res.* **15,** 337–359.

Ackerman, N. R., and Metz, C. B. (1972). Effects of multiple antibody layers on *Arbacia* eggs. *Exp. Cell Res.* **72,** 204–210.

Aketa, K. (1967). On the sperm-egg bonding as the initial step of fertilization in the sea urchin. *Embryologia* **9**, 238–245.

Aketa, K. (1973). Physiological studies on the sperm surface components responsible for sperm-egg binding in sea urchin fertilization. I. Effect of sperm-binding protein on the fertilizing capability of sperm. *Exp. Cell Res.* **80**, 439–441.

Aketa, K., and Onitake, K. (1969). Effect on fertilization of antiserum against sperm-binding protein from homo- and heterologous sea urchin egg surfaces. *Exp. Cell Res.* **56**, 84–86.

Aketa, K., Yoshida, M., Miyazaki, S., and Ohta, T. (1979). Sperm binding to an egg model composed of agrose beads. *Exp. Cell Res.* **123**, 281–284.

Alexander, N. J., and Bialy, G. (1994). Contraceptive vaccine development. *Reprod. Fertil. Dev.* **6**, 273–280.

Araki, Y., Orgebin-Crist, M. C., and Tulsiani, D. R. (1992). Qualitative characterization of oligosaccharide chains present on the rat zona pellucida glycoconjugates. *Biol. Reprod.* **46**, 912–919.

Barondes, S. H. (1988). Bifunctional properties of lectins. *Trends Biochem. Sci.* **13**, 480–482.

Baxandall, J., Perlmann, P., and Afzelius, B. A. (1964). Immuno-electron microscope analysis of the surface layers of the unfertilized sea urchin egg. I. Effects of the antisera on the cell ultrastructure. *J. Cell Biol.* **23**, 609–628.

Beebe, S. J., Leyton, L., Burks, D., Ishikawa, M., Fuerst, T., Dean, J., and Saling, P. (1992). Recombinant mouse ZP3 inhibits sperm binding and induces the acrosome rection. *Dev. Biol.* **151**, 48–54.

Bleil, J. D., and Wassarman, P. M. (1980a). Structure and function of the zona pellucida: Identification and characterization of the proteins of the mouse oocyte's zona pellucida. *Dev. Biol.* **76**, 185–202.

Bleil, J. D., and Wassarman, P. M. (1980b). Synthesis of zona pellucida proteins by denuded and follicle-enclosed mouse oocytes during culture in vitro. *Proc. Natl. Acad. Sci. U.S.A.* **77**, 1029–1033.

Bleil, J. D., and Wassarman, P. M. (1980c). Mammalian sperm-eg interaction: Identification of a glycoprotein in mouse egg zonae pellucidae possessing receptor activity for sperm. *Cell (Cambridge, Mass.)* **20**, 873–882.

Bleil, J. D., and Wassarman, P. M. (1986). Autoradiographic visualization of the mouse egg's sperm receptor bound to sperm. *J. Cell Biol.* **102**, 1363–1371.

Bleil, J. D., and Wassarman, P. M. (1988). Galactose at the non-reducing terminus of O-linked oligosaccharides of mouse egg zona pellucida glycoprotein ZP3 is essential for the glycoprotein's sperm receptor activity. *Proc. Natl. Acad. Sci. U.S.A.* **85**, 5563–5567.

Bleil, J. D., and Wassarman, P. M. (1990). Identification of a ZP3-binding protein on acrosome-intact mouse sperm by photoaffinity crosslinking. *Proc. Natl. Acad. Sci. U.S.A.* **87**, 5563–5567.

Bleil, J. D., Greve, J. M., and Wassarman, P. M. (1988). Identification of a secondary sperm receptor in the mouse egg zona pellucida: Role in maintenance of binding of acrosome-reacted sperm. *Dev. Biol.* **128**, 376–385.

Blobel, C. P., and White, J. M. (1992). Structure, function and evolutionary relationship of proteins containing a disintegrin domain. *Curr. Opin. Cell Biol.* **4**, 760–765.

Blobel, C. P., Myles, D. G., Primakoff, P., and White, J. M. (1990). Proteolytic processing of a protein involved in sperm-egg fusion correlates with acquisition of fertilization competence. *J. Cell Biol.* **111**, 69–78.

Blobel, C. P., Wolfsberg, T. G., Turck, C. W., Myles, D. G., Primakoff, P., and White, J. M. (1992). A potential fusion peptide and an integrin ligand domain in a protein active in sperm-egg fusion. *Nature (London)* **356**, 248–252.

Boldt, J., Howe, A. M., and Preeble, J. (1988). Enzymatic alteration of the ability of mouse egg plasma membrane to interact with sperm. *Biol. Reprod.* **39**, 19–27.

Boldt, J., Gunter, L. E., and Howe, A. M. (1989). Characterization of cell surface polypeptides of unfertilized, fertilized and protease-treated zona-free mouse eggs. *Gamete Res.* **23**, 91–101.

Bolwell, G. P., Callow, J. A., and Evans, L. V. (1980). Fertilization in brown algae. III. Preliminary characterization of putative gamete receptors from eggs and sperm of *Fucus serratus. J. Cell Sci.* **43,** 209–224.

Bonnell, B. S., Larabell, C. A., and Chandler, D. E. (1993). The sea urchin egg jelly coat is a three-dimensional fibrous network as seen by intermediate voltage electron microscopy and deep etching analysis. *Mol. Reprod. Dev.* **35,** 181–188.

Bonnell, B. S., Keller, S. H., Vacquier, V. D., and Chandler, D. E. (1994). Sea urchin egg jelly coat consists of globular glycoproteins bound to a fibrous fucan superstructure. *Dev. Biol.* **162,** 313–324.

Brawley, S. H. (1987). A sodium-dependent, fast block to polyspermy occurs in eggs of fucoid algae. *Dev. Biol.* **124,** 390–397.

Brawley, S. H. (1991). The fast block against polyspermy in fucoid algae is an electrical block. *Dev. Biol.* **144,** 94–106.

Brawley, S. H. (1992). Fertilization in natural populations of the dioecious brown alga *Fucus ceranoides* and the importance of the polyspermy block. *Mar. Biol. (Berlin)* **113,** 145–157.

Bronson, R. A., and Fusi, F. (1990a). Sperm-oolemmal interction: Role of the arg-gly-asp (RGD) adhesion peptide. *Fertil. Steril.* **54,** 527–529.

Bronson, R. A., and Fusi, F. (1990b). Evidence that an arg-gly-asp adhesion sequence plays a role in mammalian fertilization. *Biol. Reprod.* **43,** 1019–1025.

Bronson, R. A., Fleit, H. B., and Fusi, F. (1990). Identification of an oolemmal IgG Fc receptor: it's role in promoting binding of antibody-labelled human sperm to zona-free hamster eggs. *Am. J. Reprod. Immunol.* **23,** 87–92.

Burkly, L. C., Jakubowski, A., Newman, B. M., Rosa, M. D., Chi-Ross, G., and Lobb, R. (1991). Signaling by vascular cell adhesion molecule-1 (VCAM-1) through VLA-4 promotes CD3-dependent T cell proliferation. *Eur. J. Immunol.* **21,** 2871–2875.

Buxbaum, R. E. (1995). Biological levels. *Nature (London)* **373,** 567–568.

Callow, J. A., Callow, M. E., and Evans, L. V. (1985). Fertilization in *Fucus. In* "Biology of Fertilization" (C. B. Metz and A. Monroy, eds.), pp. 389–437. Academic Press, New York.

Carroll, D. J., and Jaffe, L. A. (1995). Proteases stimulate fertilization-like responses in starfish eggs. Submitted for publication.

Chamberlin, M. E., and Dean, J. (1990). Human homolog of the mouse sperm receptor. *Proc. Natl. Acad. Sci. U.S.A.* **87,** 6014–6018.

Chan, B. M. C., Wong, J. G. P., Rao, A., and Hemler, M. E. (1991). T cell receptor-dependent, antigen-specific stimulation of a murine T cell clone induces a transient, VLA protein-mediated binding to extracellular matrix. *J. Immunol.* **147,** 398–404.

Chandler, D. E. (1991). Multiple intracellular signals coordinate structural dynamics in the sea urchin egg cortex at fertilization. *J. Electron Microsc. Tech.* **17,** 266–293.

Cheng, A., Le, T., Palacios, M., Bookbinder, L. H., Wassarman, P. M., Suzuki, F., and Bleil, J. D. (1994). Sperm-egg recognition in the mouse: Characterization of sp56, a sperm protein having specific affinity for ZP3. *J. Cell Biol.* **125,** 867–878.

Cherr, G. N., and Clark, W. H. J. (1985). An egg envelope component induces the acrosome reaction in sturgeon sperm. *J. Exp. Zool.* **234,** 75–85.

Cherr, G. N., Lambert, H., Meizel, S., and Katz, D. F. (1986). *In vitro* studies of the golden hamster sperm acrosome reaction completion on the zona pellucida and induction by homologous soluble zonae pellucidae. *Dev. Biol.* **114,** 119–131.

Chong, L. D., Traynor-Kaplan, A. T., Bokoch, G. M., and Schwartz, M. A. (1994). The small GTP-binding protein Rho regulates a phosphatidylinositol 4-phosphate 5-kinase in mammalian cells. *Cell (Cambridge, Mass.)* **79,** 507–513.

Clark, W. H., Jr., Yudin, A. I., Lynn, J. W., Griffin, F. J., and Pillai, M. C. (1990). Jelly formation in penaeoidean shrimp eggs. *Biol. Bull. (Woods Hole, Mass.)* **178,** 295–299.

Clark, W. H., Jr., Griffin, F. J., and Wikramanayake, H. (1994). Pre-fusion events of sperm-oocyte interaction in the marine shrimp, *Sicyonia ingentis. Semin. Dev. Biol.* **5,** 225–231.

Colwin, L. H., and Colwin, A. L. (1963). Role of the gamete membranes in fertilizaiton in *Saccoglosus kowalevskii* (Enteropneusta). II. Zygote formation by gamete membrane fusion. *J. Cell Biol.* **19,** 501–518.

Colwin, L. H., and Colwin, A. L. (1967). Membrane fusion in relation to sperm-egg association. *In* "Fertilization" (C. B. Metz and A. Monroy, eds.), pp. 295–367. Academic Press, New York.

Cowan, A. E., Myles, D. G., and Koppel, D. E. (1991). Migration of guinea pig sperm membrane protein PH-20 from one localized surface domain to another does not occur by a simple diffusion-trapping mechanism. *Dev. Biol.* **144,** 189–198.

Cross, N. L. (1981). Initiation of the activation potential by an increase in intracellular calcium in eggs of the frog. *Rana pipiens. Dev. Biol.* **85,** 380–384.

Damsky, C. H., and Werb, Z. (1992). Signal transduction by integrin receptors for extracellular matrix: Cooperative processing of extracellular information. *Curr. Opin. Cell Biol.* **4,** 772–781.

Dan, J. C. (1952). Studies on the acrosome. I. Reaction to egg water and other stimuli. *Biol. Bull. (Woods Hole, Mass.)* **103,** 54–66.

Dan, J. C. (1956). The acrosome reaction. *Int. Rev. Cytol.* **5,** 365–393.

Dan, J. C. (1967). Acrosome reaction and lysins. *In* "Fertilization" (C. B. Metz and A. Monroy, eds.), pp. 237–293. Academic Press, New York.

Dan, J. C., and Wada, S. K. (1955). Studies on the acrosome. IV. The acrosome reaction in some bivalve spermatazoa. *Biol. Bull. (Woods Hole, Mass.)* **109,** 40–55.

Davis, L. S., Oppenheimer-Marks, N., Bednarczyk, J. L., McIntyre, B. W., and Lipsky, P. E. (1990). Fibronectin promotes proliferation of naive and memory T cells by signaling through both the VLA-4 and VLA-5 integrin molecules. *J. Immunol.* **145,** 785–793.

DeAngelis, P. L., and Glabe, C. G. (1985). Biochemical characterization of the interaction of sulfated fucans and the sperm adhesive protein, bindin. *J. Cell Biol.* **101,** 264a.

Denarie, J., and Cullimore, J. (1993). Lipo-oligosaccharide nodulation factors: A minireview of a new class of signaling molecules mediating recognition and morphogenesis. *Cell (Cambridge, Mass.)* **74,** 951–954.

De Santis, R., and Pinto, M. R. (1987). Isolation and partial characterization of a glycoprotein complex wiht sperm-receptor activity from *Ciona intestinalis* ovary. *Dev. Growth Differ.* **29,** 617–625.

De Santis, R., Jamunno, G., and Rosati, F. (1980). A study of the chorion and the follicle cells in relation to the sperm-egg interaction in the ascidian, *Ciona intestinalis. Dev. Biol.* **74,** 490–499.

De Santis, R., Hoshi, M., Cotelli, F., and Pinto, M. R. (1983a). Induction of acrosome reaction in *Ciona intestinalis. In* "The Sperm Cell" (J. André, ed.), pp. 274–276. Martinus Nijhoff, The Hague, The Netherlands.

De Santis, R., Pinto, M. R., Cotelli, F., Rosati, F., Monroy, A., and D'Alessio, G. (1983b). A fucosyl glycoprotein component with sperm receptor and sperm activating activities from the vitelline coat of *Ciona intestinalis* eggs. *Exp. Cell Res.* **148,** 508–513.

Dumont, J. M., and Brummett, A. R. (1985). Egg envelopes in vertebrates. *In* "Developmental Biology, A Comprehensive Synthesis: Oogenesis" (L. Browder, ed.), pp. 235–288. Plenum, New York.

Dunbar, B. S., Avery, S., Lee, V., Prasad, S., Schwahn, D., Schwoebel, E., Skinner, S., and Wilkins, B. (1994). The mammalian zona pellucida: Its biochemistry, immunochemistry, molecular biology, and developmental expression. *Reprod. Fertil. Dev.* **6(3),** 331–347.

Dunbar, B. S., Liu, C., and Sammons, D. W. (1981). Identification of the three major proteins of porcine and rabbit zonae pellucidae by high resolution two-dimensional gel electrophoresis: Comparison with serum, follicular, fluid, and ovarian cell proteins. *Biol. Reprod.* **24(5),** 1111–1124.

Elinson, R. P. (1986). A comparative approach to the block to polyspermy in amphibian eggs. *Int. Rev. Cytol.* **101,** 59–100.

Epel, D. (1989). Arousal of activity in sea urchin eggs at fertilization. *In* "The Cell Biology of Fertilization" (H. Schatten and G. Schatten, eds.), pp. 361–385. Academic Press, San Diego, CA.

Epifano, O., and Dean, J. (1994). Biology and structure of the zona pellucida: A target for immunocontraception. *Reprod. Fertil. Dev.* **6,** 319–330.

Evans, J. P., Schultz, R. M., and Kopf, G. S. (1995). Identification and localization of integrin subunits in oocytes and eggs of the mouse. *Mol. Reprod. Dev.* **40,** 211–220.

Feizi, T. (1989). Glycoprotein oligosaccharides as recognition structures. *In* "Carbohydrate Recognition in Cellular Function," (G. Bock and S. Harnett, eds.), pp. 62–79. Wiley, Chichester.

Florman, H. M., and Wassarman, P. M. (1985). O-linked oligosaccharides of mouse egg ZP3 account for its sperm receptor activity. *Cell (Cambridge, Mass.)* **41,** 313–324.

Florman, H. M., Bechtol, K. B., and Wassarman, P. M. (1984). Enzymatic dissection of the functions of the mouse egg's receptor function for sperm. *Dev. Biol.* **106,** 243–255.

Foltz, K. R. (1994). The sea urchin egg receptor for sperm. *Semin. Dev. Biol.* **5,** 243–253.

Foltz, K. R., and Lennarz, W. J. (1990). Purification and characterization of an extracellular fragment of the sea urchin egg receptor for sperm. *J. Cell Biol.* **11,** 2951–2959.

Foltz, K. R., and Lennarz, W. J. (1992). Identification of the sea urchin egg receptor for sperm using an antiserum raised against its extracellular domain. *J. Cell Biol.* **116,** 647–658.

Foltz, K. R., and Lennarz, W. J. (1993). The molecular basis of sea urchin gamete interactions at the sea urchin egg plasma membrane. *Dev. Biol.* **158,** 46–61.

Foltz, K. R., Partin, J. S., and Lennarz, W. J. (1993). Sea urchin egg receptor for sperm: Sequence similarity of binding domain and hsp70. *Science* **259,** 1421–1425.

Gabius, H. J. (1994). Non-carbohdrate binding partners/domains of animal lectins. *Int. J. Biochem.* **26,** 467–477.

Garbers, D. L. (1989). Molecular basis of fertilization. *Annu. Rev. Biochem.* **58,** 719–742.

Garbers, D. L., Kopf, G. S., Tubb, D. J., and Olson, G. (1983). Elevation of sperm adenosine 3',5'-monophosphate concentrations by a fucose sulfate-rich complex associated with eggs 1: Structural characterization. *Biol. Reprod.* **29,** 1211–1220.

Glabe, C. G. (1979). A supramolecular theory for specificity in intracellular adhesion. *J. Theor. Biol.* **78,** 417–423.

Glabe, C. G. (1985a). Interaction of the sperm adhesive protein, bindin, with phospholipid vesicles. I. Specific association of bindin with gel-phase phospholipid vesicles. *J. Cell Biol.* **100,** 794–799.

Glabe, C. G. (1985b). Interaction of the sperm adhesive protein, bindin, with phospholipid vesicles. II. Bindin induces the fusion of mixed phase vesicles that contain phosphatidylcholine and phosphatidylserine in vitro. *J. Cell Biol.* **100,** 800–806.

Glabe, C. G., and Lennarz, W. J. (1979). Species-specific sperm adhesion in sea urchins: A quantitative investigation of bindin-mediated egg agglutination. *J. Cell Biol.* **83,** 595–604.

Glabe, C. G., and Lennarz, W. J. (1981). Isolation of a high molecular weight glycoconjugate derived from the surface of S. purpuratus eggs that is implicated in sperm adhesion. *J. Supramol. Struct. Cell. Biochem.* **15,** 347–358.

Glabe, C. G., and Vacquier, V. D. (1977). Species specific agglutination of eggs by bindin isolated from sea urchin sperm. *Nature (London)* **267,** 836–837.

Glabe, C. G., and Vacquier, V. D. (1978). Egg surface glycoprotein receptor for sea urchin sperm bindin. *Proc. Natl. Acad. Sci. U.S.A.* **75,** 881–885.

Glabe, C. G., Hong, K., and Vacquier, V. D. (1991). Fusion of sperm and egg plasma membranes during fertilization. *In* "Membrane Fusion" (J. Wilschut and D. Hoekstra, eds.), pp. 627–646. Dekker, New York.

Gmachl, M., and Kreil, G. (1993). Bee venom hyaluronidase is homologous to a membrane protein of mammalian sperm. *Proc. Natl. Acad. Sci. U.S.A.* **90,** 3569–3573.

Gmachl, M., Sagan, S., Ketter, S., and Kreil, G. (1993). The human sperm protein PH-20 has hyaluronidase activity. *FEBS Lett.* **336,** 545–548.

Gould, M. C., and Stephano, J. L. (1987). Electrical responses of eggs to acrosomal protein similar to those induced by sperm. *Science* **235,** 1654–1656.

Gould, M. C., and Stephano, J. L. (1991). Peptides from sperm acrosomal protein that initiate egg development. *Dev. Biol.* **146,** 505–518.

Gould, M. C., Stephano, J. L., and Holland, L. Z. (1986). Isolation of protein from *Urechis* sperm acrosomal granules that binds sperm to eggs and initiates development. *Dev. Biol.* **117,** 306–318.

Gould, R. J., Polokoff, M. A., Friedman, P. A., Huang, T. F., Holt, J. C., Cook, J. J., and Niewiarowski, S. (1990). Disintegrins: A family of integrin inhibitory proteins from viper venoms. *Proc. Soc. Exp. Biol. Med.* **195,** 168–171.

Green, D. P. L. (1993). Mammalian fertilization as a biological machine: A working model for adhesion and fusion of sperm and oocyte. *Hum. Reprod.* **8,** 91–96.

Greve, J. M., and Wassarman, P. M. (1985). Mouse egg extracellular coat is a matrix of interconnected filaments possessing a structural repeat. *J. Mol. Biol.* **181,** 253–264.

Gwatkin, R. D. L. (1977). "Fertilization Mechanisms in Man and Mammals." Plenum, New York.

Haas, T. A., and Plow, E. F. (1994). Integrin–ligand interactions: A year in review. *Curr. Opin. Cell Biol.* **6,** 656–662.

Hagiwara, S., and Jaffe, L. A. (1979). Electrical properties of egg cell membranes. *Annu. Rev. Biophys. Bioeng.* **8,** 385–416.

Haino, K., and Kigawa, M. (1966). Studies on the egg membrane lysin of *Tegula pfeifferi:* Isolation and chemical analysis of the egg membrane. *Exp. Cell Res.* **42,** 625–633.

Haino-Fukushima, K. (1974). Studies on the egg membrane lysin of *Tegula pfeifferi:* The reaction mechanism of the lysin. *Biochim. Biophys. Acta* **352,** 179–191.

Hand, G. S. J. (1971). Stimulation of protein synthesis in unfertilized sea urchin and sand dollar eggs treated with trypsin. *Exp. Cell Res.* **64,** 204–208.

Hardy, D. M., Harumi, T., and Garbers, D. L. (1994). Sea urchin sperm receptors for egg peptides. *Semin. Dev. Biol.* **5,** 217–224.

Hedrick, J. L., and Nishihara, T. (1991). Structure and function of the extracellular matrix of anuran eggs. *J. Electron Microsc. Tech.* **17,** 319–335.

Hirsh, D., Kemphues, K. J., Stinchcomb, D. T., and Jefferson, R. (1985). Genes affecting early development in *Caenorhabditis elegans. Cold Spring Harbor Symp. Quant. Biol.* **50,** 69–78.

Hoang, K., and Foltz, K. R. (1995). Submitted for publication.

Hofmann, A., and Glabe, C. (1994). Bindin, a multifunctional sperm ligand and the evolution of new species. *Semin. Dev. Biol.* **5,** 233–242.

Hogg, N., and Landis, R. C. (1993). Adhesion molecules in cell interactions. *Curr. Opin. Immunol.* **5,** 383–390.

Honegger, T. G. (1982). Effect of fertilization and localized binding of lectins in the ascidian *Phallusia mammillata. Exp. Cell Res.* **138,** 446–451.

Honegger, T. G. (1986). Fertilization in ascidians: Studies on the egg envelope, sperm and gamete interactions in *Phallusia mammilata. Dev. Biol.* **118,** 118–128.

Hoshi, M. (1985). Lysins. *In* "Biology of Fertilization" (C. B. Metz and A. Monroy, eds.), pp. 431–462. Academic Press, New York.

Hoshi, M., De Santis, R., Pinto, M. R., Cotelli, F., and Rosati, F. (1983). Is sperm α-L-fucosidase responsible for sperm-egg binding in *Ciona intestinalis. In* "The Sperm Cell" (J. André, ed.), pp. 107–110. Martinus Nijhoff, The Hague, The Netherlands.

Hoshi, M., Okinaga, T., Kotani, K., Araki, T., and Chiba, K. (1991). Acrosome reaction-inducing glycoconjugate in the jelly coat of starfish eggs. *In* "Comparative Spermatology 20 Years After" (B. Baccetti, ed.), pp. 175–180. Raven Press, New York.

Hoshi, M., Takizawa, S., and Hirohashi, N. (1994). Glycosidases, proteases and ascidian fertilization. *Semin. Dev. Biol.* **5,** 201–208.

Huang, T. T. F., Fleming, A. D., and Yanagimachi, R. (1981). Only acrosome-reacted spermatozoa can bind to and penetrate zona pellucida: A study using the guinea pig. *J. Exp. Zool.* **217,** 287–290.

Hunnicutt, G. R., Primakoff, P., and Myles, D. G. (1993). Guinea pig PH-20 has hyaluronidase activity but this function appears to be distinct from its activity in binding acrosome-reacted sperm to the zona pelucida. *Mol. Biol. Cell* **4,** 816a.

Hynes, R. O. (1992). Integrins: Versatility, modulation, and signaling in cell adhesion. *Cell* (*Cambridge, Mass.*) **69,** 11–25.

Hynes, R. O. (1994). The impact of molecular biology on models for cell adhesion. *BioEssays* **16,** 663–669.

Ishihara, K., and Dan, J. C. (1970). Effects of chemical disruption on the biological activities of sea urchin egg jelly. *Dev. Growth Differ.* **12,** 179–187.

Iwao, Y., Miki, A., Kobayashi, M., and Onitake, K. (1994). Activation of *Xenopus* eggs by an extract of *Cynops* sperm. *Dev. Growth Differ.* **36,** 469–479.

Jaffe, L. A. (1995). Egg membranes during fertilization. *In* "Molecular Biology of Membrane Transport Disorders" (S. G. Schultz, T. Andreoli, A. Brown, D. Fambrough, J. Hoffman, and M. Welsh, eds.). Plenum, New York, (in press).

Jaffe, L. A., and Cross, N. L. (1986). Electrical regulation of sperm-egg fusion. *Annu. Rev. Physiol.* **48,** 191–200.

Jaffe, L. A., and Gould, M. (1985). Polyspermy-preventing mechanisms. *In* "Biology of Fertilization" (C. B. Metz and A. Monroy, eds.), pp. 223–250. Academic Press, New York.

Jaffe, L. A., Gould-Somero, M., and Holland, L. (1979). Ionic mechanism of the fertilization potential of the marine worm, *Urechis caupo* (Echiura). *J. Gen. Physiol.* **73,** 469–492.

Jaffe, L. A., Gould-Somero, M., and Holland, L. (1982). Studies of the mechanism of the electrical polyspermy block using voltage clamp during cross-species fertilization. *J. Cell Biol.* **92,** 616–621.

Jaffe, L. A. (1985). The role of calcium explosions, waves, and pulses in activating eggs. *In* "Biology of Fertilization" (C. B. Metz and A. Monroy, eds.), pp. 127–165. Academic Press, New York.

Jones, R. (1991). Interaction of zona pellucida glycoproteins, sulphated carbohydrates and synthetic polymers with proacrosin, the putative egg-binding protein from mammalian spermatozoa. *Development* (*Cambridge, UK*) **111,** 1155–1163.

Jones, R., Brown, C. R., and Lancaster, R. T. (1988). Carbohydrate-binding properties of boar sperm proacrosin and assessment of its role in sperm-egg recognition and adhesion during fertilization. *Development* (*Cambridge, UK*) **102,** 781–792.

Kalab, P., Visconti, P., Leclerc, P., and Kopf, G. S. (1994). p95, the major phosphotyrosine-containing protein in mouse spermatazoa, is a hexokinase with unique properties. *J. Biol. Chem.* **269,** 3810–3817.

Kannenberg, E. L., and Brewin, N. J. (1994). Host-plant invasion by *Rhizobium:* The role of cell surface components. *Trends Microbiol.* **2,** 277–283.

Katagiri, C. (1987). Role of oviducal secretions in mediating gamete fusion in anuran amphibians. *Zool. Sci.* **4,** 1–14.

Kay, E. S., and Shaprio, B. M. (1985). The formation of the fertilization membrane of the sea urchin egg. *In* "Biology of Fertilization" (C. B. Metz and A. Monroy, eds.), pp. 45–80. Academic Press, New York.

Keller, S. H., and Vacquier, V. D. (1994). The isolation of acrosome-reaction-inducing glycoproteins from sea urchin egg jelly. *Dev. Biol.* **162,** 304–312.

Kellom, T., Vick, A., and Boldt, J. (1992). Recovery of penetration ability in protease-treated zona-free eggs occurs coincident with recovery of a cell surface 94 kD protein. *Mol. Reprod. Dev.* **33**, 46–52.

Kinloch, R. A., Ruiz-Seiler, B., and Wassarman, P. M. (1990). Genomic organization and polypeptide primary structure of the zona pellucida glycoprotein hZP3, the hamster sperm receptor. *Dev. Biol.* **142**, 414–421.

Kinloch, R. A., Mortillo, S., Stewart, C. L., and Wassarman, P. M. (1991). Embryonal carcinoma cells transfected with ZP3 genes differentially glycosylate similar polypeptides and secrete active mouse sperm receptor. *J. Cell Biol.* **115**, 655–664.

Kinsey, W. H. (1986). Purification and properties of the egg plasma membrane. *In* "Methods in Cell Biology" (T. E. Schroeder, ed.), pp. 139–152. Academic Press, Orlando, FL.

Kinsey, W. H., and Lennarz, W. J. (1981). Isolation of a glycopeptide fraction from the surface of the sea urchin egg that inhibits sperm-egg binding and fertilization. *J. Cell Biol.* **91**, 325–331.

Kinsey, W. H., Decker, G. L., and Lennarz, W. J. (1980). Isolation and characterization of the plasma membrane of the sea urchin egg. *J. Cell Biol.* **87**, 248–254.

Lallier, R. (1977). The problems of sea urchin egg fertilization and its implications for biological studies. *Experientia* **33**, 1263–1406

Larabell, C. A., and Chandler, D. E. (1988a). The extracellular matrix of *Xenopus laevis* eggs: A quick-freeze, deep-etch analysis of its modification at fertilization. *J. Cell Biol.* **107**, 731–741.

Larabell, C. A., and Chandler, D. E. (1988b). In vitro formation of the "S" layer, a unique component of the fertilization envelope in *Xenopus laevis* eggs. *Dev. Biol.* **130**, 356–364.

Larabell, C. A., and Chandler, D. E. (1989). The vitelline layer and fertilization envelope of echindoerm and amphibian eggs: Visualization of a cell surface-anchored extracellular matrix. *In* "Freeze Fracture Studies of Membranes" (S. Hui, ed.), pp. 175–199. CRC Press, Boca Raton, FL.

Larabell, C. A., and Chandler, D. E. (1990). Step-wise transformation of the vitelline envelope of *Xenopus* eggs at activation: A quick-freeze, deep-etch analysis. *Dev. Biol.* **139**, 263–268.

Larabell, C. A., and Chandler, D. E. (1991). Fertilization-induced changes in the vitelline envelope of echinoderm and amphibian eggs: Self-assembly of an extracellular matrix. *J. Electron Microsc. Tech.* **17**, 294–318.

Lathrop, W. F., Carmichael, E. P., Myles, D. G., and Primakoff, P. (1990). cDNA cloning reveals the molecular structure of a sperm surface protein, PH-20, involved in sperm-egg adhesion and the wide distribution of its gene among mammals. *J. Cell Biol.* **111**, 2939–2949.

Lathrop, W. F., Myles, D. G., and Primakoff, P. (1991). Isolation and characterization of cDNA coding for the mouse homologue of guinea pig sperm protein PH-20. *J. Cell Biol.* **115**, 462a.

Lee, Y.-H., and Vacquier, V. D. (1992). The divergence of species-specific abalone sperm lysins is promoted by positive Darwinian selection. *Biol. Bull.* (*Woods Hole, Mass.*) **182**, 97–104.

Lewis, C. A., Talbot, C. F., and Vacquier, V. D. (1982). A protein from abalone sperm dissolves the egg vitelline layer by a nonenzymatic mechanism. *Dev. Biol.* **92**, 227–239.

Leyton, L., and Saling, P. (1989a). Evidence that aggregation of mouse sperm receptors by ZP3 trigers the acrosome reaction. *J. Cell Biol.* **108**, 2163–2168.

Leyton, L., and Saling, P. (1989b). 95-kD sperm proteins bind ZP3 and serve as tyrosine kinase substrates in response to zona binding. *Cell* (*Cambridge, Mass.*) **57**, 1123–1130.

Leyton, L., LeGuen, P., Bunch, D., and Saling, P. M. (1992). Regulation of mouse gamete interactions by a sperm tyrosine kinase. *Proc. Natl. Acad. Sci. U.S.A.* **89**, 11692–11695.

Li, C., and Chalfie, M. (1990). Organogenesis in *C. elegans:* Positioning of neurons and muscles in the egg-laying system. *Neuron* **4**, 681–695.

Lillie, F. R. (1914). Studies of fertilization. VI. The mechanisms of fertilization in *Arbacia*. *J. Exp. Zool.* **16**, 523–590.

Lillie, F. R. (1919). "Problems in Fertilization." Univ. of Chicago Press, Chicago.

Lin, Y., Kimmel, L., Myles, D. G., and Primakoff, P. M. (1993a). Molecular cloning of the human and monkey sperm surface protein PH-20. *Proc. Natl. Acad. Sci. U.S.A.* **90,** 10071–10075.

Lin, Y., Mahan, K., Myles, D. G., and Primakoff, P. (1993b). PH-20, a sperm membrane protein active in sperm binding to the zona pellucidae, is also a membrane-anchored hyaluronidase. *Mol. Biol. Cell* **4,** 1431a.

Lin, Y., Mahan, K., Lathrop, W. F., Myles, D. G., and Primakoff, P. (1994). A hyaluronidase activity of the sperm plasma membrane PH-20 enables sperm to penetrate the cumulus layer surrounding the egg. *J. Cell Biol.* **125,** 1157–1163.

Litscher, E. S., and Honegger, T. G. (1991). Glycoprotein constituents of the vitelline coat of *Phallusia mammillata* (Ascidiacea) with fertilization inhibiting activity. *Dev. Biol.* **148,** 536–551.

Litscher, E. S., and Wassarman, P. M. (1993). Carbohydrate-mediated adhesion of eggs and sperm during mammalian fertilization. *Trends Glycosci. Glycotechnol.* **5,** 369–388.

Loftus, J. C., Smith, J. W., and Ginsberg, M. H. (1994). Integrin-mediated cell adhesion: The extracellular face. *J. Biol. Chem.* **269,** 25235–25238.

Lopez, A., Miraglia, S. J., and Glabe, C. G. (1993). Structure/function analysis of the sea urchin sperm adhesive protein bindin. *Dev. Biol.* **156,** 24–33.

Macek, M. B., Lopez, L. C., and Shur, B. D. (1991). Aggregation of β-1,4-galactosyltransferase on mouse sperm induces the acrosome reaction. *Dev. Biol.* **147,** 440–444.

Mahajan-Miklos, S., and Cooley, L. (1994). Intracellular cytoplasm transport during *Drosophila* oogenesis. *Dev. Biol.* **165,** 336–351.

Maresh, G. A., and Dunbar, B. S. (1987). Antigenic comparisons of five species of zonae pellucidae. *J. Exp. Zool.* **244,** 299–307.

Matsuyama, Y., Yamada, A., Kay, J., Yamada, K. M., Akiyama, S. K., Schlossman, S. F., and Morimoto, C. (1989). Activation of CD4 cells by fibronectin and anti-CD3 antibody. *J. Exp. Med.* **170,** 1133–1148.

McNamee, H. P., Ingber, D. E., and Schwartz, M. A. (1993). Adhesion to fibronectin stimulates inositol lipid synthesis and enhances PDGF-induced inositol lipid breakdown. *J. Cell Biol.* **121,** 673–678.

Melton, D. W. (1994). Gene targeting in the mouse. *BioEssays* **16,** 633–638.

Metz, C. B. (1961). Use of inhibiting agents in studies on fertilization mechanisms. *Int. Rev. Cytol.* **11,** 219–253.

Metz, C. B. (1967). Gamete surface components and their role in fertilization. *In* "Fertilization" (C. B. Metz and A. Monroy, eds.), pp. 163–236. Academic Press, New York.

Metz, C. B. (1978). Sperm and egg receptors involved in fertilization. *Curr. Top. Dev. Biol.* **12,** 107–147.

Metz, C. B., and Monroy, A., eds. (1985). "Biology of Fertilization," Vols. 1-3. Academic Press, New York.

Metz, C. B., Schuel, H., and Bischoff, E. R. (1964). Inhibition of the fertilizing capacity of sea urchin sperm by papain digested, non-agglutinating antibody. *J. Exp. Zool.* **155,** 261–272.

Mikami-Takei, K., Kosakai, M., Isemura, M., Suyemitsu, T., Ishihara, K., and Schmid, K. (1991). Fractionation of jelly substance of the sea urchin egg and biological activities to induce the acrosome reaction and agglutination of spermatazoa. *Exp. Cell Res.* **192,** 82–86.

Miller, C. C., Fayrer-Hosken, R. A., Timmons, T. M., Lee, V. H., Caulde, A. B., and Dunbar, B. S. (1992). Characterization of equine zona pellucida glycoproteins by polyacrylamide gel electrophoresis and immunological techniques. *J. Reprod. Fertil.* **96,** 815–825.

Miller, D. J., and Ax, R. L. (1990). Carbohydrates and fertilization in animals. *Mol. Reprod. Dev.* **26,** 184–198.

Miller, D. J., and Shur, B. D. (1994). Molecular basis of fertilization in the mouse. *Semin. Dev. Biol.* **5,** 255–264.

Miller, D. J., Gong, X., and Shur, B. D. (1993). Sperm require β-N-acetylglucosaminidase to penetrate through the egg zona pellucida. *Development (Cambridge, UK)* **118,** 1279–1289.

Miller, D. J., Macek, M. B., and Shur, B. D. (1992). Complementarity between sperm surface β-1,4-galactosyltransferase and egg coat ZP3 mediates sperm-egg binding. *Nature (London)* **357,** 589–593.

Miller, R. L. (1985). Sperm chemo-orientation in the metazoa. *In* "Biology of Fertilization" (C. B. Metz and A. Monroy, eds.), pp. 276–337. Academic Press, New York.

Minor, J. E., Fromson, D. R., Britten, R. J., and Davidson, E. H. (1991). Comparison of the bindin proteins of *Strongylocentrotus francisanus, S. purpuratus* and *Lytechinus variegatus* sequences involved in the species specificity of fertilization. *Mol. Biol. Evol.* **8,** 781–795.

Minor, J. E., Britten, R. J., and Davidson, E. H. (1993). Species-specific inhibition of fertilization by a peptide derived from the sperm protein bindin. *Mol. Biol. Cell* **4,** 375–387.

Miyake, S., Sakurai, T., Okumura, K., and Yagita, H. (1994). Identification of collagen and laminin receptor integrins on murine T lymphocytes. *Eur. J. Immunol.* **24,** 2000–2005.

Miyazaki, S., Shirakawa, H., Nakada, K., and Honda, Y. (1993). Essential role of the inositol 1,4,5-trisphosphate receptor/Ca²⁺ release channel in Ca²⁺ waves and Ca²⁺ oscillations at fertilization of mammalian eggs. *Dev. Biol.* **158,** 62–78.

Moller, C. C., Bleil, J. D., Kinloch, R. A., and Wassarman, P. M. (1990). Structural and functional relationships between mouse and hamster zona pellucida glycoproteins. *Dev. Biol.* **137,** 276–286.

Monroy, A. (1985). Processes controlling sperm-egg fusion. *Eur. J. Biochem.* **152,** 51–56.

Monroy, A. (1986). A centennial debt of developmental biology to the sea urchin. *Biol. Bull. (Woods Hole, Mass.)* **171,** 509–519.

Monroy, A., and Rosati, F. (1979). The evolution of the cell-cell recognition system. *Nature (London)* **278,** 165–166.

Moore, A. R. (1951). Action of trypsin on the eggs of *Dendraster excentricus. Exp. Cell Res.* **2,** 284–287.

Morales, P., Cross, N. L., Overstreet, J. W., and Hanson, F. W. (1989). Acrosome-intact and acrosome-reacted human sperm can initiate binding to the zona pellucida. *Dev. Biol.* **133,** 385–392.

Mori, T., Wu, G. M., Mori, E., Shindo, Y., Mori, N., Fukuda, A., and Mori, T. (1990). Expression of class II major histocompatibility complex antigen on mouse sperm and its roles in fertilization. *Am. J. Reprod. Immunol.* **24,** 9–14.

Mori, T., Wu, G. M., and Mori, E. (1991). Expression of CD4-like structure on murine egg vitelline membrane and its signal transductive roles through p56lck in fertilization. *Am. J. Reprod. Immunol.* **26,** 97–103.

Mori, T., Gou, M. W., Yoshida, H., and Saito, S. (1992). Expression of the signal transducing regions of CD4-like and lck genes in murine egg. *Biochem. Biophys. Res. Commun.* **182,** 527–533.

Mortillo, S., and Wassarman, P. M. (1991). Differential binding of gold-labeled zona pellucida glycoproteins mZP2 and mZP3 to mouse sperm membrane compartments. *Development (Cambridge, UK)* **113,** 141–149.

Moy, G. W., and Vacquier, V. D. (1979). Immunoperoxidase localization of bindin during sea urchin fertilization. *Curr. Top. Dev. Biol.* **13,** 31–44.

Mozingo, N. M., and Chandler, D. E. (1991). Evidence for the existence of two assembly domains within the sea urchin fertilization envelope. *Dev. Biol.* **146,** 148–157.

Mozingo, N. M., and Chandler, D. E. (1993). Ultrastructural changes during fertilization envelope assembly in *Lytechinus pictus* eggs revealed by quick-freeze, deep-etch electron microscopy. *Cell Tissue Res.* **271,** 271–277.

Myles, D. G. (1993). Molecular mechanisms of sperm-egg membrane binding and fusion in mammals. *Dev. Biol.* **158,** 35–45.

Myles, D. G., and Primakoff, P. (1984). Localized surface antigens of guinea pig sperm migrate to new regions prior to fertilization. *J. Cell Biol.* **99**, 1634–1641.

Myles, D. G., Primakoff, P., and Bellve, A. R. (1981). Surface domains of the guinea pig sperm defined with monoclonal antibodies. *Cell (Cambridge, Mass.)* **23**, 433–439.

Myles, D. G., Hyatt, H., and Primakoff, P. (1987). Binding of both acrosome-intact and acrosome-related guinea pig sperm to the zona pellucida during in vitro fertilization. *Dev. Biol.* **121**, 559–567.

Myles, D. G., Kimmel, L. H., Blobel, C. P., White, J. M., and Primakoff, P. (1994). Identification of a binding site in the disintegrin domain of fertilin required for sperm-egg fusion. *Proc. Natl. Acad. Sci. U.S.A.* **91**, 4195–4198.

Nakano, E. (1969). Fishes. *In* "Fertilization" (C. B. Metz and A. Monroy, eds.), pp. 295–234. Academic Press, New York.

Nojima, Y., Humphries, M. J., Mould, A. P., Komoriya, A., Yamada, K. M., Schlossman, S. F., and Morimoto, C. (1990). VLA-4 mediates CD3-dependent CD4+T cell activation via the CS1 alternatively spliced domain of fibronectin. *J. Exp. Med.* **172**, 1185–1192.

Nuccitelli, R. (1991). How do sperm activate eggs? *Curr. Top. Dev. Biol.* **25**, 1–16.

Ohlendieck, K., and Lennarz, W. J. (1995). Role of the sea urchin egg receptor for sperm in gamete interaction. *Trends Biochem. Sci.* **20**, 29–33.

Ohlendieck, K., Dhume, S. T., Partin, J. S., and Lennarz, W. J. (1993). The sea urchin egg receptor for sperm: Isolation and characterization of the intact, biologically active receptor. *J. Cell Biol.* **122**, 887–895.

Ohlendieck, K., Partin, J. S., and Lennarz, W. J. (1994). The biologically active form of the sea urchin egg receptor for sperm is a disulfide-bonded homo-multimer. *J. Cell Biol.* **125**, 817–824.

O'Rand, M. G. (1986). Steps in the fertilization process. *In* "The Molecular and Cellular Biology of Fertilization" (J. L. Hedrick, ed.), pp. 383–393. Plenum, New York.

O'Rand, M. G. (1988). Sperm-egg recognition and barriers to interspecies fertilization. *Gamete Res.* **19**, 315–328.

O'Rand, M. G., Widgren, E. E., and Fisher, S. J. (1988). Characterization of the rabbit sperm membrane autoantigen, RSA, as a lectin-like zona binding protein. *Dev. Biol.* **129**, 231–240.

Osanai, K. (1976). Egg membrane-sperm binding in the Japanese Palolo eggs. *Bull. Mar. Biol. St. Asamushi, Tohoku Univ.* **15**, 147–155.

Paul, M. (1975). Release of acid and changes in light-scattering properties following fertilization of *Urechis caupo* eggs. *Dev. Biol.* **43**, 299–312.

Perlmann, P. (1954). Study on the effect of antisera on unfertilized sea urchin eggs. *Exp. Cell Res.* **6**, 485–490.

Perlmann, P. (1956). Response of unfertilized sea urchin eggs to antiserum. *Exp. Cell Res.* **10**, 324–353.

Perlmann, P., and Perlmann, H. (1957). Analysis of the surface structures of the sea urchin egg by means of antibodies. II. The J and A antigens. *Exp. Cell Res.* **13**, 454–474.

Pinto, M. R., De Santis, R., D'Alessio, G., and Rosati, F. (1981). Studies on fertilization in the ascidians: Fucosyl sites on vitelline coat of *Ciona intestinalis*. *Exp. Cell Res.* **132**, 289–295.

Primakoff, P. (1994). Sperm proteins being studied for use in a contraceptive vaccine. *Am. J. Reprod. Immunol.* **31**, 208–210.

Primakoff, P., and Myles, D. G. (1983). A map of the guinea pig sperm surface constructed with monoclonal antibodies. *Dev. Biol.* **98**, 417–428.

Primakoff, P., and Myles, D. G. (1990). Progress toward a birth control vaccine that blocks sperm function. *In* "Gamete Interaction: Prospects for Immunocontraception," (Alexander, N. J., ed.), pp. 89–102. Wiley-Liss, New York.

Primakoff, P., Hyatt, H., and Myles, D. G. (1985). A role for the migrating sperm surface antigen PH-20 in guinea pig sperm binding to the egg zona pellucida. *J. Cell Biol.* **101**, 2239–2244.

Primakoff, P., Hyatt, H., and Tredick-Kline, J. (1987). Identification and purification of a sperm surface protein with a potential role in sperm-egg membrane fusion. *J. Cell Biol.* **104**, 141–149.

Primakoff, P., Lathrop, W., Woolman, L., Cowan, A., and Myles, D. G. (1988a). Fully effective contraception in male and female guinea pigs immunized with the sperm protein PH20. *Nature* (*London*) **335**, 543–546.

Primakoff, P., Cowan, A., Hyatt, H., Tredick-Kline, J., and Myles, D. G. (1988b). Purification of the guinea pig sperm PH-20 antigen and detection of a site-specific endoproteolytic activity in sperm preparations that cleaves PH-20 into two disulfide-linked fragments. *Biol. Reprod.* **38**, 921–934.

Ralt, D., Goldenberg, M., Fetterolf, P., Thompson, D., Dor, J., Mashiach, S., Garbers, D. L., and Eisenbach, M. (1991). Sperm attraction to a follicular factor(s) correlates with human egg fertilizability. *Proc. Natl. Acad. Sci. U.S.A.* **88**, 2840–2844.

Ramarao, C. S., Myles, D. G., and Primakoff, P. (1994). Multiple roles for PH-20 and fertilin in sperm-egg interactions. *Semin. Dev. Biol.* **5**, 265–271.

Richardson, R. T., Yamasaki, N., and O'Rand, M. G. (1994). Sequence of a rabbit zona pellucida binding protein and localization during the acrosome reaction. *Dev. Biol.* **165**, 688–701.

Ringuiette, M. J., Chamberlin, M. E., Baur, A. W., Sobieski, D. A., and Dean, J. (1988). Molecular analysis of cDNA coding for ZP3, a sperm binding protein of the mouse zona pellucida. *Dev. Biol.* **127**, 287–295.

Rosati, F. (1985). Sperm-egg interaction in Ascidians. *In* "Biology of Fertilization" (C. B. Metz and A. Monroy, eds.), pp. 361–388. Academic Press, New York.

Rosati, F., and De Santis, R. (1980). Role of the surface carbohydrates in sperm-egg interaction in *Ciona intestinalis*. *Nature* (*London*) **283**, 762–764.

Rossignol, D. P., Roschelle, A. J., and Lennarz, W. J. (1981). Sperm-egg binding: Identification of a species-specific sperm receptor from eggs of *Strongylocentrotus purpuratus*. *J. Supramol. Struct. Cell Biochem.* **15**, 347–358.

Rossignol, D. P., Earles, B. J., Decker, G. L., and Lennarz, W. J. (1984). Characterization of the sperm receptor on the surface of eggs of *Strongylocentrotus purpuratus*. *Dev. Biol.* **104**, 308–321.

Ruiz-Bravo, N., and Lennarz, W. J. (1986). Isolation and characterization of proteolytic fragments of the sea urchin sperm receptor that retain species specificity. *Dev. Biol.* **118**, 202–208.

Ruiz-Bravo, N., and Lennarz, W. J. (1989). Receptors and membrane interactions during fertilization. *In* "The Molecular Biology of Fertilization" (H. Schatten and G. Schatten, eds.), pp. 21–36. Academic Press, New York.

Ruiz-Bravo, N., Earles, D., and Lennarz, W. J. (1986). Identification and partial characterization of sperm receptor associated with the newly-formed fertilization envelope from sea urchin eggs. *Dev. Biol.* **117**, 204–208.

Ruiz-Bravo, N., Janak, D. J., and Lennarz, W. J. (1989). Immunolocalization of the sea urchin sperm receptor in eggs and maturing ovaries. *Biol. Reprod.* **41**, 323–334.

Saling, P. M. (1991). How the egg regulates sperm function during gamete interaction: Facts and fantasies. *Biol. Reprod.* **44**, 246–251.

Saling, P. M., Waibel, R., and Lakoski, K. A. (1986). Immunological identification of sperm antigens that participate in fertilization. *Adv. Exp. Med. Biol.* **207**, 95–111.

Sánchez-Mateos, P., Campanero, M. R., Balboa, M. A., and Sánchez-Madrid, f. (1993). Co-clustering of β1 integrins, cytoskeletal proteins, and tyrosine phosphorylated substrates during integrin-mediated leukocyte aggregation. *J. Immunol.* **151**, 3817–3828.

Sander, K. (1985). Fertilization and egg cell activation in insects. *In* "Biology of Fertilization" (C. B. Metz and A. Monroy, eds.), pp. 409–430. Academic Press, New York.

Sato, M., and Osanai, K. (1986). Morphological identification of sperm receptors above egg microvilli in the polychaete, *Neanthes japonica*. *Dev. Biol.* **113**, 263–270.

Schmell, E., Earles, B. J., Breaux, C., and Lennarz, W. J. (1977). Identification of a sperm receptor on the surface of the eggs of the sea urchin *Arbacia punctulata*. *J. Cell Biol.* **72**, 35–46.

Schuel, H. (1985). Functions of egg cortical granules. *In* "Biology of Fertilization" (C. B. Metz and A. Monroy, eds.), pp. 1–43. Academic Press, New York.

Schwartz, M. A. (1992). Transmembrane signaling by integrins. *Trends Cell Biol.* **2**, 304–308.

Schwoebel, E., Prasad, S., Timmons, T. M., Cook, R., Kimur, H., Niu, E., Cheing, P., Skiner, S., Avery, S. E., Wilkins, B., and Dunbar, S. (1991). Isolation and characterization of full-length cDNA encoding the 55kDa rabbit zona pellucida protein. *J. Biol. Chem.* **266**, 7214–7219.

SeGall, G. K., and Lennarz, W. J. (1979). Chemical characterization of the component of the jelly coat from sea urchin eggs responsible for induction of the acrosome reaction. *Dev. Biol.* **71**, 33–48.

SeGall, G. K., and Lennarz, W. J. (1981). Jelly coat and induction of the acrosome rection in echinoid sperm. *Dev. Biol.* **86**, 87–93.

Sendai, Y., and Aketa, K. (1989). Involvement of wheat germ agglutinin (WGA)-binding protein in the induction of the acrosome reaction of the sea urchin Strongylocentrotus intermedias. II. Antibody against WGA-binding protein induces the acrosome reaction. *Dev. Growth Differ.* **31**, 467–473.

Shabanowitz, R. B., and O'Rand, M. G. (1988). Characterization of the human zona pellucida from fertilized and unfertilized eggs. *J. Reprod. Fertil.* **82**, 151–161.

Shalgi, R., Matityahu, A., and Nebel, L. (1986). The role of carbohydrates in sperm-egg interaction in rats. *Biol. Reprod.* **34**, 446–452.

Shapiro, B. M. (1991). The control of oxidant stress at fertilization. *Science* **252**, 533–536.

Shaw, A., McRee, D. E., Vacquier, V. D., and Stout, C. D. (1993). The crystal structure of lysin, a fertilization protein. *Science* **262**, 1864–1867.

Shaw, A., Lee, Y.-H., Stout, C. D., and Vacquier, V. D. (1994). The species-specificity and structure of abalone sperm lysin. *Semin. Dev. Biol.* **5**, 209–215.

Shimizu, T., Kinoh, H., Yamaguchi, M., and Suzuki, N. (1990). Purification and characterization of the egg jelly macromolecules, sialoglycoprotein and fucose sulfate glycoconjugate of the sea urchin *Hemicentrotus pulcherrimus*. *Dev. Growth Differ.* **32**, 473–487.

Shimizu, Y., van Seventer, G. A., Horgan, K. J., and Shaw, S. (1990). Regulated expression and binding of three VLA (β1) integrin receptors on T cells. *Nature (London)* **345**, 250–253.

Shur, B. D. (1993). Glycosyltransferases as cell adhesion molecules. *Curr. Opin. Cell Biol.* **5**, 854–863.

Sidhu, K. S., and Guraya, S. S.(1991). Current concepts in gamete receptors for fertilization in mammals. *Int. Rev. Cytol.* **127**, 253–288.

Simmons, D. L. (1993). Dissecting the modes of interactions amongst cell adhesion molecules. *Development (Cambridge, UK), Suppl.*, pp. 193–203.

Spradling, A. C. (1993). Germline cysts: Communes that work. *Cell (Cambridge, Mass.)* **72**, 649–651.

Stafford, C. J., Callow, J. A., and Green, J. R. (1993). Inhibition of fertilization in *Fucus* (Phaeophyceae) by a monoclonal antibody that binds to domains on the egg cell surface. *J. Phycol.* **29**, 325–330.

Steinhardt, R. A., Lundin, L., and Mazia, D. (1971). Bioelectrical responses of the echinoderm egg to fertilization. *Proc. Natl. Acad. Sci. U.S.A.* **68**, 2426–2430.

Stern, M. J., and DeVore, D. L. (1994). Extending and connecting signaling pathways in *C. elegans*. *Dev. Biol.* **166**, 443–459.

Summers, R. G., and Hylander, B. L. (1974). An ultrastructural analysis of early fertilization in the sand dollar *Echinarachnius parma*. *Cell Tissue Res.* **150**, 343–368.

Summers, R. G., and Hylander, B. L. (1975). Species-specificity of acrosome reaction and primary gamete binding in echinoids. *Exp. Cell Res.* **96**, 63–68.

Summers, R. G., and Hylander, B. L. (1976). Primary gamete binding. Quantitative determination of its specificity in echinoid fertilization. *Exp. Cell Res.* **100**, 190–194.

Suzuki, N. (1990). Structure and function of sea urchin egg jelly molecules. *Zool. Sci.* **7,** 355–370.

Suzuki, N., and Yoshino, K. (1992). The relationship between amino acid sequences of sperm-activating peptides and the taxonomy of echinoids. *Comp. Biochem. Physiol. B* **102,** 679–690.

Swann, K. (1993). The soluble sperm oscillogen hypothesis. *Zygote* **1,** 273–276.

Tarone, G., Russo, M. A., Hirsch, E., Odorisio, T., Altruda, F., and Silengo, L. (1993). Expression of β1 integrin complexes on the surface of unfertilized mouse oocyte. *Development (Cambridge, UK)* **117,** 1369–1375.

Tegner, M. J., and Epel, D. (1973). Sea urchin sperm-egg interactions studied with the scanning electron microscope. *Science* **179,** 685–688.

Tesarik, J., Drahorad, J., Testart, J., and Mendoza, C. (1990). Acrosin activation follows its surface exposure and precedes membrane fusion in human sperm acrosome reaction. *Development (Cambridge, UK)* **110,** 391–400.

Thillai-Koothan, P., van Duin, M., and Aitken, R. J. (1993). Cloning, sequencing and oocyte-specific expression of the marmoset sperm receptor protein, ZP3. *Zygote* **1,** 93–101.

Tilney, L. G. (1985). The acrosomal reaction. *In* "Biology of Fertilization" (C. B. Metz and A. Monroy, eds.), pp. 157–213. Academic Press, New York.

Tilney, L. G., and Jaffe, L. A. (1980). Actin, microvilli, and the fertilization cone of sea urchin eggs. *J. Cell Biol.* **87,** 771–782.

Tilney, L. G., Kiehart, D., Sardet, C., and Tilney, M. (1978). The polymerization of actin IV. The role of Ca^{++} and H^{+} in the assembly of actin and membrane fusion in the acrosomal reaction of echinoderm sperm. *J. Cell Biol.* **77,** 536–550.

Tong, Z.-B., Nelson, L. M., and Dean, J. (1995). Inhibition of zona pellucida gene expression by antisense oligonucleotides injected into mouse oocytes. *J. Biol. Chem.* **270,** 849–853.

Trimmer, J. S., and Vacquier, V. D. (1986). Activation of sea urchin gametes. *Annu. Rev. Cell Biol.* **2,** 1–26.

Trimmer, J. S., Ebina, Y., Schackman, R. W., Meinhof, C. G., and Vacquier, V. D. (1987). Characterization of a monoclonal antibody which induces the acrosome reaction of sea urchin sperm. *J. Cell Biol.* **105,** 1120–1128.

Tsuzuki, H., Yoshida, M., Onitake, K., and Aketa, K. (1977). Purification of the sperm binding factor from the egg of the sea urchin, *Hemicentrotus pulcherrimus. Biochem. Biophys. Res. Commun.* **76,** 502–511.

Tyler, A. (1948). Fertilization and immunity. *Physiol. Rev.* **28,** 180–219.

Tyler, A. (1967). Introduction: Problems and procedures of comparative gametology and syngamy. *In* "Fertilization" (C. B. Metz and C. B. Monroy, eds.), pp. 2–26. Academic Press, New York.

Tyler, A., and O'Melveny, K. (1941). The role of antifertilizin in the fertilization of sea urchin eggs. *Biol. Bull. (Woods Hole, Mass.)* **81,** 364–374.

Tyler, A., and Tyler, B. S. (1966). Physiology of fertilization and early development. *In* "Physiology of Echinodermata" (R. Boolootian, ed.), pp. 683–741. Wiley (Interscience), New York.

Urch, U. A. (1991). Biochemistry and function of acrosin. *In* "Elements of Mammalian Fertilization" (P. M. Wassarman, ed.), pp. 233–248. CRC Press, Boca Raton, FL.

Urch, U. A., and Patel, H. (1991). The interaction of boar sperm proacrosin with its natural substrate, the zona pellucida, and with polysulphated polysaccharides. *Development (Cambridge, UK)* **111,** 1165–1172.

Vacquier, V. D. (1979). The interactions of sea urchin gametes during fertilization. *Am. Zool.* **19,** 839–849.

Vacquier, V. D. (1989). Activation of sea urchin spermatazoa during fertilization. *Trends Biochem. Sci.* **11,** 77–81.

Vacquier, V. D., and Lee, Y.-H. (1993). Abalone sperm lysin: Unusual mode of evolution of a gamete recognition protein. *Zygote* **1,** 181–196.

Vacquier, V. D., and Moy, G. W. (1977). Isolation of bindin: The protein responsible for adhesion of sperm to sea urchin eggs. *Proc. Natl. Acad. Sci. U.S.A.* **74,** 2456–2460.

Vacquier, V. D., Epel, D., and Douglas, L. A. (1972a). Sea urchin eggs release protease activity at fertilization. *Nature (London)* **237,** 34–36.

Vacquier, V. D., Tegner, M. J., and Epel, D. (1972b). Protease activity establishes the block against polyspermy in sea urchin eggs. *Nature (London)* **240,** 352.

Vacquier, V. D., Tegner, M. J., and Epel, D. (1973). Protease released from sea urchin eggs at fertilization alters the vitelline layer and aids in preventing polyspermy. *Exp. Cell Res.* **80,** 111–119.

Vacquier, V. D., Carner, K. R., and Stout, C. D. (1990). Species-specific sequences of abalone lysin, the sperm protein that creates a hole in the egg envelope. *Proc. Natl. Acad. Sci U.S.A.* **87,** 5792–5796.

Vacquier, V. D., Swanson, W. J., and Hellberg, M. E. (1995). What have we learned about sea urchin sperm bindin? *Dev. Growth Differ.* **37,** 1–10.

van der Merwe, P. A., and Barclay, N. A. (1994). Transient intercellular adhesion: The importance of weak protein-protein interactions. *Trends Biochem. Sci.* **19,** 354–358.

van Seventer, G. A., Shimizu, Y., Horgan, K. J., and Shaw, S. (1990). The LFA-1 ligand ICAM-1 provides an important costimulatory signal for T cell receptor-mediated activation of resting T cells. *J. Immunol.* **144,** 4579–4586.

Vasta, G. R., Ahmed, H., Fink, N. E., Elola, M. T., Marsh, A. G., Snowden, A., and Odom, E. W. (1994). Animal lectins as self/non-self recognition molecules. Biochemical and genetic approaches to understanding their biological roles and evolution. *Ann. N.Y. Acad. Sci.* **712,** 55–73.

Ward, C. R., and Kopf, G. S. (1993). Molecular events mediating sperm activation. *Dev. Biol.* **158,** 9–34.

Ward, G. E., Brokaw, C. J., Garbers, D. L., and Vacquier, V. D. (1985). Chemotaxis of *Arbacia punctulata* spermatozoa to resact, a peptide from the egg jelly layer. *J. Cell Biol.* **101,** 2324–2329.

Wassarman, P. M. (1987a). Early events in mammalian fertilization. *Annu. Rev. Cell Biol.* **3,** 109–42.

Wassarman, P. M. (1987b). The biology and chemistry of fertilization. *Science* **235,** 553–560.

Wassarman, P. M. (1988). Zona pellucida glycoproteins. *Annu. Rev. Biochem.* **57,** 415–442.

Wassarman, P. M. (1989). Role of carbohydrates in receptor-mediated fertilization in mammals. *In* "Carbohydrate Recognition in Cellular Function" (G. Bock and S. Harnett, eds.), pp. 135–155. Wiley, Chichester.

Wassarman, P. M. (1990). Profile of a mammalian sperm receptor. *Development (Cambridge, UK)* **108,** 1–17.

Wassarman, P. M. (1991). Fertilization in the mouse. I. The egg. *In* "A Comparative Overview of Mammalian Fertilization" (B. S. Dunbar and M. G. O'Rand, eds.), pp. 151–165. Plenum, New York.

Wassarman, P. M. (1992). Mouse gamete adhesion molecules. *Biol. Reprod.* **46,** 186–191.

Wassarman, P. M., Florman, H. M., and Greve, J. M. (1985). Receptor-mediated sperm-egg interactions in mammals. *In* "Biology of Fertilization" (C. B. Metz and A. Monroy, eds.), pp. 341–360. Academic Press, New York.

Wassarman, P. M., Bleil, J. D., Fimiani, C., Florman, H., Greve, J., Kinloch, R., Moller, C., Mortillo, S., Roller, R., Salzmann, G., and Vazquez, M. (1989). The mouse egg receptor for sperm: A multifunctional zona pellucida glycoprotein. *In* "The Mammalian Egg Coat: Structure and Function" (J. Dietl, ed.), pp. 18–37. Springer-Verlag, Berlin.

Weskamp, G., and Blobel, C. P. (1994). A family of cellular proteins related to snake venom disintegrins. *Proc. Natl. Acad. Sci. U.S.A.* **91,** 2748–2751.

Whitaker, M., and Swann, K. (1993). Lighting the fuse at fertilization. *Development (Cambridge, UK)* **117,** 1–12.

White, J. M. (1990). Viral and cellular membrane fusion proteins. *Annu. Rev. Physiol.* **52,** 675–697.

Whittaker, C. A., and DeSimone, D. W. (1993). Integrin α subunit mRNAs are differentially expressed in early *Xenopus* embryos. *Development (Cambridge, UK)* **117,** 1239–1249.

Wikramanayake, A. H., and Clark, W. H. J. (1992). Molecular components involved in primary sperm-egg binding and induction of acrosomal exocytosis in the marine shrimp *Sicyonia ingentis. J. Exp. Zool.* **246,** 94–102.

Wikramanayake, A. H., and Clark, W. H. J. (1993). Two extracellular matrices from oocytes of the marine shrimp *Sicyonia ingentis* that independently mediate only primary or secondary sperm binding. *Dev. Growth Differ.* **36,** 89–101.

Wilkins, J. A., Stupack, D., Stewart, S., and Caixia, S. (1991). β1 integrin-mediated lymphocyte adherence to extracellular matrix is enhanced by phorbol ester treatment. *Eur. J. Immunol.* **21,** 517–522.

Wolfsberg, T. G., Bazan, J. F., Blobel, C. P., Myles, D. G., Primakoff, P., and White, J. M. (1993). The precursor region of a protein active in sperm-egg fusion contains a metalloprotease and a disintegrin domain: Structural, functional, and evolutionary implications. *Proc. Natl. Acad. Sci. U.S.A.* **90,** 10783–10787.

Yamada, A., Nikaido, T., Nojima, Y., Schlossman, S. F., and Morimoto, C. (1991). Activation of human CD4 T lymphocytes. *J. Immunol.* **146,** 53–56.

Yamasaki, N., Richardson, R. T., and O'Rand, M. G. (1995). Expression of the rabbit sperm protein Sp17 in Cos cells and interaction of recombinant Sp17 with the rabbit zona pellucida. *Mol. Reprod. Dev.* **40,** 48–55.

Yanagimachi, R. (1978). Sperm-egg association in mammals. *In* "Current Topics in Developmental Biology" (A. A. Moscona and A. Monroy, eds.), pp. 83–105. Academic Press, New York.

Yanagimachi, R. (1981). Mechanisms of fertilization in mammals. *In* "Fertilization and Embryonic Development In Vitro" (L. Mastroianni and J. D. Biggers, eds.), pp. 81–182. Plenum, New York.

Yanagimachi, R. (1988). Sperm-egg fusion. *In* "Current Topics in Membranes and Transport" (F. Bronner and N. Duzgunes, eds.), pp. 3–43. Academic Press, San Diego, CA.

Yoshida, M., and Aketa, K. (1978). Localization of species-specific sperm-binding factor in sea urchin eggs with immunofluorescent probe. *Acta Embryol. Exp.* **3,** 269–278.

Yoshida, M., and Aketa, K. (1979). Effect of papain-digested, univalent antibody against sperm binding factor on the fertilizability of sea urchin eggs. *Dev. Growth Differ.* **21,** 431–436.

Yoshida, M., and Aketa, K. (1982). Partial purification of the sperm-binding factor from the egg of the sea urchin, *Anthocidaris crassispina,* followed by an immunological method. *Dev. Growth Differ.* **24,** 55–63.

Yoshida, M., and Aketa, K. (1983). A 225k dalton glycoprotein is the active core structure of the sperm binding factor of the sea urchin *Anthocidaris crassispina. Exp. Cell Res.* **148,** 243–248.

Yoshida, M., and Aketa, K. (1987). Purification of the sperm-binding factor and identification of a sperm attack molecule from the egg of the sea urchin, *Hemicentrotus pulcherrimus. Gamete Res.* **18,** 1–16.

Youakim, A., Hathaway, H. J., Miller, D. J., Gong, X., and Shur, B. D. (1994). Overexpressing sperm surface β1,4-galactosyltransferase in transgenic mice affects multiple aspects of sperm-egg interactions. *J. Cell Biol.* **126,** 1573–1583.

Yurewicz, E. C., Sacco, A. G., and Subramanian, M. G. (1987). Structural characterization of the Mr=55,000 antigen (ZP3) of procine oocyte zona pellucida. *J. Biol. Chem.* **262,** 564–571.

Yurewicz, E. C., Pack, B. A., and Sacco, A. G. (1992). Porcine oocyte zona pellucida Mr 55,000 glycoproteins: Identification of O-glycosylated domains. *Mol. Reprod. Dev.* **33,** 182–188.

Yurewicz, E. C., Pack, B. A., Armant, D. R., and Sacco, A. G. (1993). Porcine zona pellucida ZP3a glycoprotein mediates binding of the biotin-labeled Mr 55,000 family (ZP3) to boar sperm membrane vesicles.*Mol. Reprod. Dev.* **36,** 382–389.

Zalokar, M. (1980). Activation of ascidian eggs with lectins. *Dev. Biol.* **79,** 232–237.

The Ultrastructure of Epithelial and Fiber Cells in the Crystalline Lens

J.R. Kuszak

Departments of Pathology and Ophthalmology, Rush-Presbyterian-St. Luke's
Medical Center, Chicago, Illinois 60612

Crystalline lenses are often simply described as inside-out stratified epithelial-like organs composed of uniform (hexagonal cross section profiles) crescent-like cells, arranged end-to-end in concentric shells around a polar axis. In this manner, as light is transmitted through lenses, their highly ordered architecture contributes to transparency by effectively transforming the multicellular organ into a series of coaxial refractive surfaces. This review will attempt to demonstrate that such a description seriously understates the structural complexity that produces lenses of variable optical quality in different species as a function of development, growth, and age. Embryological development of the lens occurs in a similar manner in all species. However, the growth patterns and effects of aging on lens fibers varies significantly among species. The terminally differentiated fiber cells of all lenses are generally hexagonal in cross section and crescent shaped along their length. But, while the fibers of all lenses are arranged in both highly ordered radial cell columns and concentric growth shells, only avian lens fibers are meridian-like, extending from pole to pole. In all other species, two types of fibers defined by different shapes are continuously formed throughout life. The majority of fibers are s-shaped, with ends that do not extend to the poles. Rather, the ends of these fibers are arranged as latitudinal arc lengths within and between growth shells. The overlap of the ends of specifically defined groups of such fibers constitutes the lens suture branches. The location, number, and extent of suture branches within and between growth shells are important considerations in lens function because the shapes of fiber ends, unlike that along fiber length, are very irregular. Consequently, as light is transmitted through sutures, spherical aberration (i.e., focal length variation) is increased. The degree of focal length variability depends on the arrangement of suture branches within and between growth shells, and this architecture varies significantly between species. The lifelong production of additional fibers at the circumference of the lens, culminating in new growth shells, neither proceeds equally around the lens equator, nor features identical fibers formed around the equator. Suture formation commences in the inferonasal quadrant, and continues sequentially in the superotemporal, inferotemporal, and finally the superonasal quadrants. During this proce ss, lens growth produces fibers of specifically defined length and shape as a function of their equatorial location. Utilizing a variation of this growth scheme, primates produce fibers that are arranged in progressively more complex suture patterns as a function of development, growth, and age that correlates with the temporal development of the zones of discontinuity seen by slit-lamp biomicroscopy. Fiber morphology also changes as a consequence of aging. This is an important consideration since the lens, owing to its inverted embryological and growth pattern, retains every cell formed throughout life. As a result fibers develop unique intercellular contacts for adhesion and communication that are specifically altered throughout life. In conclusion, the lens features many organ-specific structural specializations. While the species variations of these specializations are, in general, consistent with those of other epithelial systems, they represent specific modifications necessary for lens function.

KEY WORDS: Epithelial cells, Fiber cells, Sutures, Ultrastructure, Membrane, Intercellular junctions, Freeze-fracture, 3D CAD, Crystalline lens, TEM, SEM.

I. Introduction

One of the first detailed light microscopic studies of crystalline lens structure was conducted at the turn of the century (Rabl, 1900). Analyzing thick sections of wax-embedded lenses, Rabl demonstrated that the majority of lens cells are exceedingly long (Fig. 1a). Hence, they are referred to as fiber cells or simply fibers. It was also noted that in cross section, the hexagonally shaped fibers are arranged in ordered radial cell columns (RCCs; Fig. 1b). Thus, it was concluded that the lens is composed of concentric shells of fibers arranged end-to-end around a polar axis, which is coincident with the visual axis. Furthermore, since the fibers of each shell are precisely overlain between shells, it can also be correctly described as a series of RCCs arranged end-to-end around the visual axis. Many other investigators (Cohen, 1958, 1965; Wanko and Gavin, 1959; Smelser, 1965) have confirmed this gross structure of the lens in a variety of species and the evolution of its highly ordered architecture has been further elucidated by analysis of its development and growth (O'Rahilly and Meyer, 1959; Coulombre and Coulombre, 1963). These studies have shown that the concentric shells of fibers are in essence the strata of this unique stratified epithelial-like organ (Harding et al., 1971).

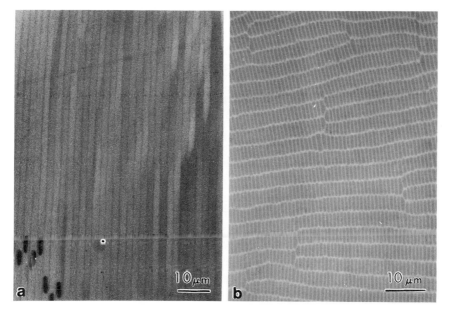

FIG. 1   (a) Light micrograph of rat lens elongating fibers thick-sectioned along the visual axis. The excessively long fiber-like cells are apparent. (b) Light micrograph of rat lens elongating fibers thick-sectioned along the equatorial axis (perpendicular to the visual axis). The arrangement of fibers in radial cell columns (RCCs) and concentric growth shells is also apparent. Note that while the fibers are generally hexagonal in cross section, occasional pentagonal cross-sectional profiles are present. Serial sectioning reveals that the pentagonal cross sections are actually pentagonal cell-to-cell fusion zones occuring for variable distances between neighboring fibers.

Thus, it has been proposed that lens architecture contributes to lens transparency by transforming this multicellular organ into a series of coaxial refractive surfaces (Trokel, 1962). Indeed, structural analysis of most cataracts, the most significant lens pathology, invariably reveals compromised lens anatomy at the sites of opacities.

Since all lenses have the gross structure described above, it has generally been accepted that any vertebrate lens is an adequate model for human cataract studies. Hence, the species selected is often dictated by experimental design. As an example, for biochemical characterization of lens crystallins, cytoskeleton, or plasma membrane, bovine, ovine, or rabbit lenses are generally employed because their large size effectively yields more material for analysis. For developmental studies, chicken lenses are often the lens of choice because of the facility in obtaining large numbers of lenses at different and well-defined stages of embryogenesis.

The efficacy of using literally any vertebrate lens model for human cataract research is an important consideration for a number of reasons: First, cataracts are the primary cause of blindness in the world (Vision Research: A National Plan 1994–1998, A Report of the National Advisory Eye Council, National Institutes of Health, National Eye Institute). Thus, cataract studies are not insignificant. Second, the availability of cataractous human lenses for study has steadily decreased as the s urgical techniques of cataract extraction have improved. Intracapsular cataract extraction, commonly used in Rabl's time, permitted almost complete retrieval of the diseased lens. This technique was replaced by extracapsular cataract extraction, a method that by design does not permit the retrieval of a completely intact cataractous lens. Though the techniques of intracapsular and extracapsular cataract surgery are still performed, today the method of choice is phacoemulsification, a technique in which the entire lens is broken apart and aspirated. Thus, the lack of adequate numbers of human lenses for study makes appropriate animal models imperative.

As additional techniques for ultrastructural analysis [e.g., scanning electron microscopy (SEM), freeze-fracture, freeze substitution, immunoelectron microscopy, three-dimensional (3D) image analysis, laser scanning confocal microscopy] were utilized, structural studies of lenses revealed that the seemingly precise architecture of the lens is, in fact, imperfect. Furthermore, although all lenses share a common gross structure, each lens has developmentally defined regions of nonuniformity that vary between species. Ostensibly subtle differences in this species-specific anatomy exert profound effects on lens optical quality. These are significant findings because the naturally occurring nonuniformities in lens architecture become more pronounced with age.

Therefore, although all crystalline lenses share a relatively common development and growth scheme, it is inappropriate to consider any vertebrate lens as completely representative of human lenses. In fact, ultrastructural analysis has also demonstrated that there are significant differences in fibers throughout the lens, and in the epithelial cells from which they are derived. Therefore, neither any one fiber nor any epithelial cell can be correctly considered as representative of, respectively, all fibers or epithelial cells in all lenses.

This review will attempt to describe the common anatomy of all crystalline lenses while highlighting some of the more significant structural differences between species. However, it will not attempt to be an all-inclusive portrait of lens morphology. Only a general description of the cytoskeleton will be given. Its comprehensive depiction will be beyond the scope of this chapter.

II. Lens Embryology: Primary Fiber Formation

A brief review of lens embryology facilitates an understanding of the unique architecture of all crystalline lenses that is paramount to the establishment and maintenance of transparency throughout life.

Lens development begins as surface ectodermal cells, overlying the bulging optic vesicle, thicken and form the lens placode. Soon thereafter, the placode is induced to invaginate into the forming optic cup (Lewis, 1907; Coulombre and Coulombre, 1963; O'Rahilly and Meyer, 1959; Saha et al., 1989; Henry and Grainger, 1990). This process continues until the placode is transformed into an inverted lens vesicle. Then, the cells approximating the retinal or posterior half of the lens vesicle begin to terminally differentiate. As a consequence of this process, these cells are transformed from cuboidal epithelial cells into long fibers. Since they are the first fibers to develop, they are referred to as primary fibers. As primary fibers form, they obliterate the lumen of the vesicle. At this stage, the lens is composed of a ball of primary fibers, overlain on its anterior half, by a monolayer of cuboidal epithelial-like cells. The entire lens is enclosed in a basement membrane-like capsule that is produced by the basal membrane of its cells.

III. Lens Growth

A. Secondary Fiber Formation

The anterior monolayer of cuboidal cells overlying the primary fibers, the lens epithelium, is essentially the germ cell layer, or stem cell population of the lens. However, rather then being distributed throughout the epithelium, the germ cells of the lens are sequestered as a narrow latitudinal band of cells, known as the germinative zone (gz), that lies just above the equator (Harding et al., 1971). Lens growth begins after completion of primary fiber development, and continues throughout life in the following manner: Cells in the gz undergo mitotic division and a number of the daughter cells are recruited to terminally differentiate into fibers. Since these are the second group of fibers to develop, they are referred to as secondary fibers. Although primary fibers are irregularly polygonal in cross section, those cells recruited to become secondary fibers undergo an initial shape change from cuboidal, with highly convoluted lateral and irregularly polygonal apical and basal membranes, to columnar, with six relatively planar lateral membrane surfaces arranged in a hexagonal profile around the cell's polar axis (Kuszak and Rae, 1982; Kuszak *et al.,* 1983). These nascent fibers are arranged in a narrow latitudinal band, known as the transitional zone (tz), that lies beneath the gz. Because additional gz cells are constantly recruited to become fibers, the tz cells are forced to migrate posteriorly (Beebe *et al.,* 1980). As migration of tz cells occurs, they simult aneously rotate 180° about their polar axis, and then bidirectionally elongate into secondary fibers as they terminally differentiate. As initial secondary fibers are formed, their anterior ends are insinuated beneath the apical membranes of the overlying lens epithelium, and above the anterior ends of the primary fiber cells. Simultaneously, the posterior ends of the same fibers are insinuated beneath the lens capsule, and above the posterior ends of the primary fibers. Fiber elongation is complete and fibers are considered mature when they are arranged end to end as a complete shell (Fig. 2a; Kuwabara, 1975).

B. Elongating Fiber Cells

Since lens growth continues throughout life, every lens has a peripheral area known as the bow region, which consists of elongating fibers arranged as incomplete shells (Fig. 2b). As additional secondary fibers develop, their anterior ends are insinuated beneath the apical membranes of the lens epithelium and above the anterior ends of the previously formed fibers, while their posterior ends are insinuated above the capsule and beneath the basal membranes of the previously formed fibers.

The elongating fibers are generally hexagonal in profile. This is the general shape that will be maintained by mature fibers. Of the six lateral membrane surfaces of fibers, two are broad and four are narrow. The two broad faces are oriented toward the lens periphery and its center. Thus, fibers have inner and outer broad faces. The four narrow faces are oriented at acute angles to the lens surface and center. As fibers elongate their middle segments are thinner (the breadth between broad faces) than their end segments. The opposite is true for mature fibers. Thus, fiber dimension

FIG. 2 Scale schematic diagram of a generic lens separated along the visual axis. Note the monolayer lens epithelium overlain onto the anterior surface of the lens. (b) If the central zone (cz) of the epithelium is lifted off of the lens, and the central core of mature fibers is further separated from the peripherally located developing population of lens cells [the pregerminative (pgz), germinative zone (gz), transitional zone (tz) epithelial cells, and elongating fibers], then the different populations of the lens defined by age, stage of the cell cycle, and state of differentiation are apparent. Reprinted from *Prog. Ret. Eye Res.* 14(2), J. R. Kuszak, The Development of Lens Sutures, 567–591, copyright (1995), with kind permission from Elsevier Science, Ltd, The Boulevard, Langford Lane, Kidlington 0X5 1GB, UK.

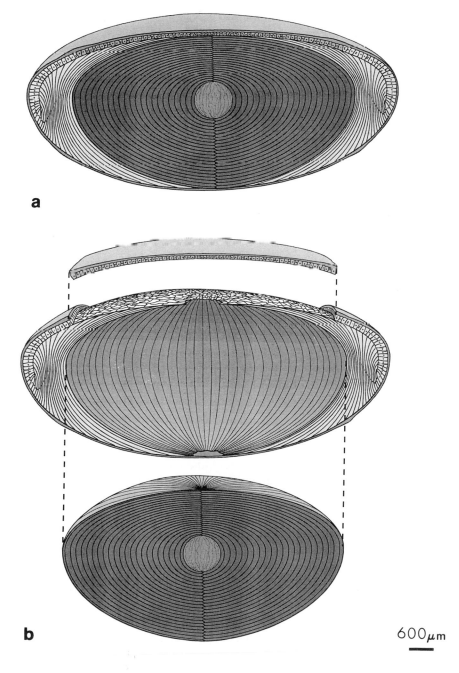

a

b

600μm

along fiber length varies as a function of development, growth, and age (Kuszak and Rae, 1982).

C. Mature Fiber Cells

The lens develops and grows essentially as a unique stratified epithelial-like organ. The lens epithelium represents the basal or germinal layer (Fig. 2a), incomplete shells of elongating fibers in the bow region represent the youngest, developing strata, and the mature fiber cells represent progressively older strata (Fig. 2b). However, because of its inverted evolution, the innermost growth shells, containing the oldest fibers of the lens, are never sloughed off as is the case with "true" stratified epithelia. Rather, the lens contains a permanent cellular record of its development and lifelong growth. It is for this reason that the lens is at once a very good system for studying the effects of aging, and a system that is highly susceptible to age-related pathologies (Kuwabara, 1975; Kuszak and Brown, 1993).

The shape of elongating and mature fibers, and their organization, convey upon the lens its typical, asymmetrical, oblate, spheroidal shape. In most species, the cross-sectional dimensions of mature secondary fibers in the equatorial plane are approximately 1.5–2.0 μm thick (the breadth between broad faces) and approximately 10–12 μm wide (the breadth between opposite pairs of narrow faces separated by and including both paired broad faces). Wider and thicker fiber dimensions have been reported in a variety of species. However, it is imperative to note that progressively para-equatorial sections through a lens produce tangential sections through the crescent fibers. In addition, the dimensions of fibers vary along their length; this parameter is a function of suture type. In this manner, reported dimensions of fibers in sections can be quite different and a comparison of fiber dimensions between species will reveal markedly dissimilar measurements. These relationships will be fur ther explained in a later section.

D. Lens Cell Populations

Although the lens consists of a single cell type, as described above, it is composed of three distinct cell populations defined by different stages of the cell cycle and age (Fig. 2; Harding et al., 1971). The majority of the epithelial cells exist as a broad polar cap known as the central zone (cz). Peripheral to this zone are, in succession, the narrow latitudinal band of epithelial cells known as the pregerminative zone (pgz), the gz, and the tz. The cz epithelial cells are generally maintained in the G_0 stage of the cell cycle, while some of the pgz and gz cells undergo mitosis, and all of the tz cells and elongating fibers represent cells actively engaged in progressively more advanced stages of terminal differentiation. Finally, the mature fibers are a terminally differentiated cell population. As the lens grows throughout life, the number of cells in the cz and mature fiber populations increase by the addition of more cells on their periphery. Thus, the cz and mature fiber populations have a range of cell ages with their oldest cells becoming more centrally located throughout life. In contrast, the gz, tz, and to a lesser degree the pgz cells, and elongating fibers, represent a continuously evolving population with a constant turnover of their number as they arise from the epithelium and then pass into the mature fiber compartment following terminal differentiation.

Consequently, the lens mass is generally described as consisting of a several age-related concentric regions (Duke-Elder, 1970; Kuszak and Brown, 1993). All of the primary fibers compose the embryonic nucleus. The fetal nucleus consists of the embryonic nucleus and all of the secondary fibers formed until birth. In a human lens, the infantile nucleus consists of the embryonic and fetal nucleus, and all of the secondary fibers formed until approximately age 4. The adult nucleus consists of the embryonic, fetal, and infantile nucleus, and all of the secondary fibers formed until sexual maturation. All of the secondary fibers formed afterward compose the cortex. The cortex is generally further subdivided into deep, intermediate, and superficial regions. The elongating fibers are usually considered to be the most superficial fibers, although arguably they are not truly fibers because they are still engaged in terminal differentiation. In a similar manner, a perinuclear region is sometimes described to exist between the deep cortex and adult nucleus. The exact boundaries of the different age-related regions are less important than the recognition of the broad range in fiber and epithelial cell age within a single lens. While the cellular age range in a mouse or rat lens is only as much as 2 years, in a human lens it can exceed 90 years. Even with allowances for the dissimilar metabolic rates between species, this is an extreme difference.

IV. Cell-to-Cell Fusion of Fibers

Given the constancy of fiber size along the equatorial plane, the arrangement of fibers into RCCs and growth shells must, by geometric definition, be imperfect. Rabl (1900) was the first to note that equatorial cross sections revealed numerous fibers with pentagonal profiles interspersed among RCCs (see Fig. 1b). However, he also noted that they were not randomly placed. In most cases the pentagonal fibers, generally larger than the typically hexagonal neighboring fibers, were found to bracket two parallel

RCCs. Rabl interpreted these fibers as being the result of two neighboring RCCs "fusing" to form a single larger RCC. In this manner, the gaps, which would eventually occur between opposed RCCs composed of fibers with constant cross-sectional dimensions, could be eliminated. With the development of scanning electron microscopy, it has been possible to examine the shape and organization of fibers in RCCs and growth shells along their entire length. This analysis reveals that larger, pentagonal fibers extending from suture to suture do not exist. Rather, opposed fibers "fuse" for variable distances or zones along their length in order to alter the existing or emerging alignment of radial cell columns, and the degree of end curvature required to produce sutural latticeworks (Kuszak *et al.*, 1985a, 1989). In this manner, fiber cell-to-cell fusion contributes to the concept of an intrinsic control of growth by acting as an "intercalation mechanism" responding to "positional information" obtained from previously formed cells (Wolpert, 1971, 1987; French *et al.*, 1976; Bryant *et al.*, 1981; Bryant and Simpson, 1984).

The density and distribution of fusion zones in lenses as a function of age, though not known quantitatively, appear to be variable, and related at least in part to suture type. As will be described in detail in a later section, the ends of mature fibers do not simply extend and overlap at the poles. Rather, specific and variably complex suture patterns are formed in different species. Suture formation involves the precise positioning of variably shaped and curved fibers into groups that combine to produce exactly offset anterior and posterior patterns of suture branches. Numerous fusion zones are found within forming growth shells during this period of construction to ensure that this rigorous architectural scheme is strictly monitored (Fig. 3). Far fewer fusion zones are seen in mature growth shells after suture construction has been completed. The initial site of fiber cell-to-cell fusion is thought to be within fiber cell gap junctions (Kuszak *et al.*, 1989). In comparison to gap junctions of other epithelia, fiber gap junctions have approximately 30% fewer IMPs/μm^2 of junctional area (Kuszak *et al.*, 1985b). This implies that the IMPs of fiber gap junctions might be more readily moved laterally within the junctional plaque area. In this manner, areas of closely apposed lipid, a prerequisite for membrane fusion, can be produced within fiber gap junctions.

In addition to a role in producing the ordered architectural scheme of the lens, fiber cell-to-cell fusion also provides the lens with patent intercellular pathways that augment intercellular communication afforded by gap junctions (Kuszak *et al.*, 1989). Although the extent of cell-to-cell fusion in a variety of species is not known quantitatively, it has been suggested that there are insufficient numbers of cell-to-cell fusion zones to transform the lens into a syncytium (Mathias and Rae, 1989). However, it should be pointed out that this appraisal was based on a morphological study that

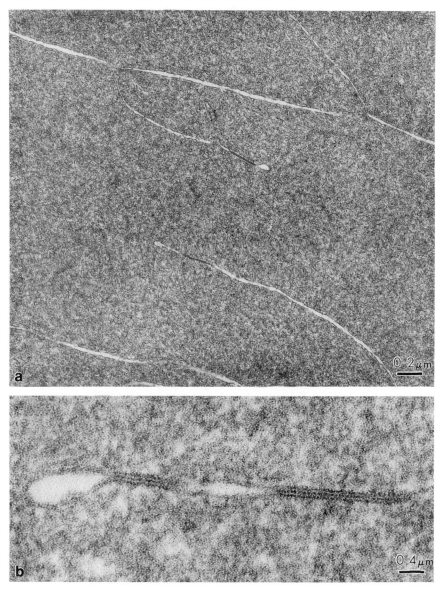

FIG. 3   (a) Note the mutual cytoplasmic region between the fused plasma membranes [membrane loops at the ends of (upper loop) or proximal to (lower loop) gap junctions] of neighboring rat elongating fibers shown by TEM thin-section. (b) Higher magnification of additional fused plasma membrane loops between neighboring elongating fibers.

underestimated the total contribution of cell-to-cell in providing patent pathways between lens fibers (Kuszak *et al.*, 1989). In quantifying the number of cell-to-cell fusion zones in a juvenile frog lens, only fusion zones with an apparent beginning and end were considered. Examples of cell-to-cell fusion that lacked either an apparent beginning or end were excluded for quantification. Further work will be necessary to quantitatively assess the extent and duration of cell-to-cell fusion in different lenses as a function of development, growth, and age in different species.

V. Lens Sutures

The overlap of mature fiber ends within and between growth shells constitutes the suture branches (Rochon-Duvigneaud, 1943; de Jong, 1981; Kuszak *et al.*, 1984, 1986, 1991, 1994; Kuszak and Brown, 1993; Sivak *et al.*, 1994). The combination of the precise arrangement of two distinct types of fibers (as regards fiber shape) and an exact variation in the length of these two fiber types within growth shells results in the formation of suture branches that collectively create anterior and posterior suture patterns. To understand the organization of suture patterns within and between growth shells it is necessary to consider the following. (1) Only birds and a few other species feature fibers that are meridian-like, that is, fibers that extend from pole to pole and are fusiform (tapering to a point at their ends). (2) All other species feature two distinct types of secondary fibers, "straight" and "S" shaped fibers. Neither of these two fiber types has ends that extend to the poles. Instead, straight fibers only have one end that reaches a pole, while neither end of an S fiber extends to a pole. (3) Although both straight and S fibers are characterized by a crescent curve along their length, they are also characterized by the additional curvature of their ends away from the polar axis. These curvatures, referred to as "fiber end curvatures," occur in opposite directions anteriorly and posteriorly, hence the name S fibers. (4) Fiber length is not merely a function of radial location.

There are four types of lens sutures with varying degrees of complexity that develop in different species: the "umbilical" suture, characteristic of certain teleosts, avian, reptilian, and salamander adult lenses; the "line" suture, characteristic of adult lenses in some teleosts, sharks, rays, frogs, and rabbit lenses; the "Y" suture, characteristic of adult lenses in such species as rat, mouse, cat, dog, sheep, and bovine; and finally the "star" suture system of primates.

A. "Umbilical" Sutures

In lenses with umbilical sutures, all secondary fibers arc meridian-like. They are crescent-shaped and fusiform, and their ends taper as they extend to confluence at the poles (Fig. 4a). The overlap of all of the anterior ends, and all of the posterior ends, constitutes, respectively, the anterior and posterior "umbilical" sutures (Fig. 4b). Theoretically, in these lenses, all fibers in any given growth shell are of equal length (intrashell fiber length variation) and the variable length of fibers in consecutive shells (intershell fiber length variation) is a simple radial or age-related function (Fig. 4c). However, as described previously, since equatorial fiber dimensions are relatively constant throughout the lens, the packing of fibers in RCCs is imperfect. Thus, in these lenses there is some intrashell fiber length variation; intershell fiber length variation is not simply an absolute radial function. For these reasons, a section coincident with the visual axis would not necessarily pass through the complete length of all fibers arranged in any RCC.

B. "Line" Sutures

The arrangement of the two types of secondary fibers in lenses with line sutures is shown in Figs. 4d–4f. As any shell of secondary fibers is formed, two pairs of straight fibers evolve around the equator to separate the lens into supero-temporal and nasal and infero-temporal and nasal quadrants. All other evolving secondary fibers develop into S fibers arranged in four distinct groups, positioned between the straight fibers. A representative anterior suture branch, formed by the two groups of fibers located in the infero-temporal and nasal quadrants of a lens, is shown in Fig. 4d. Because of opposite end curvature, the anterior ends of these two fiber groups overlap as an inferior, vertical suture branch while the posterior ends of the same fiber groups become aligned as horizontal arc lengths set at 180° to one another. The posterior ends of these fiber groups overlap with the posterior ends of the supero-temporal and nasal fiber groups while the anterior ends of these fiber groups overlap anteriorly as a superior, vertical suture branch (Fig. 4e). Note that the variable end curvature of S fibers precludes the possibility of making a thin section of a fiber along its entire length. In fact, a thick section of an S fiber with the greatest end curvature from a lens with a line suture would require a section encompassing a complete lens quadrant.

The paired anterior and posterior suture branches are oriented at 180° to create, respectively, vertical and horizontal line sutures. Essentially identical

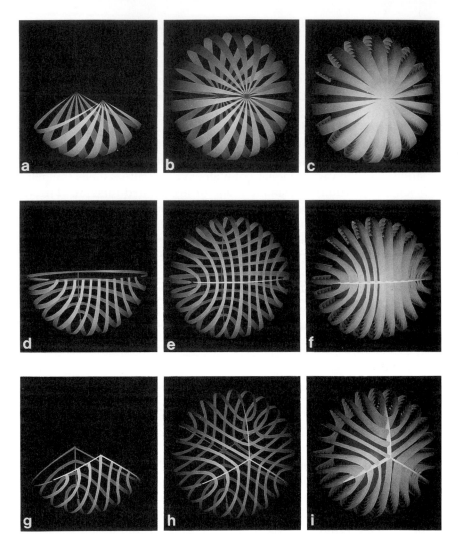

FIG. 4 A series of scale schematic 3D computer assisted drawings (CADs) showing the variable fiber types and their arrangement in individual and concentric growth shells to form "umbilical" (first row, a–c), "line" (second row, d–f), and "Y" sutures (third row, g–i). In lenses with umbilical sutures, all fibers in all growth shells are meridians. They extend from pole to pole; there is no intrashell fiber length variation, and a simple radial or age-related intershell fiber length variation. In lenses with line or Y sutures, none of the fibers in an y of the growth shells are meridians. Rather, they are either "straight" fibers, with only one end extending to a pole, or "S" shaped fibers, with their ends curving away from the polar axis in opposite directions anteriorly and posteriorly, and neither of the ends extending to a pole. Thus, because the ends of specifically defined groups of fibers overlap to form suture branches, intrashell fiber length variation is a function of suture type, and intershell fiber length variation is only a simple radial or age-related function within RCCs. Note that the identical positioning of suture branches between concentric growth shells results in the production of "continuous" suture planes extending from the embryonic nucleus (translucent central ball in c,f, and i) to the lens periphery. Reprinted from *Prog. Ret. Eye Res.* 14(2), J. R. Kuszak, The Development of Lens Sutures, 567–591, copyright (1995), with kind permission from Elsevier Science, Ltd, The Boulevard, Langford Lane, Kidlington OX5 1GB, UK.

suture patterns are created in each concentric shell (Fig. 4f). Later it will be shown that fiber ends are very irregular in size and shape. As a result, the suture branches of successive shells create two irregular planes extending from the embryonic nucleus (the primary fiber cell mass) to the lens periphery, as o pposed to the regular RCC planes created by uniformly shaped fibers. These suture planes exert a quantifiable negative effect on lens optical quality (focal length variability, i.e., spherical aberration (Kuszak et al., 1991).

C. "Y" Sutures

The arrangement of the two types of secondary fibers in lenses with Y sutures is shown in Figs. 4g–4i. As any shell of secondary fibers is formed, three pairs of straight fibers evolve around the equator to separate the lens into sextants. All other evolving secondary fibers develop into S fibers arranged as six distinct groups, positioned between the straight fibers. The anterior suture branch, formed by the two groups of fibers that bracket the infero-temporal and nasal quadrants of a lens with a line suture, is shown in Fig. 4d. The comparable two groups of fibers that bracket the infero-temporal and nasal quadrants of a lens with a Y suture are shown in Fig. 4g. Note that while the formation of a suture branch occurs in an identical manner, the opposite end curvature of Y suture lenses is less extreme. As a result, the anterior ends of six fiber groups overlap to form three anterior suture branches oriented in an upright Y pattern and three offset posterior suture branches oriented in an inverted Y pattern (Fig. 4h). In these lenses the suture branches of successive shells create six irregular planes extending from the embryonic nucleus to the lens periphery (Fig. 4i). As expected, these suture planes exert an even greater quantifiable negative effect on lens optical quality, and this effect increases as a function of age (Sivak et al., 1994).

D. "Star" Sutures

The arrangement of the two types of secondary fibers in lenses with Star sutures is shown in Fig. 5. Although the production of suture branches in these lenses occurs in a manner similar to that already demonstrated in lenses with either line or Y sutures, there is a fundamentally important difference. Consider the anterior suture branch formed by the two groups of fibers that bracket the infero-temporal and nasal quadrants of a primate lens at birth, as shown in Fig. 5a. These two groups of fibers are comparable to the groups of fibers that bracket the infero-temporal and nasal quadrants

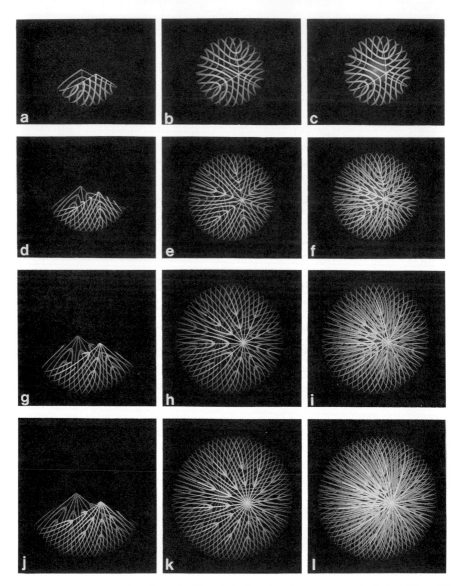

FIG. 5 A series of scale schematic 3D CADs showing the arrangement of straight and S fibers in individual and concentric growth shells to form the four generations of progressively more complex sutures ("Y" suture, first row, a–c; "simple star," second row, d–f; "star" suture, third row, g–i; and "complex star" suture, fourth row, j–l), in primate lenses. Because the defined groups of fibers that form suture branches vary from generation to generation, primate lens intrashell fiber length variation is a variable function within RCCs. Note that the nonidentical positioning of suture branches between concentric growth shells results in the production of "discontinuous" suture planes extending from the embryonic nucleus (translucent central ball in c,f, i, and l) to the lens periphery. Note also that in successive generations, as the number of suture branches increases, the end curvature of fibers and their length within and between growth shells decreases. Reprinted from *Prog. Ret. Eye Res.* 14(2), J. R. Kuszak, The Development of Lens Sutures, 567–591, copyright (1995), with kind permission from Elsevier Science, Ltd, The Boulevard, Langford Lane, Kidlington 0X5 1GB, UK.

of a lens with a Y suture, as shown in Fig. 4g. Anteriorly, they combine to form a single suture branch, while posteriorly they will contribute to 2 suture branches. Now consider the same region of superficial or younger shells in an infantile, a young, and a middle-aged adult primate lens (Figs. 5d, 5g, and 5j). In these growth shells, there are progressively more groups of fibers that combine to form increasing numbers of anterior and posterior suture branches. It is apparent that as additional fibers are added onto the lens mass, their continuous evolution is characterized by a reduction in end curvature. As a result, while the ends of the fiber groups in consecutive growth shells formed throughout gestation overlap to form typical Y suture patterns, after birth and throughout life, incremental changes in the length and shape of fibers in successive shells combine to culminate in a 6-branch star suture pattern at the end of infantile development (Fig. 5e), a more complex 9-branch star suture pattern at the end of sexual maturation (Fig. 5h), and the most complex 12-branch star suture pattern through middle and old age (Fig. 5k). As a result of the progressively more complex evolution of star sutures in primate lenses, "discontinuous" rather than continuous suture planes are formed in these lenses (Fig. 5c, 5f, 5i, and 5l). It is interesting to speculate that this key difference between the lifelong growth schemes of primate and nonprimate lenses is an evolutionary adaptation designed to reduce the negative influence of lens sutures on lens optical quality (Kuszak *et al.*, 1991; Sivak *et al.*, 1994). Indeed, laser scan analysis reveals that the more complex primate lenses are optically superior to the simpler nonprimate lenses with line and Y sutures (Kuszak *et al.*, 1994).

E. Comparative Anatomy of Lens Sutures

It is immediately apparent that the number of branches is the primary difference between the dissimilar suture patterns of different species. Less obvious are the exact variations in fiber length and shape that culminate in the production of the progressively more complex suture patterns between species.

1. Fiber Length

As shown in Figs. 4 and 5 the degree of opposite-end curvature in fibers of different species varies as a function of suture type. As a result, intrashell fiber length varies significantly in lenses with line, Y, or star sutures and intershell fiber length variation is only age-related within individual RCCs of umbilical, line, and Y sutures. The fiber that has ends extending to the midpoint of offset anterior and posterior suture branches is substantially

longer than the straight fiber whose anterior end defines the peripheral
location or origin of the same anterior suture branch and whose posterior
end defines the distal or termination of the same posterior branch. Both
of these fibers are in turn longer than the straight fiber with an anterior
end that defines the termination of the nearest neighboring anterior suture
branch and a posterior end that defines the origin of the offset posterior
suture branch. Indeed, when intrashell fiber length variation of line and Y
suture lenses is plotted against the equatorial location of fibers within any
growth shell, it can be seen that there is a bi-sinusoidal variation that is
repeated as a function of the number of suture branches (Kuszak et al.,
1986). Furthermore, as the number of branches increases, the amplitude
of this variation is reduced. However, because of the four distinct generations of star sutures in primate lenses that evolve as a function of development, growth, and age (Kuszak et al., 1994), this is not the case. Neither
intrashell nor intershell fiber length variation produce a simple bi-sinusoidal
variation as a function of the number of suture branches. Rather, primate
lens fiber length variation is more complex. In fact, throughout life, primate
lenses can feature straight fibers overlain by S fibers, and vice versa, in
successive shells (compare Figs. 5a, 5d, 5g, and 5j). Interestingly, the more
complex growth scheme of primate lenses results in fibers that are less
curved and less variable in length than as a function of age.

2. Fiber Width

The evolution of variably complex suture patterns in different species ne-
gates the simplistic notion that lens development and growth merely involve
bidirectional elongation of former cuboidal epithelial cells into fibers
around the lens equator. Two different types of fibers, straight and S shaped,
must arise in exact positions around the equator. As fiber elongation proceeds, the degree of fiber taper is determined by suture type (Kuszak et
al., 1986).

In lenses, with line sutures, the length of the suture branch formed by
the ends of all fibers within two opposed fiber groups is approximately one
third that of the width of a fiber group at the equator. Thus, in these lenses,
fiber width is reduced 31 from equator to end. In lenses with Y sutures,
the reduction is only 2:1 because the width of all fibers in a group at the
equator is only twice as great as the length of the suture branch their ends
produce. In contrast, primate lenses actually effect a widening of their fiber
ends rather than a reduction. This is because the number of fibers within
a group is progressively less than the length of the suture branches that
they produce. As described earlier, the contribution of cell-to-cell fusion
is an important component of suture development (Kuszak et al., 1989).
Fibers of exact length and variable shape along this length must evolve in

order to construct suture patterns that minimize the diffraction of light. Such a program requires a synchronized production of cytoplasmic, cytoskeletal, and membrane components that is most exact in primates. However, it should be noted that this program is progressively simplified in primates as a function of age. While the number of suture branches increases as a function of age, the intra- and intershell fiber length variation required to effect these changes is reduced. The more complex lens growth program of primates compared to nonprimates might represent an evolutionary adaptation in response to the greater lifespan of primates. If primates used a growth program entirely consistent with that of nonprimates, then progressively longer and more curved fibers would have to be produced, resulting in enlarging suture planes aligned along the visual axis as a function of age. This would undoubtedly produce lenses with inferior optical quality (Kuszak *et al.*, 1994). However, errors in the more sophisticated growth program of primates may manifest themselves as opacities (cortical cataracts) after decades of improper suture construction.

VI. Lens Ultrastructure

A. Lens Epithelium

Cells of the cz and pgz are low cuboidal cells with extensive lateral interdigitations (Figs. 6 and 7). Their large indented nuclei typically have two nucleoli and numerous nuclear pores. Nominal numbers of ribosomes, polysomes, smooth and rough endoplasmic reticulum, Golgi bodies, and small mitochondria with irregular cristae also characterize these cells. Typical lysosomes, dense bodies, and glycogen particles also characterize these cells, and they increase as a function of age.

As these epithelial cells age, they become more flattened, though the reduction in their height is insufficient to warrant referring to them as squamous. This flattening results in an increase in cellular surface area, and thus the number of cells required to reconcile the increased anterior surface area of the lens due to lifelong growth is less than the number of fibers added to the mature fiber mass.

The lens epithelial cytoskeleton is composed of actin, intermediate filaments (vimentin), microtubules, and the proteins spectrin, alpha-actinin, and myosin (Benedetti *et al.*, 1981; Aster *et al.*, 1986; Allen *et al.*, 1987). As in all eukaryotic cells, it is presumed that the various cytoskeletal elements combine to form a well-defined cytoskeleton that compartmentalizes the components of the cell's interior. An interesting feature of the lens cytoskeleton in many vertebrate lenses (most notably primate) is a prominent

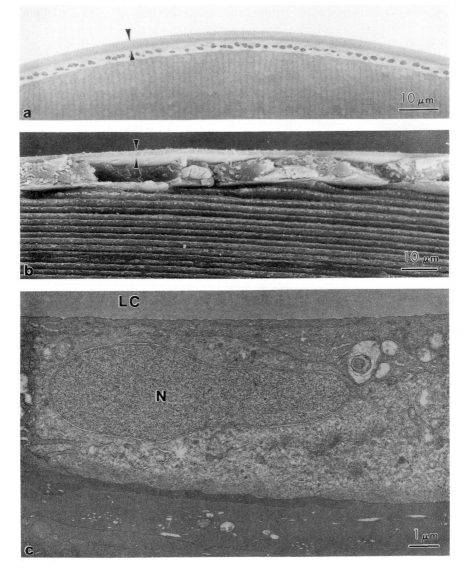

FIG. 6   (a) Light micrograph showing the cz lens epithelium overlain onto the hexagonal fibers arranged in RCCs. Note the thick basement membrane-like lens capsule (lc) produced at the basal portion of the cuboidal epithelial cells. (b) SEM micrograph showing the highly convoluted lateral membranes of cz epithelial cells oriented at acute angles beneath the lens capsule and above the elongating fibers. (c) Low-magnification TEM thin-section depicting a typical cz epithelial cell. These cells have indented nuclei, and nominal numbers of organelles (mitochondria, Golgi bodies, endoplasmic reticulum, etc.). From Spector *et al.,* 1995.

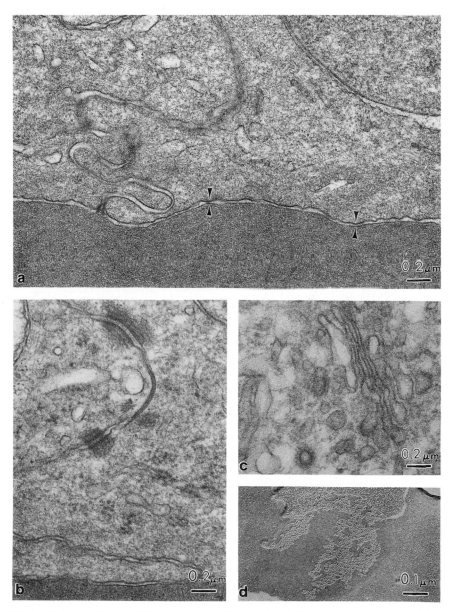

FIG. 7 High-magnification TEM micrographs showing detailed features of monkey cz epithe-lial cells. (a) Note that while some apico-lateral membrane interfaces between these cells show obvious adherens and occluding junctions (a), comparable areas between other cells (b) show either a lack of such junctions or only poorly defined examples of such junctions. (c) Note also that while there are areas of close membrane apposition at the cz epithelial-fiber interface (EFI; opposed black on white arrowheads), gap junctions are not apparent. (c) A Golgi body and clathrin-coated vesicle immediately above the apical membrane. (d) A typical gap junction between neighboring epithelial cell lateral membrane. (d) from Kuszak *et al.,* 1995. (b,c) from Kuszak, 1994.

network of actin filaments that exists in the form of polygonal arrays or "geodomes" (Rafferty and Scholz, 1984, 1989; Rafferty *et al.*, 1990; Yeh *et al.*, 1986). TEM analysis of the "geodomes" shows that they are immediately subjacent and attached to the apico-lateral and baso-lateral membranes. All of the cytoskeletal components of lens epithelial cells become more dense as the lens ages.

Lens epithelial cells are polarized with distinct apical, lateral, and basal membranes. The interfaces of these various membranes are an important component of lens physiology. The basal membranes of lens epithelial cells produce the anterior lens capsule. The posterior portion of this replicated basal lamina is produced by the basal membrane of elongating fiber cells (Parmigiani and McAvoy, 1989). The capsule is composed of fine microfibrils with the outermost portion of the capsule featuring collagen filaments intertwined with microfibrils (Rafferty, 1985). It is asymmetrical, with the anterior capsule being thicker than the posterior capsule. The thickest regions of the capsule are those to which the zonular fibers are attached.

The lateral membranes of cz lens epithelial cells are highly convoluted (see Figs. 6 and 7). These tortuous membranes are conjoined by typical gap junctions and desmosomes. The ultrastructure of lens epithelial gap junctions and tight junctions will be described in greater detail in a later section.

Gz cells can be distinguished from cz and pgz cells by their increased height and by the fact that many are actively undergoing mitosis. Beyond the germinative zone are the nascent fibers of the tz. These cells have already begun the process of terminal differentiation into fibers, as evidenced by their physical transformation from irregularly shaped cuboidal cells into hexagonally shaped columnar cells.

B. Elongating and Mature Fibers

As fibers are being formed through the terminal differentiation process, they simultaneously begin the process of eliminating most of their internal organelles including nuclei, mitochondria, golgi bodies, and smooth and rough endoplasmic reticulum (Jurand and Yamada, 1967; Papaconstantinou, 1967; Modak *et al.*, 1968; Modak and Bollum, 1972; Kuwabara and

---

FIG. 8   Stereopair micrographs showing the ball and socket lateral interdigitations that arise at the angles formed by narrow faces, and are arrayed along the length of cortical fibers in RCCs of (a) frog and (b) mouse lens fibers. Note and compare the erose nature of the narrow faces of mouse lens fibers with their broad faces, as seen in Fig. 9c.

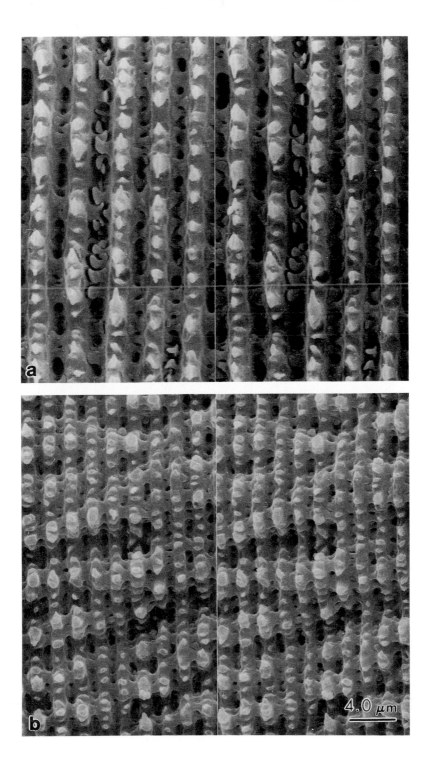

Imaizumi, 1974). This process is considered complete in mature fibers. However, as the organelles are being eliminated, there is an enormous concomitant synthesis of plasma membrane (Benedetti et al., 1976), cytoskeleton (Benedetti et al., 1981), and specialized cytoplasmic proteins, the lens crystallins (Wistow and Piatigorsky, 1988).

The basal membranes of elongating fibers interface with and produce the posterior capsule as they extend toward their posterior sutural destinations during the terminal differentiation process.

The lateral membranes of elongating fibers in all lenses are characterized by the onset of the development of unique fiber interdigitations. The production of these interdigitations begins during the latter stages of terminal differentiation and is complete by fiber maturation. As in most epithelial cells, the lateral interdigitations are believed to function as loose interlocking devices between neighboring cells. However, the lateral membrane interdigitations of fibers are more numerous, elaborate, and regularly arrayed along fiber length than in any other epithelia. Scanning electron microscopy has been unparalleled as a method to optimally demonstrate the structure of fiber lateral interdigitations (Hansson, 1970; Dickson and Crock, 1972; Farnsworth et al., 1974; Kuwabara, 1975; Harding et al., 1976; Hollenberg et al., 1976; Nelson and Rafferty, 1976; Kuszak et al., 1980, 1983; Kuszak and Rae, 1982; Hoyer, 1982; Willikens and Vrensen, 1981). There are several types of lateral membrane interdigitations between fibers that vary minimally in shape, size, and periodic arrangement between fibers of different species.

The first type is a cellular outpocketing, referred to as a ball, that consists of a globular top portion atop a short cylindrical stalk that emanates from the angle formed by two narrow faces (Fig. 8). These balls, which are regularly arrayed along the length of fibers, fit into complementary shaped sockets formed between neighboring fibers of adjacent growth shells. Smaller versions of this type of interlocking device are also found randomly emanating from broad face surfaces (Fig. 9). They fit into complementary shaped and sized sockets formed on the opposed broad faces of fibers in an opposed growth shell. An exception to this pattern is the mouse lens, which features large balls and sockets aligned along the length of broad

---

FIG. 9 Stereopair micrographs showing the flap and complementary imprint lateral interdigitations that arise at the angles formed by narrow and broad faces, and are arrayed along the length of cortical fibers in growth shells of (a) chicken, (b) bovine, and (c) mouse lens fibers. Note the variation in size and frequency of these interlocking devices. Note also the small or mini ball and socket interdigitations that exist between broad faces of successive chicken lens growth shells.

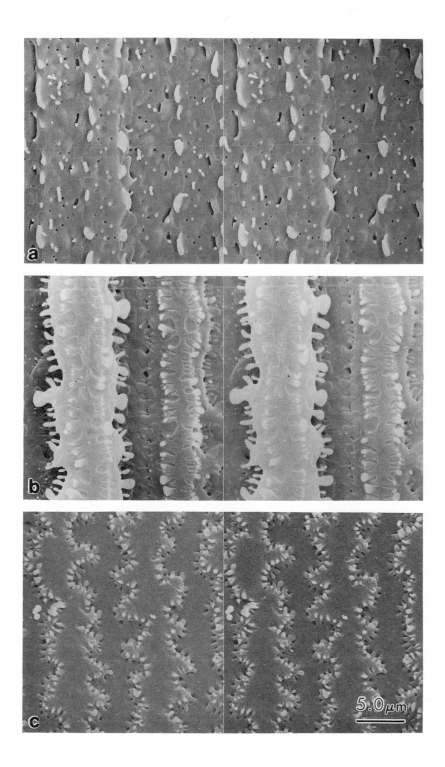

faces (Nelson and Rafferty, 1976; Lo and Reese, 1993). Gap junctions have been noted on both the neck and ball portions of these lateral interdig-itations. In fact, it has been suggested that gap junctions between fibers only occur on these devices (Lo and Reese, 1993). However, no statistical evidence to support this theory has been demonstrated.

The second type of fiber lateral membrane interlocking device is a cellular outpocketing shaped like a flap, or tongue, that emanates from the angle formed by a broad and narrow faces (Fig. 9). These devices occur along the length of fibers. They fit into complementary shaped imprints or grooves formed on the narrow faces of neighboring fibers. In primate lenses, these devices are characterized by square array membranes arrayed along their periphery (Fig. 10).

It is generally thought that the lateral interlocking devices of fibers serve to maintain fiber alignment as gross lens shape is altered during accommoda-tion and convergence. However, while this would appear to be a reasonable interpretation, it should be pointed out that the most uniform, elaborate, and extensive fiber lateral interlocking devices occur in fish lenses, which do not alter their shape to accomplish focusing.

In most lenses the lateral membranes of elongating and mature fibers are smooth or planar. In contrast, primate lenses are characterized by furrowed membranes. In primate elongating fibers, the furrows occur ran-domly on both broad and narrow faces.

C. Senescent Fiber Cells

Fiber morphology is altered as a function of maturity and age, and these variations are most significant in primates.

By the time primate fiber maturation is complete, literally their entire broad and narrow face surfaces have been transformed into polygonal domains of furrowed membrane (Fig. 11). The domains, which range in size from approximately 2.0 to 9.0 μm^2, are characterized by ridges and furrows of membrane arranged in parallel. The domains are aligned at acute angles to the length of the fiber. In nonprimate lenses, the furrowed membrane domains are less pronounced (bovine; Zampighi et al., 1989; Costello et al., 1985), chicken (Kuszak et al., 1980), frog (Sikerwar and Malhotra, 1983), and guinea pig (Kuszak et al., 1986). Small areas of smooth plasma membrane are infrequently noted between the polygonal domains. These areas are characterized by gap junctions which will be discussed in further detail in the following section. Overlying many of the ridges within the domains are microvilli that measure approximately 0.14 μm in diameter and range in length from approximately 1.25 to 4.75 μm. BEI-SEM (back-scatter electron image-scanning electron microscopy) analysis suggests that

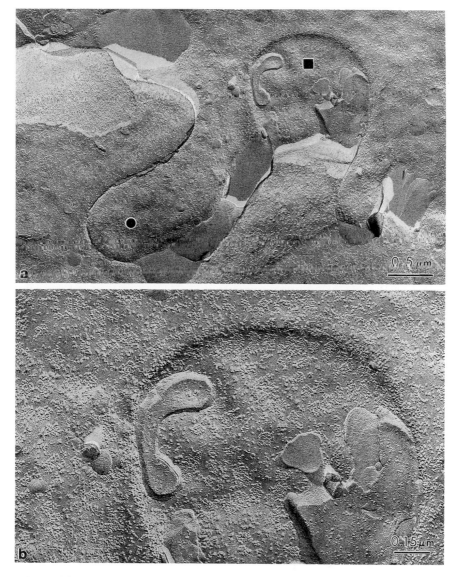

FIG. 10   (a) Freeze-etch replica of a monkey lens elongating fiber flap interlocking device (black on white circle) and a complementary imprint (black on white square). (b) At higher magnification it can be seen that patches of square array membrane are aligned along the edges of the flap and its complementary imprint.

FIG. 11   SEM micrograph showing the radical change in the surface morphology of primate lens cortical fibers as a function of development, growth, and age. The formerly smooth lateral membrane of nascent fibers is a progressively altered into polygonal domains of furrowed membrane aligned at acute angles to the length of the fibers. In addition, the balls and flaps and, respectively, the sockets and imprints that they fit into also become less regular with age.

each microvillus has an internal cytoskeleton, though the exact type of cytoskeletal elements cannot be ascertained from this technique. The furrowed and ridged membranes of the domains are characterized by square array membranes, also referred to as thin or wavy junctions in the lens (Lo and Harding, 1984; Zampighi et al., 1992).

The reason that fiber surface membrane projections increase as a function of age is unknown. In all cells the lipid composition of plasma membrane and the cytoskeleton immediately subjacent to the plasma membrane influence the conformation of plasma membrane, as well as the distribution of transmembrane proteins within the bilayer (Low and Kincade, 1985; Low and Zilversmit, 1980; He et al., 1986; Sefton and Buss, 1987; Burn, 1988; Yeagle, 1989). Biochemical studies of different-aged lens fibers suggest that the lipid composition and the integrity of cytoskeletal elements vary as a function of age (Zelenka, 1984). It is possible that as a result of cytoskeletal breakdown, and alterations in lipid composition, fiber plasma membrane becomes less planar (is transformed into polygonal domains of furrowed membrane and subsequent production of microvilli), resulting in previously unassociated small groups of square array membrane localizing to ridges and microvilli of the polygonal domains. But while these organelle changes may explain the alterations in fiber surface morphology as a function, it is appropriate to consider whether this alteration is actually a function of development, growth, and aging, occurring in a controlled temporal sequence in lenses.

A comparison of several species shows that the production of furrowed membrane domains and their associated square array membrane varies considerably between species. One reason for this disparity may be related to the longer lifespan of primates. At birth, primate lenses measure approximately 6 mm in equatorial diameter. In human lenses, this measure increases to as much as 9.5 mm by the end of the seventh decade (70+ years old). As described earlier, mature fiber thickness is relatively constant in the equatorial plane. Therefore, the projected dimensions of a lens as a function of growth can be estimated. By measuring the fiber dimensions of a young human lens (20 years old) across the equatorial plane, and then comparing the actual measures of fibers in an aged lens (70+ years old) lens) with geometric projections, it can be seen that there is actually a progressive age-related reduction in mature secondary fiber dimensions. This reduction is referred to as "fiber compaction." In primates, a programmed reconfiguration of planar fiber membrane into pleated surfaces may simply represent a means of balancing the cumulative age-related increase in lens size with the constraints of the other ocular components (e.g., the iris, cornea, pupil) that do not increase in size throughout life.

All fibers have a finely granular cytoplasm that consists primarily of soluble proteins, the lens crystallins, rare polysomes, and cytoskeletal elements. Cytoskeletal elements include actin, intermediate filaments, rare microtubules and lens-specific beaded chain filaments (Ireland and Maisel, 1983, 1984; Alcala and Maisel, 1985). The cytoskeletal elements of fibers exist as a pervading network that is anchored to the plasma membrane (Fig. 12).

Distinct structural features of the apical, lateral, and basal plasma membrane of fiber cells suggest that these cells are polarized. As described previously, the apical membranes of fibers overlap in a precise manner with the apical membranes of other fibers from the same and immediately younger and older strata to form anterior sutures. Of course, in a similar manner, the basal membranes of fibers also overlap in a very specific manner with the basal membranes of fibers of the same growth shell, as well as with those of the immediately younger and older strata, to form posterior sutures. In contrast to the uniformly hexagonal shape of fibers along their length, their ends are very irregular (Fig. 13). Thus, the overlap of suture branches in successive growth shells results in irregular planes with variable refractive surfaces rather than radial cell columns with coaxial refractive surfaces (Trokel, 1962; Kuszak *et al.*, 1991, 1994; Sivak *et al.*, 1994).

The morphology of the apico-lateral border of lens fibers is comparable to the apico-lateral border of lens epithelial cells. Both are characterized by numerous micropinocytotic and endocytotic events, square array membrane, a paucity of gap junctions, and a lack of tight junctions. Square array membrane is even more pronounced at the apico-lateral border of fibers than at that of lens epithelial cells. Micropinocytotic and endocytotic events and square array membrane are also typical of the basal membrane of lens epithelial and superficial fibers. If in fact there is a lack of efficient tight junctions between both lens epithelial and fiber cells, then this may be a distinct modification of lens cell structure as a function of differentiation. Fibers are moved progressively farther away from their source of nutrition as a function of lens development and growth. Tight junctions provide a barrier to the diffusion of substances into and from cavities lined by epithelial cells. If such junctions developed between lens cells, they would impede the passage of substances throughout the lens. However, if the presumed tight junctions between lens epithelial cells have a low transepithelial resistance (TER), than the necessary passage of substances throughout the lens would be permitted rather than impeded.

VII. Intercellular Junctions in the Lens

Because the lens is a stratified epithelial-like organ, it is reasonable to expect that intercellular junctions are present that separate the cells of this

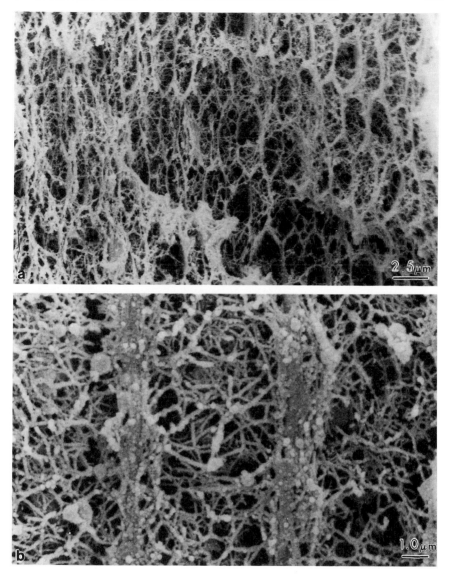

FIG. 12 SEM micrograph of DMSO-treated rabbit cortical fiber cells revealing the pervasive cytoskeleton extending from the plasma membrane (a) across their thickness and (b) along their length.

organ from other tissue (tight junctions), organize and adhere its cells together into a cohesive unit (e.g., adherens junctions and/or desmosomes), and provide cell-to-cell communication channels (gap junctions) between

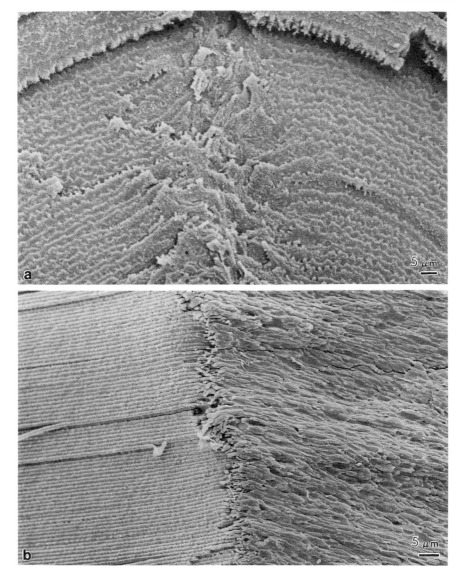

FIG. 13 (a) SEM micrograph showing the nonuniform ends of mouse fibers overlapped at a suture branch within a growth shell, and (b) aligned in an irregular suture plane (right). Note that the remainder of the fiber lengths in growth shells, shown in (a), and radial cell columns (RCCs), shown in (b, left), are very uniform. Reprinted from *Prog. Ret. Eye Res.* 14(2), J. R. Kuszak, The Development of Lens Sutures, 567–591, copyright (1995), with kind permission from Elsevier Science, Ltd, The Boulevard, Langford Lane, Kidlington OX5 1GB, UK.

all of its cells. Indeed the lens features all of these intercellular membrane specializations. However, the types, distribution, and characteristics of the intercellular junctions of the lens are highly specialized for this ocular organ.

A. Lens Epithelial Cells

Lens epithelial cells are conjoined by desmosomes, gap and tight junctions (see Figs. 7a and 7b). In thin-section micrographs, the gap junctions are characterized by pentalaminar and/or septalaminar profiles of apposed lateral membranes measuring ~16–18 μm in thickness, and separated by an ~2–3 nm gap that is permeable to lanthanum. Freeze-etch replicas of these junctions reveal intramembrane particles (IMPs) measuring ~8.5 nm in diameter that are arranged in either complete hexagonal configurations or in groups of IMPs arranged in this configuration separated by particle free aisles (Fig. 7d; Benedetti et al., 1976; Goodenough et al., 1980; Kuszak et al., 1982; Lo and Harding, 1986). Biochemical and immunoelectron microscopic analysis of these junctions shows that the IMPs of these junctions are composed of connexin 43, a member of the connexin family (Beyer et al., 1989; Musil et al., 1990).

Evidence of zonula occludens that completely encircle the apico-lateral borders of lens epithelial cells in all lenses is equivocal (see Figs. 7a and 7b). Freeze-etch analysis of mouse and rat lenses has demonstrated a few, simple strands of IMPs between apico-lateral borders of cz lens epithelial cells. Correlative physiological studies suggest that these strands provide a significant seal between these cells (Lo and Harding, 1983, 1986; Unakar et al., 1991; Zampighi et al., 1992). However, other structural and physiological studies of rat and chick lenses contradict these findings and demonstrate that there is no significant barrier to extracellular flow between lens epithelial cells (Gorthy et al., 1971; Rae and Stacey, 1979; Goodenough et al., 1980). The discrepancies between the results of these studies may be related to the age of and/or the species examined. Additional work is needed to determine the extent and physiological significance of tight junctions in the lens epithelium as a function of development, growth, age, and species.

B. Elongating Fibers

TEM thin-section analysis of elongating fibers reveals pentalaminar and/or septalaminar profiles of apposed lateral membrane with reported thicknesses of ~11–13, ~15–16, ~17–18, and ~18–20 nm (Lo and Reese, 1993; Zampighi et al., 1992). While the thin (~11–13 nm) and thick (~18–20 nm) intercellular contacts are clearly distinct junctional types, statistically significant evidence demonstrating that the ~15–16 nm and ~17–18 nm junctions are distinct entities from the thick ~18- to 20-nm junction is lacking. The thin junctions of fibers are separated by a very narrow gap (0.5–0.7 nm) that is impermeable to lanthanum, while the thick junctions are separated by a 4–5 nm gap that is permeable to lanthanum. Thus, the thick fiber junction is thicker than gap junctions of other tissues (e.g., cardiac, liver) that measure on an average ~15–16 nm. These different thicknesses imply that these intercellular contacts are composed of different size IMPs. Freeze-etch analysis of forming fibers reveals: junctional plaques with IMPs measuring ~8.5 nm in diameter packed in hexagonal arrays separated by protein-free aisles (Fig. 14a; presumed to be typical 15–16 nm thick gap junctions); junctional plaques with two distinct size IMPs measuring ~6–7 and 8.5 nm in diameter, and packed, respectively, randomly and in hexagonal arrays separated by protein-free aisles [Fig. 14b; presumed to be a mixed junctional plaque composed of both typical gap junction proteins and thick fiber junction proteins (Kuszak *et al.*, 1982)]; typical junctional formation plaques (Fig. 14c); and finally, junctional plaques with randomly packed ~7–8 nm diameter IMPs [Fig. 14d; presumed to be thick fiber junctions (Kuszak *et al.*, 1978; Goodenough, 1979; Kuszak *et al.*, 1981, 1982; Lo and Reese, 1993)].

Desmosomes are also found between elongating fibers (Lo, 1988). However, desmosomes have not been found between mature fibers. Given the progressive and dramatic structural changes that occur throughout fiber elongation, it is possible that desmosomes are necessary for the ordering of fibers during elongation, but are subsequently eliminated upon completion of differentiation.

C. Cortical and Nuclear Fiber Cells

Tight junctions and desmosomes have not been found between either cortical or nuclear fibers. However, numerous TEM thin-section, freeze-etch, biochemical, immunoelectron microscopic, and electrophysiologic studies have revealed that while gap junctions are a ubiquitous lateral membrane specialization in epithelia, their structure varies as a function of differentiation and tissue type (Spray and Bennett, 1985; Hertzberg and Johnson, 1988). Mature fibers are conjoined by typical gap junctions, and the thin and thick junctions previously described in elongating fibers. The thin and thick fiber junctions may be among the more cogent examples of gap junctions specifically modified as a function of differentiation (Yancey *et al.*, 1988; Dermietzel *et al.*, 1989). The IMPs of the thin junctions, referred to as MIP (main intrinsic protein), have an apparent molecular weight of 28 kDa that lacks significant homology with gap junction proteins of the

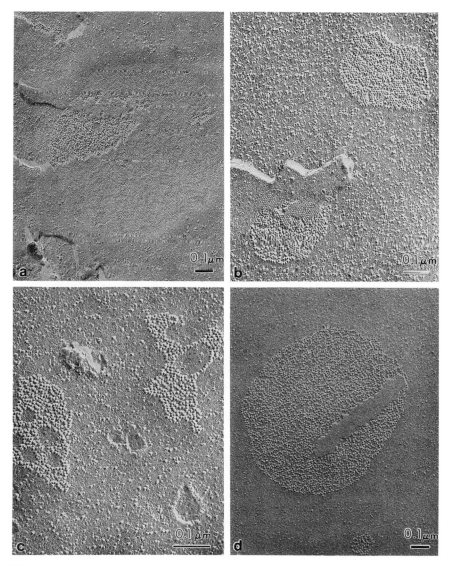

FIG. 14 Freeze-etch replicas of the various types of gap junctional contacts in elongating fibers. (a) A junction with IMPs arranged in hexagonal arrays (this junction is presumed to be comparable to a lens epithelial junction); (b) junctions with two distinct size IMPs arranged, respectively, in hexagonal and random arrays (this junction is presumed to be composed of a mixture of epithelial and fiber junction connexin proteins); (c) fiber junction formation plaques; and (d) a fiber junction.

"connexin" family (Beyer *et al.*, 1987; Ebihara *et al.*, 1989; Zhang and Nicholson, 1989). Therefore, it has been proposed that the role of these junctions in the lens may be adhesion rather than cell-to-cell communication, or a combination of both (Kuszak *et al.*, 1980, 1982; Zampighi *et al.*, 1992; Lo and Reese, 1993). Indeed, there is evidence suggesting that MIP can mediate intercellular communication. It has been immunologically localized to junctional plaques (Bok *et al.*, 1982; Fitzgerald *et al.*, 1985), and it has channel-forming properties (Zampighi *et al.*, 1985; Ehring and Hall, 1988; Ehring *et al.*, 1990); passage of dye between lentoids coupled by MIP can be blocked by antibodies to MIP (Johnson *et al.*, 1988). In contrast, the IMPs of the thick fiber junctions, referred to as MP (membrane protein) 70, have an apparent molecular weight of 70 kDa that exhibits considerable homology to connexins 32 and 43 from other tissues (liver and heart, respectively). The IMPs of nuclear fiber thick junctions have an apparent molecular weight of 38 kDa. These proteins are the endogenously proteolyzed MP70. The N-terminal sequence of both MP70 and MP38 is homologous to that of connexins 32 and 43 (Kistler *et al.*, 1988). Thus, it is proposed that these junctions provide cell-to-cell communication between cortical and nuclear fibers, respectively.

Given that there is more than one type of channel IMP in the lens, it is appropriate to ask the following questions: (1) Do different types of gap junctions mediate cell-to-cell coupling of lens cells as a function of development, growth, and age? (2) Are any of the junctions in the lens "heterogeneous" gap junctions, that is, junctions composed of more than one type of channel IMP? In other tissues it has now been demonstrated that cells in distinct stages of development are characterized by different types of gap junctions (Pozzi *et al.*, 1994) and that gap junctions composed of different connexins (heterogeneous junctions) are possible (White *et al.*, 1994; Risek *et al.*, 1994).

Though the percentage of fiber lateral membrane specialized as typical gap, thin, and thick fiber junctions is not known quantitatively in all species, it appears to vary considerably among species. Freeze-etch analysis of chick, rat, frog, and human lenses (Kuszak *et al.*, 1978, 1985b; Lo and Harding, 1986) reveal, respectively, 60, 30, 15, and 5% of fiber lateral membrane middle segments in the form of apposed junctional plaques. These junctional plaques are generally characterized by diameters of ~7–8 nm randomly arranged IMPs (Fig. 15). Thus, it is presumed that these intercellular contacts are thick junctions composed of MP70. However, when the broad and narrow faces of fibers are examined along their length it can be seen that there is an abrupt change in the presence of junctional contacts (Fitzgerald, 1986; Costello *et al.*, 1989). Anterior and posterior segments of fibers have a paucity of junctional contacts. This finding lends support to the premise that lens fibers are polarized and that their anterior and posterior

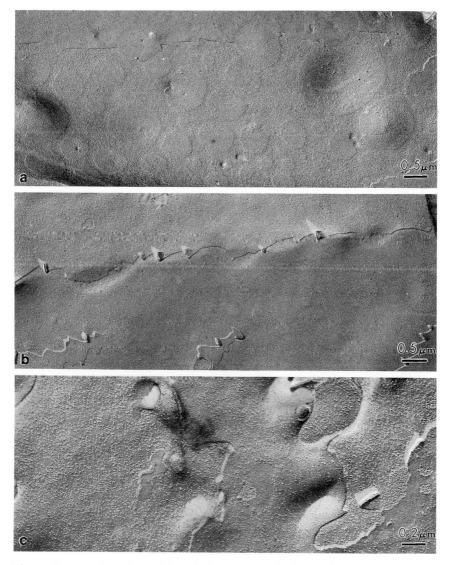

FIG. 15   Freeze-etch replicas of the middle segments of adult cortical fibers from (a) rat, (b) frog, and (c) monkey lenses. Note the wide variation in the density of fiber junctions between these comparable fibers of different species.

segments are, respectively, the apico-lateral and baso-lateral borders of these cells. The density and distribution of the junctional plaques correlate well with electrotonic coupling studies (Kuszak *et al.*, 1985b) that show lenses of different species are more or less efficiently coupled. These studies strongly suggest that these junctional plaques play a key role in mediating intercellular communication.

An explanation for the very different densities of the junctional contacts between species is unknown. However, the very different surface morphologies of cortical and nuclear fibers among species must be taken into account. While thin ~11–13 nm junctions are found in planar areas of elongating fiber membrane, they are more often found to be associated with the furrowed membrane plaques that characterize the cortical and nuclear fibers of long-lived primates, and the nuclear fibers of shorter-lived nonprimates (Zampighi et al., 1992). These junctions are presumed to be composed primarily of MIP. However, the IMPs of these junctions do not appear to be conjoined across the narrowed extracellular gap to form continuous channels between neighboring fibers. Rather, in the furrowed membrane domains, the IMPs of these junctions are only arranged on one side of the closely apposed membranes of neighboring fibers. The opposite side is particle free. This structure suggests that the role of this junction in lens physiology is not to mediate intercellular communication, but rather to increase the resistance of the extracellular space (Zampighi et al., 1992).

Given that fiber membrane and presumably fiber membrane protein synthesis is completed at the end of terminal differentiation in all lenses (Cenedella, 1993), it is interesting to consider whether or not all lenses produce equal amounts of the various intercellular junctional proteins found in lenses and whether or not all lenses alter these proteins in a similar manner as a function of development, growth, and age. Consider the following: avian lenses have relatively planar lateral fiber membrane surfaces with a high percentage of this membrane specialized presumably as "communicating junctions." In contrast, primate lenses have the vast majority of their lateral fiber membrane configured as polygonal domains of furrowed membrane with a very low percentage of this membrane specialized presumably as communicating junctions. Do avian and primate lenses produce comparable or variable amounts of MIP, MP 70, MP 38, and even MP 46, and how are the arrangements of these proteins in the plane of the membrane altered as a function of age given the very different surface morphologies of the lens fibers in these two species?

In addition, though not known quantitatively, avian lens fibers are infrequently involved in cell-to-cell fusion, while primate lens fibers routinely employ cell-to-cell fusion to effect their increasingly complex sutural architecture as a function of development, growth, and age (Kuszak et al., 1989). One consequence of cell-to-cell fusion in the elongating and/or superficial cortical fiber zone is that the cell-to-cell fusion zones effectively provide patent pathways between fibers that would permit transport of substances even too large to pass through gap junctions. Is there a relationship between the frequency of cell-to-cell fusion and the percentage of fiber membrane specialized as intercellular junctions in different species? These issues, with regard to intercellular contacts between lens fibers, remain to be resolved.

In any case, the evidence is compelling that there are epithelial–epithelial cell and fiber–fiber couplings throughout the lens. It is less certain whether or not there is coupling between epithelial cells of all the zones of the lens epithelium and the underlying elongating fibers as they terminally differentiate.

The apical membrane of elongating fibers interfaces with the apical membrane of the epithelial cells as the former extend toward their anterior sutural destinations during the terminal differentiation process. This interface is referred to as the epithelial–fiber cell interface (EFI). Dye and electrical coupling studies suggest that the movement of nutrients, ions, and essential metabolites across the EFI is mediated by gap junctions (Goodenough, 1979; Goodenough et al., 1980; Rae and Kuszak, 1983). The exchange of substances at this juncture is considered critical to the lens' survival because the underlying fibers are thought to be incapable of maintaining themselves, having eliminated most of the typical intracellular organelles as a consequence of terminal differentiation. However, as described earlier, additional new fibers are constantly being added to the lens, and these evolving fibers are continuously moving toward their sutural destinations during their elongating process. Thus, gap junctions at the EFI would have to be created, broken, and created again and again throughout fiber elongation. Correlative freeze-etch, transmission electron microscopic, dye coupling, and electrophysiological analyses have demonstrated that gap junctions are extremely rare between the apical membranes of cz zone epithelial and elongating fibers (Brown et al., 1990; Bassnett et al., 1994; Prescott et al., 1994; Kuszak et al.,1995). This is not an unexpected finding because gap junctions are typically a lateral membrane specialization (Simons and Fuller, 1985; Rodriguez-Boulan and Nelson, 1989). In addition, the type of gap junction that would exist at the EFI is unclear. As already described, the gap junctions that conjoin lens epithelial cells are not identical to those that serve the same functions between fibers. Therefore, it is reasonable to ask what type of gap junctions would mediate intercellular communication between the epithelium and elongating fibers. Would they be of the epithelial type, the fiber type, or a hybrid of the two? The EFI, however, is characterized by numerous endocytotic and micropinocytotic events. While the presence of endocytotic and micropinocytotic events cannot equally substitute for gap junctions at the EFI, these events do provide a previously unrecognized mode of transfer across the EFI.

VIII. Concluding Remarks

In conclusion, it can be seen that while all lenses develop and grow in a similar manner, structurally different types of fibers are formed; moreover, the precise organization into radial cell columns and concentric growth shells is not identical from species to species. An important consequence of this fiber variation is the production of lenses of different optical quality between species. It remains to be determined how different species direct or control the production of the different types of fibers. If all lenses have similar plasma membranes and cytoskeletal elements, then how and when do different species direct these components to be modified to effect variably complex suture patterns? If these components are actually different between species, then what are the relevant similarities? This information is particularly significant in primates that produce four distinct generations of suture patterns as a function of development, growth, and aging. It also remains to be determined how and why continued lens growth occurs asymmetrically, in (sequentially) the inferonasal, superotemporal, inferotemporal, and superonasal quadrants. It is likely that this quadrant-specific growth scheme and the development of cortical cataract opacities in humans are related.

It can also be seen that the surface morphology of lens fibers varies considerably between species. It had generally been accepted that the lateral interdigitations of fibers played a major role in the maintenance of fiber–fiber order during accommodation and convergence. However, since some species (e.g., fish) that fail to change lens shape during focusing feature elaborate systems of lateral interdigitations, it may be that their contribution to focusing is of minor significance.

The extensive series of polygonal domains of furrowed membrane in primate lenses as a function of growth and aging is unparalleled in other species. The exact reason that primate lenses employ this progressive surface modification is unknown. However, it seems likely that the long life span of this species necessitates a synchronized compaction of lens fibers. In this manner, the overall size of the lens can be controlled in relation to the rest of the eye, which does not grow significantly past sexual maturation. Thus, the furrowed membrane domains of primate lens fibers again raise the following questions: Do primate lens fibers have an inherently different ultrastructure (with regards to plasma membrane and cytoskeleton) than other species? If so are the embryonic, fetal, infantile, adult nuclear, and cortical regions of primate lenses actually composed of fibers structurally modified as a function of maturation and/or age? If so, is the reason that other species do not modify their fibers as extensively a simple consequence of their different life spans?

The extensive surface modification of primate lens fibers also raises the question of whether or not gap junctions are equally important in different species. Avian lens cortical fibers have a high percentage ($>60\%$) of their lateral membrane surfaces specialized as raised plaques of aggregated transmembrane proteins conjoined to complementary plaques on neighboring cells. It is presumed that each of these plaques is a communicating junction, suggesting that chick lenses are extensively coupled. However, it is now being suggested that many of the comparable plaques in other species are not necessarily communicating junctions. Primate lens fibers have an equally high percentage of their lateral membrane surfaces specialized as furrowed membrane domains characterized by "square array membrane" rather than communicating junctions. Does this imply that coupling is not as extensive or as necessary in primate lenses as in avian lenses? On the other hand, the contribution of cell-to-cell fusion to the development of sutures is more extensive in primate lenses that feature four distinct generations of sutures than in avian lenses that feature a branchless suture system, the umbilical suture. Yet primate and avian lenses are in many ways similar in optical quality, and both are superior to lenses with line or Y sutures.

Thus, there are many structural considerations that remain to be addressed in order to more fully comprehend the physiology and pathology of the crystalline lens. However, in conducting animal research to address these questions, it will be essential to qualify the results as a function of development, growth, age, and species in order to correctly further our understanding of human cataractogenesis.

Acknowledgments

The author acknowledges the expert technical assistance of Layne A. Novak, K.L. Peterson, James R. Kosloskey, and Joseph Dowdell, Sr. This work supported by a grant from the National Institutes of Health, National Eye Institute (No. EY-06642), The Alcon Research Institute, and The Louise C. Norton Trust.

References

Alcala, J., and Maisel, H. (1985). Biochemistry of lens plasma membrane and cytoskeleton. *In* "The Ocular Lens: Structure, Function and Pathology" (H. Maisel, ed.), pp. 169–222. Dekker, New York.

Allen, D. P., Low, P. S., Dola, A., and Maisel, H. (1987). Band 3 and ankyrin homologues are present in the eye lens: Evidence for all major erythrocyte membrane components in same non-erythroid cell. *Biochem. Biophys. Res. Commun.* **149**, 266–275.

Aster, J. C., Brewer, G. J., and Maisel, H. (1986). The 4.1-like proteins of the bovine lens: Spectrin-binding proteins closely related in structure to red blood cell protein 4.1. *J. Cell Biol.* **103**, 115–122.

Bassnett, S., Kuszak, J. R., Reinisch, L., Brown, H. G., and Beebe, D. C. (1994). Intercellular communication between epithelial and fiber cells of the eye lens. *J. Cell Sci.* **107**(4), 799–811.

Beebe, D. C., Johnson, M. C., Feagans, D. E., and Compart, P. J. (1980). The mechanism of cell elongation during lens fiber cell differentiation. *In* "Ocular Size and Shape Regulation

During Development" (S. R. Hilfer and J. B. Sheffield, eds.), pp. 79–98. Springer-Verlag, New York.

Benedetti, E. L., Dunia, I., Bentzel, C. J., Vermorken, A. J. M., Kibbelar, M. A., and Bloemendal, H. (1976). A portrait of plasma membrane specializations in eye lens epithelium and fibers. *Biochim. Biophys. Acta* **457**, 353–384.

Benedetti, E. L., Dunia, I., Ramaekers, F. C. S., and Kibbelar, M. A. (1981). Lenticular plasma membranes and cytoskeleton. In "Molecular and Cellular Biology of the Eye Lens" (H. Bloemendal, ed.), pp. 137–188. Wiley, New York.

Beyer, E. C., Paul, D. L., and Goodenough, D. A. (1987). Connexin43: A protein from rat heart homologous to a gap junction protein from liver. *J. Cell Biol.* **105**, 2621–2629.

Beyer, E. C., Kistler, J., Paul, D. L., and Goodenough, D. A. (1989). Antisera directed against connexin43 peptides react with a 43-kD protein localized to gap junctions in myocardium and other tissues. *J. Cell Biol.* **108**, 595–605.

Bok, D., Dockstader, J., and Horwitz, J. (1982). Immunocytochemical localization of the lens main intrinsic polypeptide (MIP26) in communicating junctions. *J. Cell Biol.* **92**, 213–220.

Brown, H. G., Ireland, M., and Kuszak, J. R. (1990). Ultrastructural, biochemical and immunological evidence of receptor-mediated endocytosis in the crystalline lens. *Invest. Ophthalmol. Visual Sci.* **31**, 2579–2592.

Bryant, P. J., and Simpson, P. (1984). Intrinsic and extrinsic control of growth in developing organs. *Q. Rev. Biol.* **59**(4), 387–415.

Bryant, S. V., French, V., and Bryant, P. J. (1981). Distal regeneration and symmetry. *Science* **212**, 993–1002.

Burn, P. (1988). Phosphatidylinositol cycle and its possible involvement in the regulation of cytoskeleton-membrane interaction. *J. Cell. Biochem.* **36**, 15–24.

Cenedella, R. (1993). Apparent coordination of plasma membrane component synthesis in the lens. *Invest. Ophthalmol. Visual Sci.* **34**(7), 2186–2194.

Cohen, A.I. (1958). Electron microscopic observations on the lens of the neonatal albino mouse. *Am.J. Anat.* **103**, 219–229.

Cohen, A.I. (1965). The electron microscopy of the normal human lens. *Invest. Ophthalmol. Visual Sci.* **4**, 433–446.

Costello, M. J., McIntosh, T. J., and Robertson, J. D. (1985). Membrane specializations in mammalian lens fiber cells: Distribution of square arrays. *Curr. Eye Res.* **4**, 1183–1201.

Costello, M. J., McIntosh, T. J., and Robertson, J. D. (1989). Distribution of gap junctions and square array junctions in the mammalian lens. *Invest. Ophthalmol. Visual Sci.* **30**, 975–989.

Coulombre, J. L. and Coulombre, A. J. (1963). Lens development: Fiber elongation and lens orientation. *Science* **142**, 1489–1490.

de Jong, W. W. (1981). Evolution of lens and crystallins. In "Molecular and Cellular Biology of the Eye Lens" (H. Bloemendal, ed.), pp. 221–278. Wiley, New York.

Dermietzel, R., Volker, M., Hwang, T. K., Berzborn, R. J., and Meyer, H. E. (1989). A 16kDa protein co-isolating with gap junctions from brain tissue belonging to the class of proteolipids of the vacuolar H+-ATPases. *FEBS Lett.* **253**, 1–5.

Dickson, D. H., and Crock, G. W. (1972). Interlocking patterns of primate lens fibers. *Invest. Ophthalmol. Visual Sci.* **13**, 809–815.

Duke-Elder, S. (1970). The refractive media: The lens. In "Systems of Ophthalmology" (S. Duke-Elder, ed.), pp. 309–324. Mosby, St. Louis, Mo.

Ebihara, K., Beyer, E. C., Swenson, K. I., Paul, D. L., and Goodenough, D. L. (1989). Cloning and expression of a *Xenopus* embryonic gap junction protein. *Science* **243**, 1194–1195.

Ehring, G. R., and Hall, J. E. (1988). Single channel properties of lens MIP 28 reconstituted into planar lipid bilayers. *Proc. West. Pharmacol. Soc.* **31**, 251–253.

Ehring, G. R., Zampighi, G. A., Horwitz, J., Bok, D., and Hall, J. E. (1990). Properties of channels reconstituted from the major intrinsic protein of lens fiber membrane. *J. Gen. Physiol.* **96**(3), 631–664.

Farnsworth, P. N., Fu, S. C. J., Burke, P. G., and Bahaia, I. (1974). Ultrastructure of rat lens fibers. *Invest. Ophthalmol. Visual Sci.* **13**, 274–279.

Fitzgerald, P. G. (1986). The main intrinsic polypeptide and intercellular communication in the ocular lens. *In* "Cell and Developmental Biology of the Eye. Development of Order in the Visual System" (R. S. Hilfer, and J. B. Sheffield, eds.), 9th ed., pp. 61–96. Springer-Verlag, New York.

Fitzgerald, P. G., Bok, D., and Horwitz, J. (1985). The distribution of the main intrinsic membrane polypeptide in ocular lens. *Curr. Eye Res.* **4**, 1203–1218.

French, V., Bryant, P. J., and Bryant, S. V. (1976). Pattern regulation in epimorphic fields. *Science* **193**, 969–981.

Goodenough, D. A. (1979). Lens gap junctions: A structural hypothesis for non-regulated low-resistance intercellular pathways. *Invest. Ophthalmol. Visual Sci.* **11**, 1104–1122.

Goodenough, D. A., Dick, J. S. B. I., and Lyons, J. E. (1980). Lens metabolic cooperation: A study of mouse lens transport and permeability visualized with freeze-substitution autoradiography and electron microscopy. *J. Cell Biol.* **86**, 576–589.

Gorthy, W. C., Snavely, M. R., and Berrong, N. D. (1971). Some aspects of transport and digestion in the lens of the normal young adult rat. *Exp. Eye Res.* **12**, 112–119.

Hansson, H. A. (1970). Scanning electron microscopy of lens of the adult rat. *Z. Zellforsch. Mikrosk. Anat.* **107**, 187–198.

Harding, C. V., Reddan, J. R., Unakar, N. J., and Bagchi, M. (1971). The control of cell division in the ocular lens. *Int. Rev. Cytol.* **31**, 215–300.

Harding, C. V., Susan, S., and Murphy, H. (1976). Scanning electron microscopy of the adult rabbit lens. *Ophthalmic Res.* **8**(6), 443–455.

He, H. T., Barbet, J., Chaix, J. C., and Goridis, C. (1986). Phosphatidylinisotol is involved in the membrane attachment of NCAM-120, the smallest component of the neural cell adhesion molecule. *EMBO J.* **5**, 2489–2494.

Henry, J. J. and Grainger, R. M. (1990). Early tissue interactions leading to embryonic lens formation in *Xenopus laevis. Dev. Biol.* **141**, 149–163.

Hertzberg, E. L., and Johnson, R. G. (1988). "Modern Cell Biology: Gap Junctions," 7th ed., Alan R. Liss, New York.

Hollenberg, M. J., Wyse, J. P. H., and Lewis, B. J. (1976). Surface morphology of lens fiber from eyes of normal and microphthalmic (Browman) rats. *Cell Tissue Res.* **167**, 425–428.

Hoyer, H. E. (1982). Scanning electron-microscopic study of lens fibers of the pig. *Cell Tissue Res.* **224**(1), 225–232.

Ireland, M., and Maisel, H. (1983). Identification of native actin filaments in chick lens fiber cells. *Exp. Eye Res.* **36**, 531–536.

Ireland, M., and Maisel, H. (1984). A cytoskeletal protein unique to lens fiber cell differentiation. *Exp. Eye Res.* **38**, 637–645.

Johnson, R. G., Klukas, K. A., Lu, T-H., and Spray, D. C. (1988). Antibodies to MP28 are localized to lens junctions, alter intercellular permeability and demonstrate increased expression during development. *In* "Modern Cell Biology: Gap Junctions," 7th ed., (E. L. Hertzberg and R. G. Johnson, eds.), pp. 81–98. Alan R. Liss, New York.

Jurand, A., and Yamada, T. (1967). Elimination of mitochondria during Wolffian lens degeneration.. *Exp. Cell Res.* **46**, 636–638.

Kistler, J., Christie, D., and Bullivant, S. (1988). Homologies between gap junction proteins in lens, heart and liver. *Nature (London)* **331**, 721–723.

Kuszak, J. R., and Brown, H. G. (1993). Embryology and anatomy of the crystalline lens. *In* "The Principles and Practice of Opthalmology" (D. A. Albert and F. A. Jacobiec, eds.), pp. 82–96. Saunders, Philadelphia.

Kuszak, J. R., and Rae, J. L. (1982). Scanning electron microscopy of the frog lens. *Exp. Eye Res.* **35**, 499–519.

Kuszak, J. R., Novak, L. A., and Brown, H. G. (1995). An ultrastructural analysis of the epithelial–fiber interface (EFI) in primate lenses. *Exp. Eye Res.* **61** (in press).

Kuszak, J. R., Maisel, H., and Harding, C. V. (1978). Gap junctions of chick lens fiber cells. *Exp. Eye Res.* **27,** 495–498.

Kuszak, J. R., Alcala, J., and Maisel, H. (1980). The surface morphology of embryonic and adult chick lens-fiber cell. *Am. J. Anat.* **159,** 395–410.

Kuszak, J. R., Alcala, J., and Maisel, H. (1981). Biochemical and structural features of chick lens gap junctions. *Exp. Eye Res.* **33,** 157–166.

Kuszak, J. R., Rae, J. L., Pauli, B. U., and Weinstein, R. S. (1982). Rotary replication of lens gap junctions. *J. Ultrastruct. Res.* **81,** 249–256.

Kuszak, J. R., Macsai, M. S., and Rae, J. L. (1983). Stereo scanning electron microscopy of the crystalline lens. *Scanning Electron Microsc.* **III,** 1415–1426.

Kuszak, J. R., Bertram, B. A., Macsai, M. S., and Rae, J. L. (1984). Sutures of the crystalline lens: A review. *Scanning Electron Microsc.* **III,** 1369–1378.

Kuszak, J. R., Macsai, M. S., Bloom, K. J., Rae, J. L., and Weinstein, R. S. (1985a). Cell-to-cell fusion of lens fiber cells in situ: Correlative light,scanning electron microscopic, and freeze-fracture studies. *J. Ultrastruct. Res.* **93,** 144–160.

Kuszak, J. R., Shek, Y. H., Carney, K. C., and Rae, J. L. (1985b). A correlative freeze-etch and electrophysiological study of communicating junctions in crystalline lenses. *Curr. Eye Res.* **4,** 1145–1153.

Kuszak, J. R., Bertram, B. A., and Rae, J. L. (1986). The ordered structure of the crystalline lens. *In* "Cell and Developmental Biology of the Eye.Development of Order in the Visual System" (S. R. Hilfer and J. B. Sheffield, ed.), 9th ed., pp. 35–60. Springer-Verlag, New York.

Kuszak, J. R., Ennesser, C. A., Bertram, B. A., Imherr-McMannis, S., Jones-Rufer, L. S., and Weinstein, R. Ś. (1989). The contribution of cell-to-cell fusion to the ordered structure of the crystalline lens. *Lens Eye Toxic. Res.* **6,** 639–673.

Kuszak, J. R., Sivak, J. G., and Weerheim, J. A. (1991). Lens optical quality is a direct function of lens sutural architecture. *Invest. Ophthalmol. Visual Sci.* **32**(7), 2119–2129.

Kuszak, J. R., Peterson, K. L., Herbert, K. L., and Sivak, J. G. (1994). The inter-relationship of lens anatomy and optical quality. II. Primate lenses. *Exp. Eye Res.* **59,** 521–535.

Kuwabara, T. (1975). The maturation of the lens cell: A morphologic study. *Exp. Eye Res.* **20,** 427–443.

Kuwabara, T., and Imaizumi, M. (1974). Denucleation process of the lens. *Invest. Ophthalmol. Visual Sci.* **13,** 973–981.

Lewis, W. H. (1907). Experiments on the origin and differentiation of the optic vesicle in amphibia.. *Am. J. Anat.* **7,** 259–278.

Lo, W.-K. (1988). Adherens junctions in the ocular lens of various species: Ultrastructural analysis with an improved fixation. *Cell Tissue Res.* **254,** 31–40.

Lo, W.-K., and Harding, C. V. (1983). Tight junctions in the lens epithelia of human and frog: freeze-fracture and protein tracer studies.. *Invest. Ophthalmol. Visual Sci.* **24,** 396–402.

Lo, W.-K., and Harding, C. V. (1984). Square arrays and their role in ridge formation in human lens fibers. *J. Ultrastruct. Res.* **86,** 228–254.

Lo, W.-K., and Harding, C. V. (1986). Structure and distribution of gap junctions in lens epithelium and fiber cells. *Cell Tissue Res.* **244,** 253–263.

Lo, W.-K., and Reese, T. S. (1993). Multiple structural types of gap junctions in mouse lens. *J. Cell Sci.* **106,** 227–235.

Low, M. G., and Kincade, P. W. (1985). Phosphatidylinisotol is the membrane-anchoring domain of the Thy-1 glycoprotein. *Nature* (*London*) **318,** 62–64.

Low, M. G., and Zilversmit, D. B. (1980). Role of phosphatidylinisotol in attachment of alkaline phosphate to membranes. *Biochemistry* **19,** 3913–3918.

Mathias, R. T., and Rae, J. L. (1989). Cell to cell communication in the lens. *In* "Cell Interactions and Gap Junctions" (N. Sperelakis and W. C. Cole, eds.), pp. 29–50. CRC Press, New York.

Modak, S. P., and Bollum, F. J. (1972). Detection and measurement of single-strand breaks in nuclear DNA in fixed lens sections. *Exp. Cell Res.* **75,** 544–561.

Modak, S. P., Morris, G., and Yamada, T. (1968). DNA synthesis and mitotic activity during early development of chick lens. *Dev. Biol.* **17,** 544–561.

Musil, L. S., Beyer, E. C., and Goodenough, D. A. (1990). Expression of the gap junction protein connexin43 in embryonic chick lens: Molecular cloning, ultrastructural localization and post-translational phosphorylation. *J. Membr. Biol.* **116,** 163–175.

Nelson, K. J., and Rafferty, N. S. (1976). A scanning electron microscopic study of lens fibers in healing mouse lens. *Exp. Eye Res.* **22,** 335–346.

O'Rahilly, R., and Meyer, D. B. (1959). The early development of the eye in the chick *Gallus domesticus* (stages 8 to 25). *Acta Anat.* **36,** 20–58.

Papaconstantinou, J. (1967). Molecular aspects of lens cell differentiation. *Science* **156,** 338–346.

Parmigiani, C. M., and McAvoy, J. W. (1989). A morphometric analysis of the development of the rat lens capsule. *Curr. Eye Res.* **8,** 1271–1277.

Pozzi, A., Kiang, D. T., Risek, B., Gilula, N. B., and Kumar, N.M. (1994). Characterization of three different gap junction gene products during different stages of mammary gland activity. *Mol. Biol. Cell* **5,** 326a.

Prescott, A., Duncan, G., Van Marle, J., and Vrensen, G. (1994). A correlated study of metabolic cell communication and gap junction distribution in the adult frog lens. *Exp. Eye Res.* **58**(6), 737–746.

Rabl, C. (1900). Concerning the structure and development of the lens. *Zt. Wiss. Zool.* **67,** 1–138.

Rae, J. L., and Kuszak, J. R. (1983). The electrical coupling of epithelium and fibers in the frog lens. *Exp. Eye Res.* **36,** 317–326.

Rae, J. L., and Stacey, T. R. (1979). Lanthanum and procion yellow as extracellular markers. *Exp. Eye Res.* **28,** 1–21.

Rafferty, N. S. (1985). Lens Morphology. *In* "The Ocular Lens: Structure, Function and Pathology" (H. Maisel, ed.), pp. 1–60. Dekker, New York.

Rafferty, N. S., and Scholz, D. L. (1984). Polygonal arrays of microfilaments in epithelial cells of the intact lens. *Curr. Eye Res.* **3,** 1141–1149.

Rafferty, N. S., and Scholz, D. L. (1989). Comparative study of actin filament patterns in lens epithelial cells. Are these determined by the mechanism of lens accommodation? *Curr. Eye Res.* **8,** 569–579.

Rafferty, N. S., Scholz, D. L., Goldberg, M., and Lewyckyj, M. (1990). Immunocytochemical evidence for an actin-myosin system in lens epithelial cells. *Exp. Eye Res.* **51,** 591–600.

Risek, B., Klier, G. F., and Gilula, N. B. (1994). Molecular composition and structural organization of connexons in heterogeneous gap junctions. *Mol. Biol. Cell* **5,** 326a

Rochon-Duvigneaud, A. (1943). "Les yeux et la vision des Vertebres." Masson, Paris.

Rodriguez-Boulan, E., and Nelson, W. J. (1989). Morphogenesis of the polarized epithelial cell phenotype. *Science* **245,** 718–725.

Saha, M. S., Spann, C. L., and Grainger, R. M. (1989). Embryonic lens induction: More than meets the optic vesicle. *Cell Differ. Dev.* **28,** 153–172.

Sefton, B. M., and Buss, J. E. (1987). The covalent modification of eukaryotic proteins with lipid. *J. Cell Biol.* **104,** 1449–1453.

Sikerwar, S. S., and Malhotra, S. K. (1983). A structural characterization of gap junctions isolated from mouse liver. *Cell Biol. Int. Rep.* **7,** 897–903.

Simons, K., and Fuller, S. D. (1985). Cell surface polarity in epithelia. *Annu. Rev. Cell Biol.* **1,** 243–288.

Sivak, J. G., Herbert, K. L., Peterson, K. L., and Kuszak, J. R. (1994). The inter-relationship of lens anatomy and optical quality. I. Non-primate lenses. *Exp. Eye Res.* **59,** 505–520.

Smelser, G. K. (1965). Embryology and morphology of the lens. *In* "Symposium on the Lens" (J. E. Harris, ed.), pp. 22–34. Mosby, St. Louis, MO.

Spray, D. C., and Bennett, M. V. L. (1985). Physiology and pharmacology of gap junctions. *Annu. Rev. Physiol.* **47,** 281–303.

Trokel, S. (1962). The physical basis for transparency of the crystalline lens. *Invest. Ophthalmol. Visual Sci.* **1,** 493–501.

Unakar, N. J., Johnson, M. J., and Hynes, K. (1991). Permeability studies in neonatal rat lens epithelium. *Lens Eye Toxic. Res.* **8**(1), 75–99.

Wanko, T., and Gavin, M. A. (1959). EM of the lens. *J. Biophys. Biochem. Cytol.* **6,** 97–102.

White, T. W., Bruzzone, R., Wolfram, S., Paul, D. L., and Goodenough, D. A. (1994). Selective interactions among the multiple connexin proteins expressed in the vertebrate lens: The second extracellular domain is a determinant of compatibility between connexins. *J. Cell Biol.* **125**(4), 879–892.

Willikens, B., and Vrensen, G. (1981). The three-dimensional organization of lens fibers in the rabbit. A scanning electron microscopic reinvestigation. *Albrecht von Graefes Arch. Klin. Exp. Ophthalmol.* **216,** 275–289.

Wistow, G. J., and Piatigorsky, J. (1988). Lens crystallins: The evolution and expression of proteins for a highly specialized tissue. *Annu. Rev. Biochem.* **57,** 479–504.

Wolpert, L. (1971). Positional information and pattern formation. *Curr. Top. Dev. Biol.* **6,** 183–224.

Wolpert, L. (1987). The development of the pattern of growth. *In* "Pediatrics and Growth" (D. Baltrop, ed.), pp. 15–22. Blackwell, Oxford.

Yancey, S. B., Koh, K., Chung, J., and Revel, J. P. (1988). Expression of the gene for main intrinsic polypeptide (MIP): Separate distributions of MIP and beta-crystallin gene transcripts in rat lens development. *J. Cell Biol.* **106,** 705–714.

Yeagle, P. L. (1989). Regulation of membrane function through composition, structure, and dynamics. *Ann. N.Y. Acad. Sci.* **568,** 29–34.

Yeh, S., Scholz, D. L., Liou, W., and Rafferty, N. S. (1986). Polygonal arrays of actin filaments in human lens epithelial cells. *Invest. Ophthalmol. Visual Sci.* **27,** 1535–1540.

Zampighi, G. A., Hall, J. E., and Kreman, M. (1985). Purified lens junctional protein forms channels in planar lipid films. *Proc. Natl. Acad. Sci. U.S.A.* **82,** 8468–8472.

Zampighi, G. A., Hall, J. E., Ehring, G. R., and Simon, S. A. (1989). The structural organization and protein composition of lens fiber junctions. *J. Cell Biol.* **108,** 2255–2275.

Zampighi, G. A., Simon, S. A., and Hall, J. E. (1992). The specialized junctions of the lens. *Int. Rev. Cytol.* **136,** 185–225.

Zelenka, P. S. (1984). Lens lipids. *Curr. Eye Res.* **3,** 1337–1359.

Zhang, J. T., and Nicholson, B. J. (1989). Sequence and tissue distribution of a second protein of hepatic gap junctions, Cx26 as deduced from its cDNA. *J. Cell Biol.* **109,** 3391-3401.

INDEX

A

Abalone, sperm-binding proteins, 262–263
ABP-120 mutants, computer-assisted motion
analysis, 91–93
Abscisic acid, fibrillin activation, 203
Acceleration, cell, 2D analysis, 62–63
Acrosin, role in sperm binding, 268
Acrosomal reaction
carbohydrate role, 281–282
changes concomitant with, 252
Acyl lipids, chromoplast
in *Capsicum* and *Narcissus,* 190
characterization, 188
metabolism, 219–220
ADP, chromoplast, 207–208
Aging, lens epithelial cells, 312, 323
Amitrole, induction of chromoplast
differentiation, 204
Amoeboid motility, *see* Cell motility,
amoeboid
Amphibians, egg surface sperm-binding
proteins, 263–264
Amphotericin
osmotic conductance (P_{os}), 3–5
osmotic flow, bimodal analysis, 30–34
Apoptosis
calcium role, 114–115
cell death mechanisms, 112–113
DNA fragmentation, 112–113
endonucleases in, 112–114
defined, 107
gene activation during, 116–117
genetic regulation, 119–120
growth factor regulation, 117–119
hormonal regulation, 117–119

morphological characteristics, 109–111
signal transduction during, 115–116
techniques for examining cell death,
120–123
Aquaporins, CHIP28, *see* CHIP28
Ascidians, sperm-binding proteins, 258–259
Asterina miniata, sperm-binding proteins,
279–280
ATP, chromoplast, 207–208
Axis tilt
computation, 63
two-dimensional analysis, 63

B

Baculovirus p35 protein, effects on cell death,
129
Bat star *(Asterina miniata),* sperm-binding
proteins, 279–280
bcl-2 gene, role in cell death, 119–120
Behavioral cycles, amoeboid cells, 79–80
Bimodal theory
drag coefficients, 19–20
empirical, 23–24
fixed-section pores
P_f/P_d ratios, 26–30
P_{os} values, 30–34
kinetics, 23
leaky pores, 21–26
osmotic conductance, 24–26
pore radius prediction, 26–30
reflexion coefficient, 24–26
semipermeable pores, 19–21
thermodynamics, 21–22
variable-section pores, 34–36

351